Trends in Mathematics is a series devoted to the publication of volumes arising from conferences and lecture series focusing on a particular topic from any area of mathematics. Its aim is to make current developments available to the community as rapidly as possible without compromise to quality and to archive these for reference.

Proposals for volumes can be sent to the Mathematics Editor at either

Birkhäuser Verlag
P.O. Box 133
CH-4010 Basel
Switzerland

or

Birkhäuser Boston Inc.
675 Massachusetts Avenue
Cambridge, MA 02139
USA

Material submitted for publication must be screened and prepared as follows:

All contributions should undergo a reviewing process similar to that carried out by journals and be checked for correct use of language which, as a rule, is English. Articles without proofs, or which do not contain any significantly new results, should be rejected. High quality survey papers, however, are welcome.

We expect the organizers to deliver manuscripts in a form that is essentially ready for direct reproduction. Any version of TeX is acceptable, but the entire collection of files must be in one particular dialect of TeX and unified according to simple instructions available from Birkhäuser.

Furthermore, in order to guarantee the timely appearance of the proceedings it is essential that the final version of the entire material be submitted no later than one year after the conference. The total number of pages should not exceed 350. The first-mentioned author of each article will receive 25 free offprints. To the participants of the congress the book will be offered at a special rate.

Geometric Group Theory

Geneva and Barcelona Conferences

Goulnara N. Arzhantseva
Laurent Bartholdi
José Burillo
Enric Ventura
Editors

Birkhäuser
Basel · Boston · Berlin

Editors:

Goulnara N. Arzhantseva
Université de Genève
Section de mathématiques
2–4, rue du Lièvre
Case postale 64
1211 Genève 4
Switzerland
e-mail : Goulnara.Arjantseva@math.unige.ch

José Burillo
Departament de Matemàtica Aplicada IV
Universitat Politècnica de Catalunya
Escola Politècnica Superior de Castelldefels
Avda. Del Canal Olímpic s/n
08860 Castelldefels (Barcelona)
Spain
e-mail: burillo@ma4.upc.edu

Laurent Bartholdi
École Polytechnique Fédérale
de Lausanne (EPFL)
Institut de Mathématiques B (IMB)
1015 Lausanne
Switzerland
e-mail : laurent.bartholdi@epfl.ch

Enric Ventura
Department of Applied Mathematics III
Escola Politècnica Superior
d'Enginyeria de Manresa
Universitat Politècnica de Catalunya
Av. Bases de Manresa 61–73
08242 Manresa / Barcelona
Spain
e-mail: enric.ventura@upc.edu

2000 Mathematical Subject Classification: 20-xx, 22Dxx

Library of Congress Control Number: 2007928997

Bibliographic information published by Die Deutsche Bibliothek. Die Deutsche Bibliothek lists
this publication in the Deutsche Nationalbibliografie; detailed bibliographic data is available in
the Internet at http://dnb.ddb.de

ISBN 978-3-7643-8411-1 Birkhäuser Verlag AG, Basel - Boston - Berlin

© 2007 Birkhäuser Verlag AG
Basel · Boston · Berlin
P.O. Box 133, CH-4010 Basel, Switzerland
Part of Springer Science+Business Media
Printed on acid-free paper produced from chlorine-free pulp. TCF ∞
Cover Design: Alexander Faust, CH-4051 Basel, Switzerland
Printed in Germany
ISBN 978-3-7643-8411-1   e-ISBN 978-3-7643-8412-8

9 8 7 6 5 4 3 2 1   www.birkhauser.ch

Contents

Preface vii

U. Baumgartner
Totally Disconnected, Locally Compact Groups as
Geometric Objects. A survey of work in progress 1

J. Burillo, S. Cleary und D. Wiest
Computational Explorations in Thompson's Group F 21

T. Ceccherini-Silberstein and M. Coornaert
On the Surjunctivity of Artinian Linear Cellular Automata
over Residually Finite Groups 37

Y. de Cornulier and A. Mann
Some Residually Finite Groups Satisfying Laws 45

R.J. Flores
Classifying Spaces for Wallpaper Groups 51

V. Guirardel and G. Levitt
A General Construction of JSJ Decompositions 65

Y. de Cornulier and P. de la Harpe
Décompositions de Groupes par Produit Direct
et Groupes de Coxeter 75

A. Ould Houcine
Limit Groups of Equationally Noetherian Groups 103

A. Juhász
Solution of the Conjugacy Problem and Malnormality
of Subgroups in Certain Relative Small Cancellation
Group Presentations 121

A. Juhász
Solution of the Membership Problem for Magnus Subgroups
in Certain One-Relator Free Products 169

M. Lustig
Conjugacy and Centralizers for Iwip Automorphisms
of Free Groups 197

A. Miasnikov, E. Ventura and P. Weil
Algebraic Extensions in Free Groups 225

Preface

We are pleased to present "Geometric Group Theory, Geneva and Barcelona", a selection of research articles contributed by speakers and participants of two strongly related recent international conferences. The first one, "Asymptotic and Probabilistic Methods in Geometric Group Theory", was held at the University of Geneva (Switzerland), from June 20th to 25th, 2005. The second one, "Barcelona Conference in Group Theory", was held at the Centre de Recerca Matemàtica in Barcelona (Catalonia, Spain), from June 28th to July 2nd, 2005.

This volume contains a selection of twelve refereed articles highlighting several active areas of research in Geometric and Combinatorial Group Theory. The specific material presented here will be of interest to researchers in this domain; owing to the broad nature of group theory, it should also appeal to graduate students and mathematicians throughout mathematics, by providing a direct introduction to active themes of research.

We are very grateful to the organisms that supported these two conferences. The Geneva event was organized by Goulnara Arzhantseva, Laurent Bartholdi, Alexander Ol'shanskii, Mark Sapir and Efim Zelmanov. It was part of the Swiss National Science Foundation research project "Geometry of groups and asymptotic invariants", within the SNF Professorship of Goulnara Arzhantseva, no. PP002-68627, and was also sponsored by the University of Geneva, the foundation "L'Enseignement mathématique", the École Polytechnique Fédérale de Lausanne, the Swiss Mathematical Society ("Journées de Printemps"), and the American National Science Foundation.

The Barcelona event was organized by Noel Brady, José Burillo and Enric Ventura. It was part of the "Centre de Recerca Matemàtica" 2004–2005 research programme on "The Geometry of the Word Problem", which contributed administrative, financial and institutional support to the organization of the Barcelona Group Theory Conference. Thanks are due as well to the Universitat Autònoma de Barcelona and Universitat Politècnica de Catalunya for their support.

Above all, we wish to thank the contributors to the present volume, and the anonymous referees who ensured the high quality of its contents. Our thanks also go to Thomas Hempfling at Birkhäuser for his assistance in the typesetting and preparation of this volume. Without these joint efforts, this book would never have appeared.

Geneva, Lausanne
and Barcelone,
March 2007

The editors,
G. Arzhantseva, L. Bartholdi,
J. Burillo, E. Ventura.

Geometric Group Theory

Trends in Mathematics, 1–20

Totally Disconnected, Locally Compact Groups as Geometric Objects

A survey of work in progress

Udo Baumgartner

Abstract. This survey outlines a geometric approach to the structure theory of totally disconnected, locally compact groups. The content of my talk at Geneva is contained in Section 3.

Mathematics Subject Classification (2000). 22D05, 22D45.

Keywords. totally disconnected group, automorphism group, scale function, flat subgroup, eigenfactor, flat rank, rank of $\mathsf{CAT}(0)$-space, space of directions, Tits metric, contraction group.

1. Introduction

Locally compact groups usually arise as automorphism groups of finite dimensional geometric objects. This explains their omnipresence in several areas of mathematics and neighboring disciplines such as physics as well as the interest in understanding their structure.

To a large extent, the study of the structure of locally compact groups can be reduced to the study of the subclasses of connected and totally disconnected locally compact groups, since for every locally compact group G the connected component of the identity is a closed, normal subgroup with totally disconnected quotient. By the solution of Hilbert's fifth problem, connected locally compact groups are projective limits of Lie groups. This fact leads to a thorough understanding of connected locally compact groups.

Only recently has a fundamental breakthrough by Willis ([Wil94, Wil01, Wil04]) made a systematic study of totally disconnected, locally compact groups feasible. Several results about connected, locally compact groups have been extended to arbitrary locally compact groups. This paper will focus on a different

---

This article originates from the Geneva conference.

aspect of the structure theory of totally disconnected, locally compact groups: the effort to understand the invariants invented by Willis in a geometric context.

This geometric approach may be justified by the observation that locally compact groups usually arise as automorphism groups of finite dimensional geometric objects. For the automorphism group to be totally disconnected, the geometric object should be discrete, and an example of such a discrete geometric object is a connected, locally finite simplicial complex.

This survey is organized as follows:

Section 2 will introduce a geometric framework for Willis' invariants of totally disconnected, locally compact groups, which will be used in the sequel.

Section 3 will introduce flat subgroups of a totally disconnected, locally compact group and explain their geometric aspect.

Section 4 will introduce an asymptotic notion which provides a framework for formalizing relative position of flat subgroups 'at infinity'.

Finally, in Section 5 we will hint at a group theoretic translation of the asymptotic notions; this last section will be rather sketchy.

Conventions: 0 is a natural number. If G is a group acting on a set X and F is a subset of X, then $G_{\{F\}}$ denotes the subgroup of G stabilizing F, while G_F denotes the subgroup of G fixing F pointwise. Conjugation by a group element g is understood to be the map $x \mapsto gxg^{-1}$. The relations $\subset$, $\lhd$ *etc.* always imply strict inclusion. Any automorphism of a topological group will be assumed to be a homeomorphism. The modular function Δ_G of a locally compact topological group G is defined by the equation $\mu(\alpha(M)) = \Delta_G(\alpha)\mu(M)$ where μ is a left Haar measure on G. We use e for the unit element of a group and 1 for the trivial group.

2. Fat points; the space of compact, open subgroups

Let G be a totally disconnected, locally compact group. In this section we introduce a metric space $\mathcal{B}(G)$ with an isometric action by the group of automorphisms of G.

We will think of the space $\mathcal{B}(G)$ as a coarse version of any metric space with a nice isometric G-action. This view will be vindicated by yielding useful information for interesting classes of examples, such as those discussed in later sections. We do not dare to suggest that for a general totally disconnected, locally compact group G, any metric space X with a sufficiently nice isometric G-action must have the same large scale geometry as an orbit of G in $\mathcal{B}(G)$. (However, we encourage anyone who can prove such a statement to please come forward.)

The set of points of $\mathcal{B}(G)$ is $\{V : V$ a compact, open subgroup of $G\}$. This set is not empty, because every totally disconnected, locally compact group has a base of neighborhoods of the identity consisting of compact, open subgroups by Theorem 7.7 in [HR79, Chapter II]. Inspired by the case where G is a group of automorphisms of a locally finite cell complex, we will think of the points in $\mathcal{B}(G)$

as the pointwise fixators of bounded subsets of some space X with a nice G-action. Hence, informally, $\mathcal{B}(G)$ will be a set of **fat points**.

Next, we wish to measure the discrepancy between compact, open subgroups of G. One can check that the function d defined by

$$d(V, W) := \log(|V : V \cap W| \cdot |W : W \cap V|)$$

is a metric on the set $\mathcal{B}(G)$ using the decomposition $d(V, W) = d_+(V, W) + d_-(V, W)$ where we put $d_+(A, B) := \log(|A : A \cap B|)$ and $d_-(A, B) := d_+(B, A)$. Only the triangle inequality for d is not obvious. Using elementary properties of subgroup-indices, one establishes the *triangle inequality for d_+*, which holds in the form

$$d_+(A, C) \le d_+(A, B) + d_+(B, C). \tag{1}$$

The triangle inequality for d_- and d follow from this.

The function d is a natural distance on $\mathcal{B}(G)$ and the group of automorphisms of G acts by isometries on the metric space $(\mathcal{B}(G), d)$. Note that the set of values of the function d is a discrete, well-ordered subset of the reals. In particular, $\mathcal{B}(G)$ is a discrete metric space.

Suppose for a moment, that G acts isometrically on some metric space X. To get a feeling for when $d(G_A, G_B)$ is a reasonable measurement for the distance between bounded subsets A and B of X, observe that for $x, y \in X$ we have

$$d(G_x, G_y) = \log(|G_x.y| \cdot |G_y.x|). \tag{2}$$

3. Quasiflats; flat subgroups and eigenfactors

In this section we present results from the paper [BRW05] and the unfinished work [BMW04]. Roughly speaking, both papers establish connections between quasi-flats in metric spaces with a nice isometric G-action and quasi-flats in the space $\mathcal{B}(G)$. The first two subsections are preparatory and give geometric interpretations of concepts developed in [Wil94] and [Wil04]. The first formulation of these concepts in the geometric language used below was given in [BW06].

3.1. Geodesics, tidy subgroups and the scale function

Orbits of points in $\mathcal{B}(G)$ under the powers of a single automorphism of G are quasi-isometric to the real line. We will discuss consequences of that fact in this subsection. Later we will discuss groups of automorphisms whose orbits are quasi-isometric to Euclidean spaces of higher dimension.

We first introduce a discrete version of the concept of geodesic. Call a finite, one- or two-sided infinite sequence of points in $\mathcal{B}(G)$ a *geodesic* if the distance between any two terms equals the sum of the distances of the adjacent terms in between. A finite geodesic sequence will be called a *geodesic segment*, while a one-sided (respectively two-sided) infinite geodesic sequence will be called a *geodesic ray* (respectively a *geodesic line*). Note that a geodesic line may be a constant sequence.

Let O be a point of $\mathcal{B}(G)$. The sequence $\left(\alpha^n(O)\right)_{n\in\mathbb{N}}$ will be called *the ray generated by* $\boldsymbol{\alpha}$ *based at* O while the two-sided infinite sequence $\left(\alpha^n(O)\right)_{n\in\mathbb{Z}}$ will be called *the line generated by* $\boldsymbol{\alpha}$ *based at* O.

It turns out that for each automorphism α of G there is a compact, open subgroup O of G such that the line generated by α based at O is a geodesic. It is reasonable to suspect that a compact, open subgroup satisfying the following condition has this property: An element O of $\mathcal{B}(G)$ is called *tidy for* α if the *displacement function* of α, denoted by $d_\alpha\colon \mathcal{B}(G) \to \mathbb{R}$ and defined by $d_\alpha(V) = d(\alpha(V), V)$, attains its minimum at O. Since the set of values of the metric d on $\mathcal{B}(G)$ is a well-ordered discrete subset of $\mathbb{R}$, every $\alpha \in \mathsf{Aut}(G)$ has a subgroup tidy for α. Suppose that O is tidy for α. The integer

$$s_G(\alpha) := |\alpha(O)\colon \alpha(O)\cap O|,$$

which is also equal to $\min\{|\alpha(V)\colon \alpha(V)\cap V|\colon V \in \mathcal{B}(G)\}$, is called *the scale of* α.

The following lemma proves our intuition correct, that the set of subgroups tidy for α is the set of basepoints for geodesics in $\mathcal{B}(G)$ generated by α.

Lemma 1 (see Lemma 2 in [BW06]).

1. *The line generated by α based at $O \in \mathcal{B}(G)$ is a geodesic line (geodesic ray) if and only if O is tidy for α;*
2. *If O is tidy for α, then*

$$d_+(\alpha^n(O), O) = \log(s_G(\alpha^n)), \quad d_-(\alpha^n(O), O) = \log(s_G(\alpha^{-n}))$$

   and $d(\alpha^n(O), O)$ grow linearly.
3. *The element O in $\mathcal{B}(G)$ is tidy if and only if $|\alpha(O)\colon \alpha(O)\cap O| = s_G(\alpha)$ (iff $|O\colon \alpha(O)\cap O| = s_G(\alpha^{-1})$).*

We give a brief indication towards a proof of this result. The second statement follows from the definition of the scale and $s_G(\alpha^n) = s_G(\alpha)^n$ for n in $\mathbb{N}$, which also implies that the line generated by α based at a tidy subgroup for α is a geodesic. The third statement uses that a subgroup V minimizes $d_+(\alpha(V), V)$ if and only if it minimizes $d_+(\alpha^{-1}(V), V) = d_-(\alpha(V), V)$. That a basepoint for a geodesic ray generated by α is tidy for α is slightly more complicated.

The definition of tidy subgroup given above is not the original one and often not the most useful characterization of a tidy subgroup. We finish this section by introducing the original definition of a tidy subgroup V for an automorphism α in terms of a factoring of V into a product of factors V_+ and V_- such that α magnifies V_+ and shrinks V_-. The generalization of this factorization will prove useful when examining so-called flat subgroups in the following subsection.

Given an automorphism, α, and a compact, open subgroup V of G, define subgroups V_+, V_- of V and V_{++}, V_{--} of G by:

$$V_\pm = \bigcap_{n\geq 0} \alpha^{\pm n}(V) \quad \text{and} \quad V_{\pm\pm} = \bigcup_{n\geq 0} \alpha^{\pm n}(V_\pm)\,.$$

The automorphism α magnifies V_+ and shrinks V_-, that is, $\alpha(V_+) \geqslant V_+$ and $\alpha^{-1}(V_-) \geqslant V_-$. It is shown in [Wil01, Theorem 3.1] that a compact, open subgroup V satisfies condition 3 in Lemma 1, and hence is tidy if and only if

(T1) $V = V_+ V_-\ (= V_- V_+)$

(T2) V_{--} (and V_{++}) are closed.

In terms of these subgroups the scale $s_G(\alpha)$ is given by $|\alpha(V_+)\colon V_+|$.

3.2. Eigenfactors of tidy subgroups, flat subgroups and their orbits

This subsection collects the necessary requisites on flat subgroups. A thorough account can be found in [Wil04].

An element of $\mathcal{B}(G)$ is called *tidy for a subset* $\mathcal{M}$ of $\mathsf{Aut}(G)$ if and only if it is tidy for every element of $\mathcal{M}$. A subgroup $\mathcal{H}$ of $\mathsf{Aut}(G)$ is called *flat* if and only if there is an element of $\mathcal{B}(G)$ which is tidy for $\mathcal{H}$. We will call a subgroup H of G flat if and only if the group of inner automorphisms induced by H is flat.

Let $\mathcal{H}$ be a group of automorphisms of G. By Lemma 1, $\mathcal{H}$ is flat if and only if there is a point O in the space of compact, open subgroups $\mathcal{B}(G)$, such that any line through O generated by an element of $\mathcal{H}$ is a geodesic.

In this subsection we will improve on this description of the orbit $\mathcal{H}.O$ of such a common tidy subgroup O under the flat group $\mathcal{H}$. It will be shown in Proposition 6, that $\mathcal{H}.O$ with the restriction of the metric on $\mathcal{B}(G)$ is isometric to a free abelian group A with the metric on A induced by some norm on the vector space $\mathbb{R} \otimes A$, which can be given explicitly and does not depend on the choice of O.

Since any two orbits of a group of isometries, like $\mathcal{H}$ acting on $\mathcal{B}(G)$, are quasi-isometric, we obtain that all orbits of flat subgroups in $\mathcal{B}(G)$ are quasi-flats. (It is not known, whether every group of automorphisms $\mathcal{Q}$ of G, whose orbits in $\mathcal{B}(G)$ are quasi-flats, is necessarily flat, though this has been verified in [BW06, Proposition 5] in the case where the flat rank of $\mathcal{Q}$, to be defined in Definition 7, is equal to 0.)

These results on flat groups of automorphisms will be deduced from a refinement of the factorization **(T1)** of a tidy subgroup V into its positive and negative parts V_+ and V_-. The refined factors in such a factorization of a common tidy subgroup for a flat group of automorphisms are called eigenfactors due to a connection with eigenspaces and eigenvalues. The measurement of the amount of expansion or contraction of an eigenfactor caused by the application of an automorphism defines a refinement of the scale function, associated to that eigenfactor.

Definition 2. *Let $\mathcal{H}$ be a flat subgroup of $\mathsf{Aut}(G)$ and O be tidy for $\mathcal{H}$.*

1. *A subgroup, V, of O satisfying:*
   a) *for every α in $\mathcal{H}$ either $\alpha(V) \geqslant V$ or $\alpha(V) \leqslant V$*
   b) *$V = \bigcap\{\alpha(O)\colon \alpha \in \mathcal{H} \text{ and } \alpha(V) \geqslant V\}$*
   will be called a O-eigenfactor for $\mathcal{H}$.
2. *The O-eigenfactor $\bigcap\{\alpha(O)\colon \alpha \in \mathcal{H}\}$ will be denoted by O_0. It is a subgroup of every other O-eigenfactor and is invariant under all elements of $\mathcal{H}$.*

6 U. Baumgartner

3. *The function* $s_V \colon \mathcal{H} \to \mathbb{N} \setminus \{0\}$ *defined by* $s_V(\alpha) := |\alpha(V) : \alpha(V) \cap V|$ *is called the* scale relative to the O-eigenfactor V.

The basic properties of the eigenfactors and associated relative scales are stated in the following lemma.

Lemma 3. *Let* $\mathcal{H}$ *be a flat subgroup of* $\mathsf{Aut}(G)$ *and* O *be tidy for* $\mathcal{H}$. *Let* V *be a* O-*eigenfactor for* $\mathcal{H}$ *different from* O_0.
Put $t_V := \min\{|\alpha(V) : V| \colon \alpha(V) > V\}$ *and choose* α_V *such that* $t_V = |\alpha_V(V) : V|$. *Then:*

1. *There is a surjective homomorphism* $\rho_V \colon \mathcal{H} \to \mathbb{Z}$, *such that* $\alpha(V) = \alpha_V^{\rho_V(\alpha)}(V)$ *for every* α *in* $\mathcal{H}$.
2. $s_V(\alpha) = t_V^{\rho_V(\alpha)}$ *for every* α *in* $\mathcal{H}$.
3. $V = \bigcap\{\alpha(O) \colon \alpha \in \mathcal{H} \text{ and } \rho_V(\alpha) \geq 0\}$.

The relative scale functions on $\mathcal{H}$ are independent of the tidy subgroup used to define them, and the function ρ_V is independent of the tidy subgroup O in which V is a O-eigenfactor. Hence we are able to derive the following result; see [Wil04, Theorems 6.12 and 6.14].

Theorem 4. *Let* $\mathcal{H}$ *be a flat subgroup of* $\mathsf{Aut}(G)$. *There is a set of surjective homomorphisms* $\Phi := \Phi(\mathcal{H}, G) \subseteq \mathsf{Hom}(\mathcal{H}, \mathbb{Z})$ *and positive integers* t_ρ *for* ρ *in* $\Phi(\mathcal{H}, G)$ *such that for each* α *in* $\mathcal{H}$ *we have*

1. $\rho(\alpha) \neq 0$ *for only finitely many* ρ *in* $\Phi(\mathcal{H}, G)$;
2. $s_G(\alpha) = \prod_{\rho \in \Phi, \, \rho(\alpha) \geq 0} t_\rho^{\rho(\alpha)}$.

The map
$$(\rho)_{\rho \in \Phi(\mathcal{H}, G)} \colon \mathcal{H} \to \mathbb{Z}^{(\Phi(\mathcal{H}, G))}$$
is a homomorphism; factoring by its kernel $\mathcal{H}(1)$, and applying Theorem 4 together with part (3) of Lemma 1, we obtain the following theorem.

Theorem 5. *Suppose that* $\mathcal{H}$ *is a flat group of automorphisms of* G *and let* O *be tidy for* $\mathcal{H}$. *The set*
$$\mathcal{H}(1) := \bigcap_{\rho \in \Phi(\mathcal{H}, G)} \ker(\rho) = \{\alpha \in \mathcal{H} \colon s_G(\alpha) = 1 = s_G(\alpha^{-1})\} = \mathcal{H}_{\{O\}}$$
is a normal subgroup of $\mathcal{H}$ *and* $\mathcal{H}/\mathcal{H}(1)$ *is free abelian.*

Let $\mathcal{H}$ be a flat group of automorphisms and let O be tidy for $\mathcal{H}$. By Theorem 5, the orbit $\mathcal{H}.O$ is in bijection with the free abelian group $\mathcal{H}/\mathcal{H}_{\{O\}} = \mathcal{H}/\mathcal{H}(1)$. It remains to calculate the metric on the orbit $\mathcal{H}.O$ to establish our claim that orbits of flat groups of automorphisms are quasi-flats.

Since the metric on $\mathcal{B}(G)$ is invariant under automorphisms of G, the metric on $\mathcal{H}.O$ we seek is determined by the distances of all points in $\mathcal{H}.O$ from O. The formula for the scale function restricted to a flat subgroup given in Theorem 4 implies the explicit formula for this distance from O given in the next result.

Proposition 6. *Let $\mathcal{H}$ be a flat group of automorphisms and let O be tidy for $\mathcal{H}$. Put $\|\alpha\mathcal{H}(1))\|_{\mathcal{H}} := d(\alpha(O), O)$ for α in $\mathcal{H}$. Then*

$$\|\alpha\mathcal{H}(1)\|_{\mathcal{H}} = \sum_{\rho \in \Phi(\mathcal{H}, G)} \log(t_\rho)\, |\rho(\alpha)|.$$

Hence the function $\| \cdot \|_{\mathcal{H}}$ extends to a norm on the vector space $\mathbb{R} \otimes \mathcal{H}/\mathcal{H}(1)$ and each orbit of a flat group of automorphisms of finite flat rank in $\mathcal{B}(G)$ is quasi-isometric to $\mathbb{R} \otimes \mathcal{H}/\mathcal{H}(1)$.

If $\mathcal{H}$ is a flat group of automorphisms, then the cardinality of the $\mathbb{Z}$-rank of the quotient $\mathcal{H}/\mathcal{H}(1)$ is an important invariant of $\mathcal{H}$ which can be used to define another invariant for a group of automorphisms $\mathcal{A}$ of a totally disconnected, locally compact group G as follows.

Definition 7. *Let $\mathcal{H}$ be a flat group of automorphisms of a totally disconnected, locally compact group G . The* flat rank *of $\mathcal{H}$, denoted* flat-rk$(\mathcal{H})$, *is the $\mathbb{Z}$-rank of $\mathcal{H}/\mathcal{H}(1)$. If $\mathcal{A}$ is a group of automorphisms of G, then its flat rank is defined to be the supremum of the flat ranks of all flat subgroups of $\mathcal{A}$. The flat rank of the group G itself is the flat rank of the group of inner automorphisms of G.*

The flat rank of a flat group of automorphisms can be any cardinality as shown by the example discussed on page 32 in [Wil04].

However, we will be mainly interested in automorphism groups of finite flat rank. If $\mathcal{H}$ is a flat group of automorphisms of G of finite flat rank then the set $\Phi(\mathcal{H}, G)$ is finite as well thanks to the following result.

Theorem 8. *Let $\mathcal{H}$ be a flat subgroup of $\mathsf{Aut}(G)$ of finite flat rank and let O be tidy for $\mathcal{H}$. Then O has only finitely many O-eigenfactors and O is the product, in some order, of its O-eigenfactors $O_0,\ldots, O_q$.*

$$O = O_0 \cdot O_1 \cdots O_q,$$

For further information on flat groups see [Wil04].

3.3. Flat rank of automorphism groups of buildings

This subsection summarizes the papers [BRW05] and [BMW04]. These papers contribute to our understanding of how geometric properties of a metric space X influence topological properties of a totally disconnected, locally compact group G acting isometrically on X.

Both papers look at the flat rank only. As the reader might anticipate, a good correspondence between geometric properties of X and topological properties of G is only to be expected if the action of G on X is sufficiently rich. The statements made here will be rather elementary; they will however be important due to the importance of the examples to which they apply. By that rationale the main results of this section are Theorems 15, 16 and 19.

Let G be a totally disconnected, locally compact group. By Proposition 6, orbits of flat groups of automorphisms of G in the space of compact, open subgroups are quasi-flats. If G acts on a metric space X and the G-action is sufficiently rich, then we expect X to be very similar to $\mathcal{B}(G)$ and we expect that the orbits of a flat subgroup of G in X are quasi-flats as well. The next proposition shows that this conclusion can be drawn if the map $X \to \mathcal{B}(G)$ which assigns to a point its stabilizer is a quasi-isometric embedding.

Proposition 9. *Let G be a totally disconnected, locally compact group. Suppose that G acts on a metric space X such that G-stabilizers of points in X are compact, open subgroups of G. Assume that the map $X \to \mathcal{B}(G)$ which assigns to each point its stabilizer is a quasi-isometric embedding. Let H be a flat subgroup of G of finite flat rank n. Then, for any point x in X the inclusion of the orbit of x under H in X defines an n-quasi-flat in X.*

To prove this result, one argues as follows. The orbit of x under H is quasi-isometric to the orbit of its stabilizer G_x under H acting by conjugation. The latter orbit is quasi-isometric to the H-orbit of a tidy subgroup, say O, for H. But $H.O$ is isometric to the subset $\mathbb{Z}^n$ of $\mathbb{R}^n$ with the norm $\| \cdot \|_H$, and $\mathbb{Z}^n$ is quasi-dense. Since the identity map between $\mathbb{R}^n$ equipped with $\| \cdot \|_H$ and $\mathbb{R}^n$ with the Euclidean norm is bi-Lipschitz the result follows.

Assume now that X is also a complete $\mathsf{CAT}(0)$-space. Then Proposition 9 applied to the flat subgroup generated by an element g in G and the fixed point theorem for isometric actions on such spaces ([BH99, II.2, Corollary 2.8(1)]) can be used to reformulate existence of fixed points for g on X in terms of values assumed by the scale function.

Corollary 10. *Let G be a totally disconnected, locally compact group. Suppose that G acts isometrically on a complete $\mathsf{CAT}(0)$–space X with compact, open point stabilizers. Assume that the map $X \to \mathcal{B}(G)$, which assigns a point its stabilizer, is a quasi-isometric embedding. Then an element g of G has a fixed point in X if, and only if, $s_G(g) = 1 = s_G(g^{-1})$.*

The correspondence between flat subgroups and quasi-flats described in Proposition 9 is useful if the maximal dimension of quasi-flats in X is finite. An important class of metric spaces which have this property, are complete, cocompact $\mathsf{CAT}(0)$-spaces.

We adopt the following definition for the rank of a complete $\mathsf{CAT}(0)$-space. For alternative definitions see [Gro93, pp. 127–133].

Definition 11. *The* rank *of a complete $\mathsf{CAT}(0)$-space X, denoted $\mathrm{rk}(X)$, is the maximal dimension of an isometrically embedded Euclidean space in X.*

Recall that a metric space X is called *cocompact* if and only if the isometry group of X acts cocompactly on X ([BH99, p. 202]). With the terminology introduced above, we can state the following theorem, which follows from [Kle99, Theorem C], Proposition 9 and Corollary 10.

Theorem 12. *Let G be a totally disconnected, locally compact group. Suppose that G acts isometrically on a complete, locally compact, cocompact $\mathsf{CAT}(0)$-space X*

with compact, open point stabilizers such that the map $X \to \mathcal{B}(G)$ which assigns a point its stabilizer is a quasi-isometric embedding. Then flat-rk$(G) \le$ rk(X); in particular the flat rank of G is finite. We have flat-rk$(G) = 0$ if and only if every element of G fixes a point in X.

The rank of a Gromov-hyperbolic $\mathsf{CAT}(0)$-space is 1. This leads to the following special case of Theorem 12.

Corollary 13. *Let G, X and $X \to \mathcal{B}(G)$ be as above. Assume further that X is Gromov-hyperbolic. Then flat-rk$(G) = 1$, unless every element of G has a fixed point in X, in which case flat-rk$(G) = 0$.*

As a consequence of Theorems 15 and 16 stated below, Corollary 13 applies to the case where G is a topological Kac-Moody group (a concept recalled in the paragraph preceding Theorem 15) with Gromov-hyperbolic Weyl group.

The following, similar result for flat groups of automorphisms is shown in [BMW04]. To prove it, one argues that Proposition 6 implies that a hyperbolic orbit for $\mathcal{A}$ in $\mathcal{B}(G)$ precludes $\mathcal{A}$ having rank bigger than 1.

Theorem 14. *Let $\mathcal{A}$ be a group of automorphisms of the totally disconnected locally compact group G. Suppose that $\mathcal{A}$ has a hyperbolic orbit in the space of compact open subgroups of G. Then the flat rank of $\mathcal{A}$ is at most 1.*

Theorem 14 applies for example to the group $\mathcal{A}$ of type preserving automorphisms of a Bourdon-building, X say. We refer the reader to [Bou97] for the definition of Bourdon-buildings. It can be shown that the orbit of the fixator of any chamber of X under $\mathcal{A}$ is hyperbolic and is not bounded unless X is thin. It follows that the flat rank of $\mathcal{A}$ is 1 if X is thick.

For the previous results of this subsection to be interesting, we need examples of metric spaces X with an action of a totally disconnected, locally compact group G on X such that the map $X \to \mathcal{B}(G)$ which assigns a point its stabilizer is a quasi-isometric embedding.

The following Theorem 15 shows that there are many such examples. The most interesting examples for Theorem 15 are given by topological Kac-Moody groups, which were introduced in [RR06]. For the convenience of the reader we recall the definition of a topological Kac-Moody group. Let Λ be the group of rational points of an abstract Kac-Moody group over a finite field k. The associated topological Kac-Moody group G is the closure of Λ in the automorphism group of the Davis-realization $|X|_0$ of the positive building of Λ. Since only spherical residues are realized in the Davis-realization and the field k is finite, $|X|_0$ is a locally finite, connected cell complex. Therefore $\mathsf{Aut}(|X|_0)$ and its subgroup G are totally disconnected, locally compact groups.

Theorem 15. *Let $(\mathcal{C}, S)$ be a locally finite building with Weyl group W. Denote by δ and by X the W-distance function and the $\mathsf{CAT}(0)$-realization of $(\mathcal{C}, S)$, respectively. Let G be a closed subgroup of the group of automorphisms of $(\mathcal{C}, S)$. Assume* *that the G-action is transitive on ordered pairs of chambers at given δ-distance. Then the following statements hold.*

1. *The map $\varphi\colon X \to \mathcal{B}(G)$ mapping a point to its stabilizer, is a quasi-isometric embedding.*
2. *For any point $x \in X$, the image of the orbit map $g \mapsto g.x$, restricted to a flat subgroup of flat rank n in G, is an n-dimensional quasi-flat of X.*
3. *We have:* $\mathrm{flat\text{-}rk}(G) \le \mathrm{rk}(X)$.
4. *If X contains an n-dimensional flat, so do all of its apartments.*

As a consequence, we obtain: $\mathrm{flat\text{-}rk}(G) \le \mathrm{rk}(|W|_0)$, *where* $\mathrm{rk}(|W|_0)$ *is the maximal dimension of isometrically embedded flats in the* $\mathsf{CAT}(0)$*-realization* $|W|_0$ *of* W.

We have already seen above that part (1) of Theorem 15 implies parts (2) and (3). To obtain a proof of part (1), one uses the transitivity condition on the action. This condition implies that if $q_s + 1$ denotes the common cardinality of s-residues, (c, c') is a pair of chambers, and $s_1 \cdots s_l$ is the type of some minimal gallery connecting c to c', then $|G_c : G_c \cap G_{c'}| = \prod_{j=1}^{l} q_{s_j}$.

Thus we see that distance between stabilizers of chambers is just gallery-distance between corresponding chambers, 'weighted' by the cardinalities of the panels crossed. One then compares the Davis-realization and the metric space of chambers with the gallery-distance to obtain part (1).

To obtain (4), the equality of ranks of flats in the building and in any apartment, it suffices to show that for each n in $\mathbb{N}$ there is an apartment A_n of $(\mathcal{C}, S)$ which contains an isometric copy of the ball of radius n around $\mathbf{0}$ in $\mathbb{R}^d$ (Lemma 9.34 in chapter II of [BH99]). Starting with a flat F inside the building, and a ball of radius n inside it, one shows that there are two chambers c_n and c'_n such that the minimal galleries connecting c_n and c'_n cover B_n. The choice of the chambers c_n and c'_n is made using the pattern of hyperplanes that is formed by the intersections of walls with F. By combinatorial convexity, any apartment A_n containing c_n and c'_n contains B_n. The final statement of Theorem 15 is an immediate consequence of the others.

While the assumptions in Theorem 15 are quite weak, it is more difficult to obtain lower bounds for the flat rank of a totally disconnected, locally compact group G, because to do so, we need to construct a flat subgroup of G of the appropriate rank. Nevertheless, under conditions somewhat more restrictive than those of Theorem 15, we can obtain a lower bound for the flat rank; see Theorem 16 below. The class of groups satisfying the assumptions of Theorem 16 is contained in the class of groups satisfying the assumptions of Theorem 15 and contains all topological Kac-Moody groups.

Theorem 16. *Let G be a group with a locally finite twin root datum of associated Weyl group W. We denote by $\overline{G}$ the geometric completion of G, i.e., the closure of the G-action in the full automorphism group of the positive building of G. Let A be an abelian subgroup of W. Then A lifts to a flat subgroup $\widetilde{A}$ of $\overline{G}$ such that* $\mathrm{flat\text{-}rk}(\widetilde{A}) = \mathrm{rank}_{\mathbb{Z}}(A)$.

As a consequence, we obtain: $\text{alg-rk}(W) \leq \text{flat-rk}(G)$, *where* $\text{alg-rk}(W)$ *is the maximal $\mathbb{Z}$-rank of abelian subgroups of W.*

One takes the inverse image of A in the subgroup N of G which is defined by the root datum as the group $\widetilde{A}$ above. Hence $\widetilde{A}$ is a finite extension of a finitely generated abelian group. Generalizing one of the main results of [Wil04], one shows that such a group is flat. Using Corollary 10, one sees that $A(1)$ is the set of elements of A admitting a fixed point. One concludes that $A(1)$ is the torsion subgroup of A, which implies $\text{flat-rk}(\widetilde{A}) = \text{rank}_{\mathbb{Z}}(A)$. Theorem 16 follows.

The question arises, whether the upper and lower bounds derived in Theorems 15 and 16 are equal. In [CH] Pierre-Emmanuel Caprace and Frédéric Haglund recently answered this question in the affirmative.

Theorem 17 (see Corollary C in [CH]). *Let (W, S) be a Coxeter system with S finite. Then*

$$\text{alg-rk}(W) = \text{rk}(|W|_0).$$

As pointed out to us by Caprace, a result in Krammer's PhD thesis, [Kra94, Theorem 6.8.3], provides a combinatorial control on the abelian subgroups of a Coxeter group. As a consequence, the algebraic rank of a Coxeter group of finite rank can be computed from its Coxeter diagram. Combining this with Theorems 15 and 16 gives a method to compute the flat rank of any topological Kac-Moody group.

For groups of rational points of semisimple algebraic groups over local fields a result similar to Theorem 16 can be proved. That result implies the following Corollary.

Corollary 18. *Let $\mathbf{G}$ be an algebraic semi-simple group over a local field k, with affine Weyl group W. Then* $\text{flat-rk}(\mathbf{G}(k)) = k\text{-rk}(\mathbf{G}) = \text{alg-rk}(W)$.

By this Corollary, the sequence $\left(\text{PSL}_{d+1}(\mathbb{Q}_p)\right)_{d \geq 1}$ consists of topologically simple, compactly generated, locally compact, totally disconnected groups of flat rank $d \geq 1$. Theorem 16 enables us to exhibit a sequence of non-linear groups with the same properties.

Theorem 19. *For every integer $n \geq 1$ there is a non-linear, topologically simple, compactly generated, locally compact, totally disconnected group of flat rank n.*

These examples are provided by topological Kac-Moody groups.

4. Asymptotic information; space of directions

In this section we summarize the results of the paper [BW06], which examines the asymptotic behavior under iteration of automorphisms of a totally disconnected, locally compact group G.

That approach defines a framework to discuss relative positions of flat groups of automorphisms of G, and we will learn how flat groups of automorphisms of G

look 'at infinity'. A meaningful direct description of the relative position of flat subgroups inside $\mathcal{B}(G)$ would be desirable, but does not yet exist.

The approach taken is similar to standard techniques employed in the definition of the Tits-boundary of a $\mathsf{CAT}(0)$-space. We construct a complete metric G-space, the space of directions of G, from asymptote-classes of automorphisms, with the asymptote relation defined via a straightforward asymptote relation on the rays generated by the automorphisms in question. This space of directions reproduces well-known structures at infinity in familiar cases, as will be seen in Subsection 4.4.

It would be desirable to define a coarser topology on the set underlying the space of directions in analogy to the cone topology on the geodesic boundary of a $\mathsf{CAT}(0)$-space but at the time of writing it is not clear how such a topology might be defined.

4.1. Asymptotic rays and asymptotic automorphisms

We start by defining the asymptote relation on sequences of points in $\mathcal{B}(G)$. In the sequel, we will be interested only in sequences which are rays generated by some automorphism of G.

Definition 20. *Two sequences $\left(V_n\right)$ and $\left(W_n\right)$ in the space $\mathcal{B}(G)$ are said to be asymptotic, written $\left(V_n\right) \asymp \left(W_n\right)$, if there is a constant M such that $d(V_n, W_n) \leq M$ for all $n \in \mathbb{N}$.*

It is clear that $\asymp$ defines an equivalence relation on sequences in $\mathcal{B}(G)$. The rays generated by an automorphism α based at any two basepoints are asymptotic. The group of automorphisms of G acts on sequences of elements in $\mathcal{B}(G)$ with images of asymptotic sequences again asymptotic. The automorphism β maps the ray based at O generated by α to the ray based at $\beta(O)$ generated by $\beta\alpha\beta^{-1}$.

We transfer the concept of asymptotic rays to automorphisms of the totally disconnected, locally compact group G as follows.

Definition 21. *Two automorphisms, α and β of G are called* asymptotic, *written $\alpha \asymp \beta$, if there exist positive integers k_α, k_β and O_α, $O_\beta \in \mathcal{B}(G)$ such that the ray based at O_α generated by α^{k_α} is asymptotic to the ray based at O_β generated by β^{k_β}.*

For two automorphisms α and β, being asymptotic is independent of the choice of basepoints O_α and O_β, from which it follows that $\asymp$ is an equivalence relation on the set of automorphisms of G. The action of the automorphism group on itself by conjugation preserves this equivalence relation.

The positive powers of a given automorphism are pairwise asymptotic. Further, it can be shown, using the triangle inequality for d_+ (formula (1)) and Theorem 7.7 from [Möl02], that if $\alpha \asymp \beta$, as witnessed by the rays $\left(\alpha^{nk_\alpha}(O_\alpha)\right)_n$ and $\left(\beta^{nk_\beta}(O_\beta)\right)_n$ being asymptotic, then

$$s_G(\alpha^{k_\alpha}) = s_G(\beta^{k_\beta}) \quad \text{and} \quad s_G(\alpha^{-k_\alpha}) = s_G(\beta^{-k_\beta}) \tag{3}$$

$$\text{and thus} \quad k_\alpha \, d(\alpha(O_\alpha), O_\alpha) = k_\beta \, d(\beta(O_\beta), O_\beta)$$

for O_α tidy for α and O_β tidy for β. In particular, if α is asymptotic to β and α has a fixed point, then β has a fixed point and if $s_G(\alpha) \neq 1$ then $s_G(\beta) \neq 1$ also.

Definition 22. *An automorphism α of G will be said to* move towards infinity *if for any pair of elements $V \subseteq W$ in $\mathcal{B}(G)$ there is an integer $n \in \mathbb{N}$ such that $\alpha^n(V) \not\subseteq W$.*

It can be seen that an automorphism α moves towards infinity if and only if $s_G(\alpha) \neq 1$. Hence the notion of moving to infinity is compatible with the relation $\asymp$, and we can make the following definition.

Definition 23. *Let A be a group of automorphisms of G. The quotient set of the set of automorphisms in A which move to infinity by the relation $\asymp$ will be called the* set of directions of A *and will be denoted ∂A.*

We will write ∂G for the set of directions of the group of inner automorphisms of G and call it the set of directions of G.

If α is an automorphism moving to infinity, then $\partial \alpha$ will denote its $\asymp$-class and will be called the direction of $\boldsymbol{\alpha}$. *If for two automorphisms α and β, which move to infinity, we have $\partial \alpha = \partial \beta$, then we say that $\boldsymbol{\alpha}$ and $\boldsymbol{\beta}$ have the same direction.*

4.2. Pseudo-metric on the set of directions

In this subsection we define a pseudo-metric δ on the set of directions in analogy to the Tits-metric on geodesic rays in a $\mathsf{CAT}(0)$-space. This pseudo-metric will be derived from a pseudo-metric, also called δ, on the set of automorphisms moving towards infinity, which we introduce first.

The discrepancy $\delta(\alpha, \beta)$ between two automorphisms α and β moving towards infinity will be measured by watching the limiting behavior of the distances between images of basepoints V and W under powers α^n and β^m which are 'equally far out' as n and m go to infinity. As for the distance d between compact open subgroups, the function δ will be obtained from two functions δ_+ and δ_- with δ_+ measuring the position of the image of W under powers of β with respect to $\alpha^n(V)$ as n goes to infinity while δ_- will do the same with the roles of α and β reversed.

Let α and β be two automorphisms of G moving towards infinity, that is, such that $s_G(\alpha) \neq 1$ and $s_G(\beta) \neq 1$. Choose $V, W \in \mathcal{B}(G)$ and define

$$\delta_{+n}^{V,W}(\alpha, \beta) := \min \left\{ \frac{d_+(\alpha^n(V), \beta^k(W))}{n \log(s_G(\alpha))} : k \in \mathbb{N}, \ s_G(\beta^k) \leq s_G(\alpha^n) \right\} \qquad (4)$$

This allows us to define the quantity

$$\delta_+(\alpha, \beta) := \limsup_{n \to \infty} \delta_{+n}^{V,W}(\alpha, \beta) \leq 1 \,.$$

It does not depend on the pair (V, W) chosen; see [BW06, p. 406].

The following result establishes the triangle inequality for δ_+, among other properties; see Lemma 15 in [BW06].

Lemma 24. *Let α, β and γ be automorphisms of G moving to infinity. Then we have*

1. $0 \le \delta_+(\alpha, \beta) \le 1$;
2. $\delta_+(\alpha, \alpha) = 0$ *and, provided α^{-1} moves to infinity as well* $\delta_+(\alpha, \alpha^{-1}) = 1$;
3. $\delta_+(\alpha, \gamma) \le \delta_+(\alpha, \beta) + \delta_+(\beta, \gamma)$.

Corollary 25. *Put $\delta_-(\alpha, \beta) := \delta_+(\beta, \alpha)$. The function $\delta(\alpha, \beta) := \delta_+(\alpha, \beta) + \delta_-(\alpha, \beta)$ is a pseudo-metric on the set of automorphisms moving towards infinity.*

The function δ is compatible with the relation $\asymp$ and therefore induces a pseudo-metric on the set of directions, also denoted δ, as the next lemma shows.

Lemma 26. *Let α and β be two automorphisms moving towards infinity. Suppose that $\alpha \asymp \beta$. Then $\delta(\alpha, \beta) = 0$.*

To see that Lemma 26 holds, choose tidy base points O_α and O_β for the rays generated by α and β respectively. By definition of the asymptote relation, $\left(\alpha^{nk_\alpha}(O_\alpha)\right)_n$ and $\left(\beta^{nk_\beta}(O_\beta)\right)_n$ are asymptotic for some k_α, k_β and equation (3) tells us that $s_G(\alpha^{k_\alpha}) = s_G(\beta^{k_\beta})$.

Choosing the index k as the largest multiple of k_β with the property $s_G(\beta^k) \le s_G(\alpha^n)$ in the defining equation (4) for $\delta_{+n}^{O_\alpha, O_\beta}(\alpha, \beta)$ and using the existence of a bound for $d(\alpha^{nk_\alpha}(O_\alpha), \beta^{nk_\beta}(O_\beta))$, one sees that $\delta_{+n}^{O_\alpha, O_\beta}(\alpha, \beta)$ goes to 0 as n goes to infinity, hence $\delta_+(\alpha, \beta) = 0$. Reversing the roles of α and β gives $\delta_-(\alpha, \beta) = 0$ also.

In general δ does not induce a metric on the set of directions of a group of automorphisms. Still, there is a coarser notion of $o(n)$-asymptotic rays giving rise to the notion of $o(n)$-asymptotic automorphisms such that two automorphisms moving to infinity have zero pseudo-distance if and only if they are $o(n)$-asymptotic. We will not introduce this notion here.

Definition 27. *Let A be a group of automorphisms of a totally disconnected, locally compact group. Then the completion, $\overline{\partial A}$, of the metric space $\partial A / \delta^{-1}(0)$ is the space of directions of A.*

The next result follows immediately from the definitions.

Proposition 28. *The action of A on ∂A defined by*

$$\beta.\partial\alpha = \partial(\beta\alpha\beta^{-1})$$

induces an action by isometries on the space of directions $\overline{\partial A}$.

Lemma 24 and the definition of δ imply that the diameter of the space of directions of a group of automorphisms is at most 2.

4.3. Directions in flat groups

Let $\mathcal{H}$ be a flat group of automorphisms of the totally disconnected, locally compact group G and let O be tidy for $\mathcal{H}$. To determine the directions generated by automorphisms in $\mathcal{H}$, and pseudo-distances between them, we may base rays at O.

We have learned in Subsection 3.2 that the orbit $\mathcal{H}.O$ is, via the identification with $\mathcal{H}/\mathcal{H}(1)$, in natural bijection with a free abelian group, and we therefore expect directions generated by automorphisms in $\mathcal{H}$ to correspond to rays in the free abelian group $\mathcal{H}/\mathcal{H}(1)$. This is indeed so, see Proposition 29. It need not be the case that every automorphism in $\mathcal{H} \setminus \mathcal{H}(1)$ moves to infinity however, because $s_G(\alpha) \neq 1$ does not imply $s_G(\alpha^{-1}) \neq 1$, as the example of α being multiplication by p^{-1} on $\mathbb{Q}_p$ with p prime shows.

It follows from Theorem 4 that the set of automorphisms in $\mathcal{H}$ not moving to infinity can be described as $\{\alpha \subset \mathcal{H} : \rho(\alpha) \leq 0 \text{ for all } \rho \subset \Phi(\mathcal{H}, G)\}$. Since the elements of $\Phi(\mathcal{H}, G)$ define $\mathbb{Z}$-linear forms on $\mathcal{H}/\mathcal{H}(1)$, it follows that the set of automorphisms in $\mathcal{H}$ not moving to infinity are those inside the polyhedral cone $\bigcap_{\rho \in \Phi(\mathcal{H},G)} \rho^{-1}(] - \infty, 0])$. This fact can be made more precise and can be used to compute the set and space of directions of a flat group of automorphisms $\mathcal{H}$. The result is as follows.

Proposition 29. *The bijection $\mathcal{H}.O \to \mathbb{Z}^{(\mathrm{flat\text{-}rk}(\mathcal{H}))}$ induces a bijection between $\partial\mathcal{H}$ and the set of directions of rays in $\mathbb{Z}^{(\mathrm{flat\text{-}rk}(\mathcal{H}))}$ starting at the origin and avoiding the polyhedral cone defined by $\bigcap_{\rho \in \Phi(\mathcal{H},G)} \rho^{-1}(] - \infty, 0])$.*

The map δ restricted to directions defined by automorphisms in $\mathcal{H}$ is a metric. The space of directions of $\mathcal{H}$ is homeomorphic to the intersection of the unit sphere in $\mathbb{R} \otimes \mathbb{Z}^{(\mathrm{flat\text{-}rk}(\mathcal{H}))} = \mathbb{R}^{(\mathrm{flat\text{-}rk}(\mathcal{H}))}$ with the complement of the (closed!) polyhedral cone defined by $\bigcap_{\rho \in \Phi(\mathcal{H},G)} (\mathrm{id}_{\mathbb{R}} \otimes \rho)^{-1}(] - \infty, 0])$.

The proof involves an explicit formula for the function δ_+ on $\mathcal{H} \times \mathcal{H}$ and a comparison of the metric δ with the metric on the unit sphere, which relies on an estimate of $\delta(x, y)$ that applies when x and y are close to the cone defined by $\bigcap_{\rho \in \Phi(\mathcal{H},G)} (\mathrm{id}_{\mathbb{R}} \otimes \rho)^{-1}(] - \infty, 0])$. For details see Subsection 4.2 of [BW06].

Whenever there is an automorphism α in $\mathcal{H}$ such that both α and α^{-1} move to infinity, the cone $\bigcap_{\rho \in \Phi(\mathcal{H},G)} \rho^{-1}(] - \infty, 0])$ will be empty and hence the space of directions of $\mathcal{H}$ a sphere. If $\mathcal{H}$ is an unimodular group of automorphisms, then any automorphism α in $\mathcal{H} \setminus \mathcal{H}(1)$ has this property, since for any automorphism α the number $s_G(\alpha)/s_G(\alpha^{-1})$ equals the value of the modular function of G at α. It follows that the space of directions of a unimodular flat group of automorphisms is homeomorphic to a sphere (empty if $\mathcal{H} = \mathcal{H}(1)$).

The group of non-zero p-adic numbers acting on the p-adics by multiplication is an example of a flat group of automorphisms whose space of directions is not a whole sphere; indeed, in this case the space of directions consist of one point only.

4.4. Examples of spaces of directions

Keeping with the spirit of this survey, we only discuss examples of geometric origin in this subsection. These examples may generate the impression that it is fairly easy to guess the space of directions of a totally disconnected, locally compact group. However, we want to caution the reader that spaces of directions need not be finite dimensional, the pseudo-metric δ need not be a metric on the set of directions and it is possible to obtain the set of Borel subsets of a standard measure space with sets identified if they are equal almost everywhere and metric $\delta(B_1, B_2)$ equal to the measure of the symmetric difference of B_1 and B_2 as space of directions of some group; see [BW06], Subsection 5.2.

The easiest interesting examples of spaces of directions of totally disconnected, locally compact groups are automorphism groups of locally finite trees. The following proposition describes this case. In the special case where G is the full automorphism group of a homogenous tree X, which is not a line, Proposition 30 says that ∂G is the set ends of X with the discrete topology.

Proposition 30. *Let X be a locally finite tree and let G be a closed subgroup of* $\mathrm{Aut}(X)$. *Then the following holds.*

1. *The map which assigns to each element of non-trivial scale its attracting end defines a bijection between ∂G and the set of those ends ϵ of X, which have the property that there are elements in G with common attracting end ϵ and different repelling ends.*
2. *Automorphisms of X with distinct directions have pseudo-distance 2.*

To obtain Proposition 30, one argues as follows. First, if h and h' are two hyperbolic elements (on non-trivial scale) with the same attracting end, they are seen to have the same direction, by applying integer powers of h and h' to the stabilizer of a vertex which lies on the intersection of the axes of h and h'.

It may be seen that the set of ends which the first claim puts in bijection with the set of directions of G, are exactly those whose stabilizer contains elements, necessarily hyperbolic, with non-trivial scale. More precisely, the scale of a hyperbolic automorphism h may be seen to be equal to $\prod q_i$, where the numbers $q_i + 1$ are the branching indices, with respect to the tree consisting of the axes of all hyperbolic elements in G fixing the attracting end of h, of a set of representatives of the vertices on the axis of h for the $\langle h \rangle$-action.

Suppose then that h and h' are two elements of non-trivial scale in G with different attracting ends ϵ and ϵ' respectively. If two vertices v and w on the line $]\epsilon', \epsilon[$ are chosen sufficiently far apart, then, using formula (2) on page 3, we may see that $d(G_{h^n.v}, G_{h'^n.w}) \geq n \log(s_G(h) \cdot s_G(h'))$. Both claims follow from this inequality.

Another class of interesting totally disconnected, locally compact groups are the groups of rational points of semisimple (or, more generally, reductive) algebraic groups over local fields. Proposition 31 below describes the space of directions in that case; its statement simplifies considerably if $\mathbf{G}$ is connected and semisimple (in this case $[\mathbf{G}^0, \mathbf{G}^0]$ equals $\mathbf{G}$, and the conjugation action equals the natural action by left translation).

Proposition 31. *Let k be a local field and let $\mathbf{G}$ be a reductive k-group. Consider the action π of $\mathbf{G}(k)$ on the Bruhat-Tits building $\Delta([\mathbf{G}^0, \mathbf{G}^0], k)$ of $[\mathbf{G}^0, \mathbf{G}^0]$ over k induced by the conjugation action of $\mathbf{G}(k)$ on $[\mathbf{G}^0, \mathbf{G}^0](k)$.*

1. *An element g of $\mathbf{G}(k)$ moves to infinity if and only if $\pi(g)$ is a hyperbolic automorphism of $\Delta([\mathbf{G}^0, \mathbf{G}^0], k)$.*

2. *Two automorphisms g and h moving to infinity have the same direction (meaning conjugation by g is asymptotic to conjugation by h) if and only if $\pi(g)$ and $\pi(h)$ define the same limit point in the building at infinity, $\Delta([\mathbf{G}^0, \mathbf{G}^0], k)_\infty$. This is the case if and only if any pair of axes for $\pi(g)$ and $\pi(h)$ are asymptotic.*

3. *The set of limit points of hyperbolic automorphisms induced by elements of $\mathbf{G}(k)$ equals the set of endpoints of rays in $\Delta([\mathbf{G}^0, \mathbf{G}^0], k)$ which have rational direction vectors with respect to the translation lattice of some affine apartment. In particular, the set of directions of $\mathbf{G}(k)$ is dense in the building at infinity with respect to the Tits metric.*

4. *The function δ is a metric on the set of directions of $\mathbf{G}(k)$ and the space of directions of $\mathbf{G}(k)$ is homeomorphic to $\Delta([\mathbf{G}^0, \mathbf{G}^0], k)_\infty$, with the Tits metric.*

This result is obtained as follows. First, it suffices to prove the result in the special case where $\mathbf{G}$ is Zariski-connected, semisimple and adjoint. Assuming that, $G := \mathbf{G}(k)$ embeds as a closed subgroup of its building $X := \Delta(\mathbf{G}, k)$ and its action thereon is strongly transitive.

Therefore, the action of G on X satisfies the conditions listed in Theorem 15. We conclude that the map $X \to \mathcal{B}(G)$ which maps a point to its stabilizer is a quasi-isometric embedding as stated in part (1) of that result.

The space X admits only semisimple isometries; hence part 1 of Proposition 31 follows from Corollary 10.

Using that the map $X \to \mathcal{B}(G)$ which maps a point to its stabilizer is a quasi-isometric embedding, part 2 of Proposition 31 follows from the definitions.

To obtain part 3 of Proposition 31, we need to determine the endpoints of axes of elements in G acting by hyperbolic isometries. This is simple for elements belonging to some k-split torus in $\mathbf{G}$, $\mathbf{S}$ say. The group $\mathbf{S}(k)$ leaves the corresponding affine apartment invariant and acts there as a subgroup of finite index in the translation lattice. The endpoints of axes of elements belonging to the group of rational points of some k-split torus account for all points at infinity mentioned in part 3 of Proposition 31. The proof of that part is obtained by using the Jordan-Chevalley decomposition to show that every group element moving to infinity has the same direction as some element of that special form.

Finally, the quasi-distance between directions of inner automorphisms, g and h say, which we may assume to lie in a common k-split torus of $\mathbf{G}$, may be compared to the angle spanned by axes for g and h emanating from a common point, thus proving part 4 of Proposition 31.

5. Parabolics and contraction groups

In the paper [BW04] asymptotic information of a different kind has been studied. For an automorphism α of G define a subgroup of G by

$$P_\alpha := \{x \in G \colon \{\alpha^n(x) \colon n \in \mathbb{N}\} \text{ is bounded}\}$$

and a normal subgroup of P_α by

$$U_\alpha := \{x \in G \colon \alpha^n(x) \xrightarrow[n \to \infty]{} e\}\,.$$

We call P_α and U_α respectively the *parabolic* subgroup and the *contraction* group associated to α. If the automorphism α is inner and is conjugation by g, we relax notation and write P_g and U_g. If V is tidy for α then $P_\alpha \geqslant V_{--} \geqslant U_\alpha$.

If $\mathbf{G}$ is an algebraic semisimple group, and k is a local field, P_g is the group of rational points of a k-parabolic subgroup and U_g is the group of rational points of its unipotent radical. Furthermore all k-parabolic subgroups of $\mathbf{G}$ can be represented in that way. For these groups, parabolic subgroups are the stabilizers of points in the spherical building at infinity, the structure at infinity of the symmetric space of $\mathbf{G}(k)$.

We have the following general related result, shown in [BW06].

Proposition 32. *Let G be a totally disconnected, locally compact group and let α be an automorphism of G. Then*

1. *$P_{\alpha^{-1}}$ stabilizes the $\asymp$-class of α;*
2. *Any element of $U_{\alpha^{-1}}$ fixes all but finitely many elements of the sequence $(\alpha^n(O))_{n \in \mathbb{N}}$ for any compact open subgroup O.*

The value of the scale function of a group G at an automorphism α can be reinterpreted as the value of the scale function of various subgroups at α. This can be used to establish the connection of the scale function to eigenvalues, mentioned earlier; see Proposition 34.

Proposition 33. *Let $N \trianglelefteq H$ be α-stable closed subgroups of P_α and let V be tidy for α in the ambient group G. Then writing q for the canonical map $H \to H/N$ and $\overline{\alpha}$, respectively $\alpha_|$, for the induced automorphisms on H/N and N, we have*

1. $s_H(\alpha^{-1}) = \Delta_H(\alpha^{-1})$
2. $s_H(\alpha^{-1}) = s_{H/N}(\overline{\alpha}^{-1}) s_N(\alpha_|^{-1})$
3. $s_G(\alpha^{-1}) = s_{P_\alpha}(\alpha^{-1}) = s_{V_{--}}(\alpha^{-1}) = s_{\overline{U}_\alpha}(\alpha^{-1})\,.$

The combination of (3) and (1) above implies that $s_G(g) = \Delta_{\overline{U}_{g^{-1}}}(g)$. This enables us to compute the scale function of the group of rational points of a semisimple algebraic group $\mathbf{G}$ over a local field of positive characteristic.

Proposition 34. *Let k be a nonarchimedean local field and let $\mathbf{G}$ be a Zariski-connected reductive k-group. For any element g of $\mathbf{G}(k)$ its scale $s_{\mathbf{G}(k)}(g)$ equals the product of the multiplicative valuations of those eigenvalues of $Ad(g)$, whose valuation is greater than 1 (counted with their proper multiplicities).*

Let α be an automorphism and V a subgroup tidy for α. Then the action of α on the group $\overline{U}_\alpha$ and its supergroup V_{--}, can be characterized geometrically.

Theorem 35. *Let G be a totally disconnected, locally compact group, α an automorphism of G of infinite order and V a subgroup tidy for α. Then there is a continuous representation $\rho\colon V_{--} \rtimes \langle\alpha\rangle \to Aut(T)$ onto a closed subgroup of the automorphism group of a homogeneous tree T of degree $s_G(\alpha^{-1}) + 1$.*

1. *The action of $V_{--} \rtimes \langle\alpha\rangle$ on T via ρ: fixes an end, $-\infty$; is transitive on $\partial T \setminus \{-\infty\}$; and the quotient graph of T by this action is a loop.*
2. *The stabilizer of each end in $\partial T \setminus \{-\infty\}$ is a conjugate of $(V_+ \cap V_-) \rtimes \langle\alpha\rangle$. The kernel of ρ is the largest compact normal α-stable subgroup of V_{--}.*
3. *The image of V_{--} under ρ is the set of elliptic elements in $\rho(V_{--} \rtimes \langle\alpha\rangle)$.*

The representation ρ restricts to give a representation of $\overline{U}_\alpha \rtimes \langle\alpha\rangle$ on T, about which more can be said.

Theorem 36. *Let G be a totally disconnected locally compact metric group and α an automorphism of G of infinite order. Let ρ be the representation of $\overline{U}_\alpha \rtimes \langle\alpha\rangle$ on the tree T as above.*

1. *The action of U_α is transitive on $\partial T \setminus \{-\infty\}$ and is simply transitive if and only if U_α is closed.*
2. *If U_α is closed, then ρ is a topological isomorphism onto its image.*

Acknowledgement

I wish to thank the referee, who saved the day by pointing out a serious mistake.

References

[BH99]   Martin Bridson and André Haeflinger. *Metric Spaces of non-positive Curvature*, volume 319 of *Grundlehren der mathematischen Wissenschaften*. Springer Verlag, 1999.

[BMW04] Udo Baumgartner, Rögnvaldur G. Möller, and George A. Willis. Groups of flat rank at most 1. preprint, 2004.

[Bou97]   M. Bourdon. Immeubles hyperboliques, dimension conforme et rigidité de Mostow. *Geom. Funct. Anal.*, 7:245–268, 1997.

[BRW05] Udo Baumgartner, Bertrand Rémy, and George A. Willis. Flat rank of automorphism groups of buildings.
Preprint, available at `http://arxiv.org/abs/math.GR/0510290`, 2005.
To appear in *Transformation Groups*.

[BW04]   Udo Baumgartner and George A. Willis. Contraction groups and scales of automorphisms of totally disconnected locally compact groups. *Israel J. Math.*, 142:221–248, 2004.

[BW06]   Udo Baumgartner and George A. Willis. The direction of an automorphism of a totally disconnected locally compact group. *Math. Z.*, 252:393–428, 2006.

[CH]   Pierre-Emmanuel Caprace and Frédéric Haglund. On geometric flats in the CAT(0) realization of Coxeter groups and Tits buildings. Available at `http://www.arxiv.org/abs/math.GR/0607741`.

[Gro93] M. Gromov. Asymptotic invariants of infinite groups. In *Geometric group theory, Vol. 2 (Sussex, 1991)*, volume 182 of *London Math. Soc. Lecture Note Ser.*, pages 1–295. Cambridge Univ. Press, Cambridge, 1993.

[HR79]   Edwin Hewitt and Kenneth A. Ross. *Abstract Harmonic Analysis; Volume I*, volume 115 of *Grundlehren der mathematischen Wissenschaften*. Springer Verlag, second edition, 1979.

[Kle99] Bruce Kleiner. The local structure of length spaces with curvature bounded above. *Math. Z.*, 231(3):409–456, 1999.

[Kra94] Daan Krammer. *The conjugacy problem for Coxeter groups*. PhD thesis, Universiteit Utrecht, Faculteit Wiskunde & Informatica, 1994.

[Möl02] Rögnvaldur G. Möller. Structure theory of totally disconnected locally compact groups via graphs and permutations. *Canadian Journal of Mathematics*, 54:795–827, 2002.

[RR06]   Bertrand Rémy and Mark Ronan. Topological groups of Kac-Moody type, right-angled twinnings and their lattices. *Comment. Math. Helv.*, 81(1):191–219, 2006.

[Wil94] George A. Willis. The structure of totally disconnected locally compact groups. *Math. Ann.*, 300:341–363, 1994.

[Wil01] George A. Willis. Further properties of the scale function on a totally disconnected locally compact group. *J. Algebra*, 237:142–164, 2001.

[Wil04] George A. Willis. Tidy subgroups for commuting automorphisms of totally disconnected locally compact groups: An analogue of simultaneous triangularisation of matrices. *New York Journal of Mathematics*, 10:1–35, 2004. available at `http://nyjm.albany.edu:8000/j/2004/Vol10.htm`.

Udo Baumgartner
School of Mathematical and Physical Sciences
The University of Newcastle
University Drive, Building V
Callaghan, NSW 2308, Australia
e-mail: `Udo.Baumgartner@newcastle.edu.au`

Geometric Group Theory
Trends in Mathematics, 21–35
© 2007 Birkhäuser Verlag Basel/Switzerland

Computational Explorations in Thompson's Group F

José Burillo, Sean Cleary and Bert Wiest

Abstract. Here we describe the results of some computational explorations in Thompson's group F. We describe experiments to estimate the cogrowth of F with respect to its standard finite generating set, designed to address the subtle and difficult question whether or not Thompson's group is amenable. We also describe experiments to estimate the exponential growth rate of F and the rate of escape of symmetric random walks with respect to the standard generating set.

1. Introduction

Richard Thompson's group F has attracted a great deal of interest over the last years. The group F is a finitely presented group which arises quite naturally in different contexts, and allows several different, but fairly simple, descriptions – for instance by a presentation, as a diagram group [14], as a group of homeomorphisms of the unit interval, as the geometry group of associativity [7], and as the fundamental group of a component of the loop space of the dunce hat. Cannon, Floyd and Parry [3] give an excellent introduction to F.

The interest in this group stems partly from F's unusual properties, and partly from the fact that some of the basic questions about this group are still open, in particular those related to its cogrowth and growth. It seems clear is that F lies very close to the borderline between different regimes.

Probably the most famous open question is whether or not F is amenable. Also, it is known that F has exponential growth, but the growth rate is unknown. Similarly, the rate of escape of random walks in F is unknown.

The question of amenability is especially intriguing since F is either an example of a finitely presented non-amenable group without free non-abelian subgroups, or an example of a finitely presented amenable but not elementary amenable group. Though there are finitely presented examples of groups for each of these phenomena

---

This article originates from the Barcelona conference.

22 J. Burillo, S. Cleary and B. Wiest

from Grigorchuk [11] and Sapir and Olshanskii [17], those groups were constructed explicitly for those purposes, whereas F is a more "naturally occurring" example to consider – so either answer would be remarkable.

The aim of this paper is to contribute new empirical evidence to the quest to understand cogrowth, growth, and escape rate. This evidence was obtained using large computer simulations.

The structure of this paper is as follows. In Section 2 we recall briefly the definition and those properties of the group F that will be needed in the paper. Moreover, we give the definition of amenability which will be used in our experiments (there are other, equivalent, definitions which are probably more well known). In Section 3 we describe the algorithms used in our computations relating to amenability. In Section 4 we present the results of our computer experiments, with the aim of obtaining evidence for or against the amenability of F. In Section 5, we describe two computational approaches to estimate the exponential growth rate of F with respect to the standard two-generator generating set, and in Section 6, we describe the results of some computations to measure the average distance from the origin of increasingly-long random walks, known as the rate of escape.

2. Background on Thompson's group F and amenability

Richard J. Thompson's group F is usually defined as the group of piecewise-linear orientation-preserving homeomorphisms of the unit interval, where each homeomorphism has finitely many changes of slope ("breakpoints") which all are dyadic integers and and whose slopes, when defined, are powers of 2. F admits an infinite presentation given by

$$\langle x_1, x_2, x_3, \ldots \mid x_j x_i = x_i x_{j+1} \text{ if } i < j \rangle$$

which is convenient for its symmetry and simplicity, while there is a finite presentation given by

$$\langle x_0, x_1 \mid [x_0 x_1^{-1}, x_0^{-1} x_1 x_0], [x_0 x_1^{-1}, x_0^{-2} x_1 x_0^2] \rangle .$$

Brin and Squier [1] showed that F has no free non-abelian subgroups, and thus the question of the amenability of F is potentially connected to the conjecture of Von Neumann that a group is amenable if and only if it had no free non-abelian subgroups. The conjecture has since been solved negatively, but the problem of the amenability of F is of independent interest and it has been open for at least 25 years.

The usefulness of the infinite presentation is the fact that F admits a normal form based on the infinite set of generators. The relators of the infinite presentation can be used to reorder generators of a given word into an expression of the following form:

$$x_{i_1}^{r_1} x_{i_2}^{r_2} \cdots x_{i_n}^{r_n} x_{j_m}^{-s_m} \cdots x_{j_2}^{-s_2} x_{j_1}^{-s_1}$$

with

$$i_1 < i_2 < \cdots < i_n \qquad j_1 < j_2 < \cdots < j_m.$$

This normal form is unique if one requires the following extra condition: if the generators x_i and x_i^{-1} both appear, then either x_{i+1} or x_{i+1}^{-1} must appear as well. Indeed, if neither x_{i+1} nor x_{i+1}^{-1} appeared, then the relator could be applied so as to obtain a shorter word representing the same element. The uniqueness of this normal form can be used to solve the word problem in short time: given a word in the infinite set of generators, find the normal form, which can be done in quadratic time, and the element is the identity if and only if the normal form is empty. This unique normal form is most helpful when the task at hand is to decide whether two words represent the same element of F. If one wishes simply to test whether a given word represents the trivial element of F, it is enough to reorder the generators, but without checking the extra condition for uniqueness.

For an introduction to F and proofs of its basic properties see Cannon, Floyd and Parry [3]. Also, for an excellent introduction to amenability, the interested reader can consult Wagon [18], Chapters 10 to 12.

There are several equivalent definitions of amenability, especially for finitely generated groups. The standard definition is given by the existence of a finitely-additive left-invariant probability measure on the set of subsets of G. If the group is finitely generated, a celebrated characterization due to Følner [8] in terms of the existence of sets with small boundary, has given a special interest to this concept from the point of view of geometric group theory, making it easier to see that amenability is a quasi-isometry invariant.

The numerical criterion we will use extensively in this paper is due to Kesten [15, 16] and it uses the concept of cogrowth.

Definition 2.1. *Let G be a finitely generated group and let*

$$1 \to K \to F_m \to G \to 1$$

be a presentation for G. The cogrowth *of G is the growth of the subgroup K inside F_m. In particular, the* cogrowth function *of G is*

$$g(n) = \#(B(n) \cap K),$$

where $B(n)$ is the ball of radius n in F_m, and the cogrowth rate *of G is*

$$\gamma = \lim_{n \to \infty} g(n)^{1/n}.$$

Kesten's cogrowth criterion for amenability states basically that a group is amenable when it has a large proportion of freely reduced words, for every length n, representing the trivial element; that is, when the cogrowth is large.

Theorem 2.2 (Kesten). *Let G be a finitely generated group, and let X be a finite set of generators, with cardinal m. Let γ be its cogrowth rate. Then G is amenable if and only if $\gamma = 2m - 1$.*

This can also be interpreted in terms of random walks. If the group is non-amenable (that is, if there are very few nontrivial words representing the trivial element of the group), then the probability of a random walk in the group ending at 1 is small. Since our random walks are taken to be non-reduced, we consider the $(2m)^L$ non-reduced words of length L in m generators, and let $T(L)$ be the set of these words which represent the identity in the group G. Then, define

$$p(L) = \frac{\#T(L)}{(2m)^L},$$

that is, we define $p(L)$ to be the proportion of words which are equal to the identity in G. Then, a rewriting of Kesten's criterion for non-reduced words can be given by

Theorem 2.3 (Kesten). *A group is amenable if and only if*

$$\limsup_{L\to\infty} p(L)^{1/L} = 1$$

Roughly speaking, a group is amenable if the probability of a random walk of length L returning to 1 decreases more slowly than exponentially with L. This form of the criterion will be used in the subsequent sections to try to study numerically the amenability of F.

3. Algorithms and programs

The direct approach at finding the numbers $p(L)$ exactly for Thompson's group F fails even at quite small values of L due to the fact that the number of words grows exponentially, so the computation times get large easily. For instance, for a length as small as 14 the number of total words is 4^{14}=268,435,456, out of which there are 1,988,452 representing the neutral element, for a value $p(14)^{1/14} = 0.704423677$. It would be hard to decide whether the sequence approaches 1. A number of improvements can be made to ease the calculation so it becomes more feasible to estimate whether the sequence tends to 1.

First, we take samples of words of a given length instead of the all words of a given length. The number 4^L grows impracticably large even for small values of L, so sampling becomes a necessity. Since the number $p(L)$ is basically a proportion (or a probability), it can be approximated by Monte Carlo methods. One can always take a random non-reduced word in the two generators x_0 and x_1 and check if it is the identity by solving the word problem quickly using the normal form. Repeating this process one can find a reasonably good approximation of the number $p(L)$.

A further improvement can be implemented by taking only *balanced* words. We observe that, since the two relators in G are commutators, a word which represents the identity has to be *balanced*: it has to have total exponent zero in both generators x_0 and x_1. So we consider not *all* random words, but only balanced ones. We remark that the abelianization of F is $\mathbb{Z}^2$, generated by x_0 and x_1, so being balanced is in fact equivalent to representing the trivial element of $\mathbb{Z}^2 = F_{\mathrm{ab}}$. Now we let $C(L)$ be the set of balanced words among the 4^L non-reduced words of length L in F_2, and define

$$\widehat{p}(L) = \frac{\#T(L)}{\#C(L)},$$

the proportion of words representing the identity of F among balanced words of length L. We have

$$\sqrt[L]{p(L)} \;=\; \sqrt[L]{\frac{\#T(L)}{4^L}} = \sqrt[L]{\frac{\#T(L)}{\#C(L)}} \cdot \sqrt[L]{\frac{\#C(L)}{4^L}}$$

$$=\; \sqrt[L]{\widehat{p}(L)} \cdot \sqrt[L]{\frac{\#C(L)}{4^L}}.$$

Moreover, the last factor $\sqrt[L]{\frac{\#C(L)}{4^L}}$ tends to 1 as L tends to infinity, because $\mathbb{Z}^2$ is amenable. Thus F is amenable if and only if we have $\limsup_{L\to\infty} \widehat{p}(L)^{1/L} = 1$.

So in order to decide whether F is amenable, we shall try to find good approximations of $\widehat{p}(L)$, the proportion of words representing 1_F among balanced words of length L, and this for values of L which are as large as possible. Obviously, the algorithm for creating random balanced words must be designed in such a way that all balanced words of length L have the same chance of appearing. The practical advantage of approximating $\widehat{p}(L)$ rather than $p(L)$ is that $\widehat{p}(L)$ is much larger (roughly by a factor $\pi L/2$), so much smaller sample sizes are required.

Yet another improvement, which substantially increases the efficiency of the algorithm, can be made by using a "divide and conquer" strategy. The underlying observation is that if L is even, then the probability that a random word of length L represents the trivial element of F is equal to the probability that two random words of length $L/2$ represent the same element. Thus, the idea of the algorithm is to create a large number N of random words of length $L/2$ (in our implementations, values for N between 15,000 and 200,000 were generally used). Each of the N words is immediately brought into normal form, and these normal forms are stored. In order to decide if two words represent the same element of F, we simply compare their normal forms. Therefore we can consider all $N(N-1)/2$ unordered pairs of words in normal form, and we count how many identical pairs we see. This number, divided by $N(N-1)/2$, is an approximation for the proportion $p(L)$. However, the description just provided is an oversimplification, because as described above, we would like to restrict our sample to *balanced* words. Here we describe the estimation algorithm more precisely:

Each iteration of the algorithm has the following steps. In a preliminary step, we create one random *balanced* word of length L. Then we focus our attention on the first half (the first $L/2$ letters) of this word and we count which element in the quotient $F_{\mathrm{ab}} = \mathbb{Z}^2$ this first half represents – that is, we count the exponent sums of the letters x_0 and x_1 for the first half of the word.

In the second step, we create N random words of length $L/2$ which all represent this same element of the abelianization $F_{\mathrm{ab}} = \mathbb{Z}^2$, in such a way that all possible words of length $L/2$ with the given x_0-balance and x_1-balance have the same chance of appearing. As soon as it is created, each random word is transformed into normal form, and this normal form is stored.

In the third step, we count the proportion of identical pairs among all $N(N-1)/2$ unordered pairs of stored words in normal form.

In this way, each iteration of the algorithm gives an approximation to the true value of $\widehat{p}(L)$. Performing a few thousand iterations, and taking the mean of the proportions obtained in each step, one obtains an approximation to $\widehat{p}(L)$.

The expected value for the result of this algorithm is indeed $\widehat{p}(L)$, which we interpret as the probability that two random words of length $L/2$ represent the same element of F, under the condition that they represent the same element of $F_{\mathrm{ab}} = \mathbb{Z}^2$. It is immediate from the construction of the algorithm that for any pair $(k, l) \in \mathbb{Z}^2$, the proportion of words representing (k, l) in F_{ab} among all words constructed by the algorithm is what it should be – namely the probability that the first half of a balanced random word of length L represents (k, l) in $F_{\mathrm{ab}} = \mathbb{Z}^2$.

Then, having fixed some pair (k, l) in $\mathbb{Z}^2$, we restrict our attention to those iterations of the algorithm that deal with words with x_0-balance k and x_1-balance l (and length $L/2$). We have to prove that the expected value for the proportion of identical pairs of words in our algorithm is what it should be – namely the probability that a pair of random words, chosen with uniform probability from the set pairs of words of length $L/2$ representing the element (k, l) of $F_{\mathrm{ab}} = \mathbb{Z}^2$, represent the same element of F. That is, we have to prove that our taking words in batches of N and comparing all couples in that batch, rather than taking independent samples of pairs of words, does not distort the result. That, however, follows immediately from the fact that in our algorithm, all pairs of words of length $L/2$ with x_0-balance k and x_1-balance l, appear on average with the same frequency (they have uniform probability). The fact that our $N(N-1)/2$ samples are not independent has no impact on the expected value. It does have an impact on the variation, that is, on the size of the error bars, but even this negative impact becomes negligible when we have, on average, less than one identical pair per batch of N words, as we typically have.

The authors have implemented the last two algorithms in computer programs written in FORTRAN and C. These programs were run for several weeks on the "Wildebeest" 132-processor Beowulf cluster at the City University of New York. The results of these implementations will be shown in the next section.

4. Computational results concerning amenability

The results for the computations of trivial words for F are represented in Table 1. This table contains the following information. For lengths $L = 20, 40, \ldots, 300, 320$, it gives in the second and third columns the sample size (the number of words that were tested) and the number of words among them that were found to represent the trivial element of F; thus the quotient of these two quantities is an approximation of $\widehat{p}(L)$. The fourth column contains the Lth root of this proportion. The last column contains the 20th root of the quotient of the proportions obtained for length L and for length $L - 20$.

length L	sample size	trivial	$\sqrt[L]{\widehat{p}(L)}$	$\sqrt[20]{\dfrac{\widehat{p}(L)}{\widehat{p}(L-20)}}$
20	$2.000 \cdot 10^{7}$	$1\,364\,638$	0.8744	
40	$2.000 \cdot 10^{7}$	$82\,922$	0.8718	0.8693
60	$2.000 \cdot 10^{7}$	$6\,341$	0.8744	0.8794
80	$2.500 \cdot 10^{11}$	$7\,255\,725$	0.8776	0.8873
100	$3.125 \cdot 10^{11}$	$938\,587$	0.8806	0.8928
120	$8.750 \cdot 10^{12}$	$2\,961\,321$	0.8832	0.8966
140	$1.312 \cdot 10^{13}$	$551\,480$	0.8857	0.9009
160	$1.238 \cdot 10^{13}$	$67\,542$	0.8879	0.9030
180	$2.420 \cdot 10^{13}$	$18\,618$	0.8900	0.9067
200	$1.425 \cdot 10^{14}$	$16\,040$	0.8918	0.9084
220	$1.572 \cdot 10^{15}$	$26\,596$	0.8934	0.9096
240	$2.063 \cdot 10^{16}$	$55\,941$	0.8950	0.9125
260	$2.716 \cdot 10^{16}$	$12\,162$	0.8964	0.9139
280	$7.566 \cdot 10^{15}$	599	0.8976	0.9139
300	$1.343 \cdot 10^{16}$	196	0.8993	0.9221
320	$5.856 \cdot 10^{16}$	148	0.9003	0.9161

TABLE 1. Cogrowth estimates for F.

In order to clarify the last two columns we remark that the sequences $\sqrt[L]{\widehat{p}(L)}$ and $\sqrt[20]{\widehat{p}(L)/\widehat{p}(L-20)}$ have the same limits – for instance if we had $\widehat{p}(L) \simeq \mathrm{const} \cdot a^{L}$ then we would obtain

$$\lim_{L \to \infty} \sqrt[L]{\widehat{p}(L)} = \lim_{L \to \infty} \sqrt[20]{\widehat{p}(L)/\widehat{p}(L-20)} = a$$

The difference between the two sequences is that the second one converges much more quickly, but it is also more sensitive to statistical errors related to insufficient sample size.

In summary, the question of amenability comes down to the question whether the numbers in the last two columns converge to 1, or to a smaller number. The numbers in the second to last column converge more slowly, but they are more reliable.

Before we can establish any conclusions, it would be interesting to compare these results with the corresponding results for groups which are known to be amenable or not. As test groups we will take the free group on two generators as a nonamenable example, and the group $\mathbb{Z} \wr \mathbb{Z}$ ($\mathbb{Z}$ wreath $\mathbb{Z}$). The latter group is amenable since it is abelian-by-cyclic, and it appears as a subgroup of F in multiple ways [14, 4]. The group $\mathbb{Z} \wr \mathbb{Z}$ admits the presentation

$$\left\langle a, t \mid \left[a^{t^{i}}, a^{t^{j}} \right], i, j \in \mathbb{Z} \right\rangle,$$

L	$\mathbb{Z} \wr \mathbb{Z}$			F_2		
	sample	trivial	$\sqrt[L]{\widehat{p}(L)}$	sample	trivial	$\sqrt[L]{\widehat{p}(L)}$
20	$2.475 \cdot 10^7$	1 802 935	0.8772	$1.000 \cdot 10^7$	655 940	0.8727
40	$2.475 \cdot 10^7$	247 710	0.8913	$1.000 \cdot 10^7$	30 685	0.8653
60	$1.980 \cdot 10^7$	34 658	0.8996	$2.000 \cdot 10^7$	2 888	0.8630
80	$2.475 \cdot 10^7$	9 669	0.9066	$3.000 \cdot 10^7$	230	0.8631
100	$1.980 \cdot 10^7$	2 079	0.9125	$4.000 \cdot 10^8$	159	0.8630
120	$1.095 \cdot 10^8$	3 485	0.9173	$6.975 \cdot 10^{11}$	14 167	0.8628
140	$9.950 \cdot 10^7$	1 035	0.9213	$8.000 \cdot 10^{11}$	819	0.8626
160	$4.990 \cdot 10^8$	1 847	0.9248	$2.400 \cdot 10^{12}$	136	0.8629
180	$2.997 \cdot 10^9$	4 141	0.9278			
200	$4.740 \cdot 10^{10}$	26 919	0.9306			
220	$8.636 \cdot 10^{10}$	20 625	0.9330			
240	$1.859 \cdot 10^{11}$	19 469	0.9352			
260	$4.249 \cdot 10^{11}$	20 112	0.9372			
280	$5.734 \cdot 10^{11}$	12 735	0.9390			
300	$5.844 \cdot 10^{11}$	6 256	0.9407			
320	$4.050 \cdot 10^{12}$	21 229	0.9422			

TABLE 2. Cogrowth estimates for $\mathbb{Z} \wr \mathbb{Z}$ and the free group of rank 2.

and being two-generated it appears to be a good match to compare with F. The results for these two groups are in Table 2.

A graphical representation of comparing these estimates of cogrowth in the three groups F, $\mathbb{Z} \wr \mathbb{Z}$ and $F(2)$ is given in Figures 1.

Do these pictures suggest that F is amenable or non-amenable? It is difficult to discern convergence to 1 or something less than 1 with this data, and it is clear by considering other amenable groups such as iterated wreath products like $\mathbb{Z} \wr \mathbb{Z} \wr \mathbb{Z}$ that the convergence to 1 could be exceptionally slow.

5. Computational results concerning the growth of F

Another family of open questions about Thompson's group F center on the growth of F with respect to its standard generating set $\{x_0, x_1\}$. To study the growth of a group with respect to a generating set, we consider g_n, the number of distinct elements of F of length n and we form the spherical growth series, $g(x) = \sum g_n x^n$. If we consider balls of radius n and the number of elements b_n whose length is less than or equal to n, we have the growth series $b(x) = \sum b_n x^n$. Thompson's group has exponential growth as the submonoid generated by x_0, x_1, and x_1^{-1} is the free product of the subgroup generated by x_1 and the submonoid generated by x_0 (see Cannon, Floyd and Parry [3]). Burillo [2] computed the exact growth function for

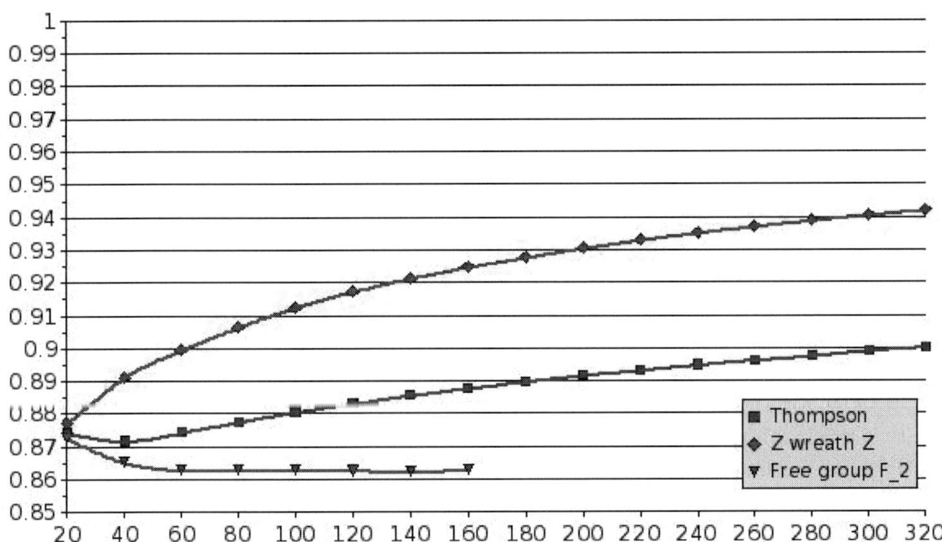

FIGURE 1. Comparing cogrowth estimates $\sqrt[L]{\widehat{p}(L)}$ for three groups.

positive words in F with respect to the standard two generator generating set $\{x_0, x_1\}$ which gives a lower bound for the growth rate of words in the full group as the largest root of $x^3 - 2x^2 - x + 1$, which is about 2.24698. Guba [12] used the normal forms for elements of F developed by Guba and Sapir [13] to sharpen the lower bound of the growth function to $\frac{1}{2}(3 + \sqrt{5})$ which is about 2.61803. Guba conjectures that 2.7956043 is an upper bound by considering the ratio of the ninth and eighth terms in the spherical growth series of F. But the exact growth function of F remains unknown – it is not even known if the growth function is rational, though Cleary, Elder and Taback [5] show that there are infinitely many cone types, which may be evidence that the growth of the full language of geodesics is not rational.

Here, we use a computational approach to estimate the growth function of F. We use two methods both based upon taking random samples of words via random walks. Both of these methods estimate the number of words in successive n-spheres of F. For the first method, we take an element of length n and consider its "inward" and "outward" valence in the Cayley graph. Since the relators of F with respect to the standard finite presentation are all of even length, application of a generator x to an element w of F will either increase or reduce the length by 1. The *inward valence* of w is the number of generators which reduce the word length and the *outward valence* of w is the number of generators which increase word length. If the length of w is n, then the outward valence gives the number of words adjacent to w which lie on the $n + 1$ sphere. By taking an average of the outward valence of a large number of elements in the n sphere, we can estimate the ratio of the number of elements in the $n + 1$ sphere to the number of elements in the n sphere. Thus we can estimate the rate of growth, as the limit of these ratios (for $n \to \infty$) will be the exponential growth rate for the group.

For the second method, we consider a variation of this approach where instead of looking at the words at distance 1 from w, we look at the words at distance 2 from w and see how many of those words lie in the $n + 2$ sphere. This gives an estimate of the ratio of the number of elements in the $n + 2$ sphere to the number of elements in the n sphere, and in the limit, we expect the square root of these ratios to approach the exponential growth rate for the group.

We expect both methods to yield overestimates of the true growth rate, but the error should be larger for the first method than for the second one. The raw outward valence method is expected to overestimate because it may count elements in the $n + 1$ sphere which are adjacent to more than one element in the n sphere multiple times. An extreme example of this are "dead-end" elements in F, characterized by Cleary and Taback [6]. These dead-end elements have the property that right multiplication by any generator reduces word length. The "outward valence" method includes these dead-end elements in the count of growth – if the randomly selected element in the n sphere is one of the 4 elements in the n sphere which is adjacent to a particular dead-end element in the $n + 1$ sphere, it will contribute to the average outward valence at least 1. For the distance two method, however, such elements will not contribute to the growth as there will be no words adjacent to the dead-end element which lie in the $n + 2$ ball.

To compute the length of an element of F, we use Fordham's method [10] for measuring word length of elements of F with respect to $\{x_0, x_1\}$. This remarkable method amounts to building the reduced tree pair diagram associated to an element of F, classifying each internal node of the trees diagram into one of seven possible types, and then pairing the nodes and summing a weight function of those node type pairs to get the exact length of the element.

We note that selecting a random element of the n sphere for a predetermined value of n is not feasible given current understanding of the metric balls in F – we do not even know the number of such elements, as in fact that is what we are trying to estimate. So we construct elements by taking random walks in the group with respect to the standard generating set of a predetermined length n, and then measure the length l of the element obtained. We then compute its outward valence by measuring the lengths of elements adjacent to it in the Cayley graph and we also count the number of elements at distance two from it which lie in the $l + 2$ sphere. Thus, we obtain simultaneously estimates of outward valence for elements in a range of balls. Furthermore, we can record the length l of a word obtained by a random walk of length n and use that to estimate crudely the rate of escape of a random walk in F, as described in the next section. The results of the computations concerning growth are presented in Table 3 and Figure 2.

As we can see from the data, and as expected, the estimates using the distance two method are lower than the estimate from the outward valence method. Moreover, for the first experiment, the values lie between the proven lower bound

Lengths	Words	Average outward valence	Average num. at dist. 2	Growth estimate from dist 2
$0 - 19$	5723	2.8440	7.8363	2.7993
$20 - 39$	629964	2.7334	7.3239	2.7063
$40 - 59$	1017998	2.7128	7.2521	2.6930
$60 - 79$	602694	2.6781	7.0389	2.6531
$80 - 99$	612613	2.6698	7.0041	2.6465
$100 - 119$	514665	2.6564	6.9256	2.6317
$120 - 139$	392069	2.6512	6.9074	2.6282
$140 - 159$	272564	2.6407	6.8529	2.6178
$160 - 179$	234893	2.6331	6.8057	2.6088
$180 - 199$	281806	2.6275	6.7779	2.6034
$200 - 219$	283764	2.6299	6.7897	2.6057
$220 - 239$	164359	2.6336	6.8234	2.6122
$240 - 259$	48750	2.6341	6.8431	2.6159
$260 - 279$	7326	2.6403	6.8756	2.6221
$280 - 299$	521	2.6430	6.8829	2.6235
$300 - 319$	17	2.6470	6.8235	2.6122

TABLE 3. Average outward valence of words arising from random walks.

of 2.618... and the conjectured upper bound of 2.763..., for words of length 20 and more. However, other aspects of the computational results are more surprising. Both functions appear to have a minimum at length about 190. Moreover, for the second experiment, the values obtained lie below the proven lower bound for words of length between 140 and 260, and lie in the expected range before and after that. This data suggests that the rate of growth is close to the proven lower bound or that random walks are not an unbiased method for estimating growth by average outward valence. Of course, since we do not know the growth function, it is difficult to effectively pick a random element, so perhaps random walks tends to bias toward those which have lower outward valence than is representative. The role of "dead-end" elements of outward valence 0 may play a role in this bias and we describe estimates of densities of dead-end elements in the next section. It may be that random walks get stuck near dead-end elements and other low outward valence items and thus random walks may select these elements at a greater proportion than uniform.

Finally, we mention that we have also computed first twelve terms of the exact spherical growth function of F to obtain:

$$g(x) \;=\; 1 + 4x + 12x^2 + 36x^3 + 108x^4 + 314x^5 + 906x^6 + 2576x^7$$
$$+ 7280x^8 + 20352x^9 + 56664x^{10} + 156570x^{11} + \cdots$$

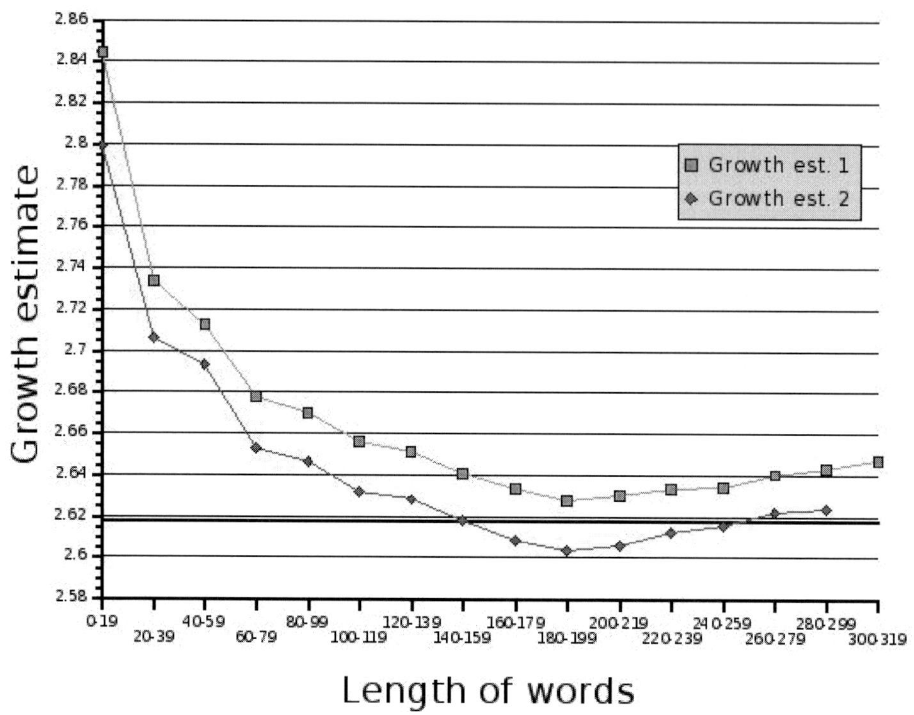

FIGURE 2. Estimates for the exponential growth rate from the data in Table 3.

Guba [12] had already calculated the first ten terms of this sequence and noticed that the ratios of successive terms of this series appear to decrease and form a natural conjectural upper bound to the growth function. The two additional successive quotients arising from our additional terms continue the decreasing pattern and are $2.7841981\ldots$ and $2.7631300\ldots$ and lie well above the experimental estimates of growth described above.

6. Rate of escape of random walks and dead-ends in F

Here we note that as a side effect of the computations described in the previous section to estimate growth, we obtain two pieces of data which are interesting in their own right.

First, since the random elements used to estimate growth are constructed by random walks and we measure their exact lengths using Fordham's method, we are able to see how quickly these random walks leave the origin. Since these are symmetric random walks, there is of course the possibility of backtracking to get non-freely reduced words, so we do not expect a random walk of length 100

Length of random walk	Number of walks	Average length	Standard deviation	Rate of escape
100	4764000	41.18	8.34	0.4118
200	3242898	76.01	12.33	0.3800
300	2700000	109.3	15.51	0.3545
400	1500000	141.8	18.33	0.3544
500	600000	173.8	20.82	0.3476
600	1500000	205.3	23.08	0.3421
700	900000	236.5	25.14	0.3379
800	900000	267.6	27.14	0.3345
900	300000	298.5	29.02	0.3316
1000	300000	329.0	30.86	0.3290

TABLE 4. Distance from origin (word length) as a function of random walk length.

to actually reach the sphere of radius 100 with non-negligible probability. Our estimates of the rate of escape of random walks of lengths 100 to 1000 are shown in Table 4 and the rate of escape seems to be decreasing in this range.

Second, since we compute the outward valence of words to estimate the growth, we can look for words of outward valence zero- these are exactly the "dead-end" elements discovered by Fordham [9] and characterized by Cleary and Taback [6]. Though dead-end elements can occur in any group (with respect to generating sets contrived for that purpose) groups with dead-end elements with respect to natural generating sets are much less common. Geodesic rays from the identity towards infinity cannot pass through dead-end elements, and thus the existence of many dead-end elements tends to reduce the growth of the group. Table 5 shows the observed incidence of dead ends during the course of the growth estimation calculations in Section 5. We see that there are significant numbers of dead ends but that the fraction decreases as the lengths of elements increases.

Acknowledgment

We would like to thank the Centre de Recerca Matemàtica for their hospitality and support, and the first and second authors are grateful for support from NSF grant #0305545.

Range of lengths	Number of words	Number of dead-ends	Fraction
0 – 39	634927	665	0.001047
40 – 79	1620692	1386	0.0008552
80 – 119	1127278	625	0.0005544
120 – 159	665245	239	0.0003593
160 – 199	561502	149	0.0002654
200 – 239	825785	162	0.0001962
240 – 279	689500	114	0.0001653
280 – 319	393643	39	0.00009907
320 – 359	128254	11	0.00008577
360 – 399	20926	1	0.00004779
400 – 439	1193	0	0
440 – 479	21	0	0

TABLE 5. Fractions of dead-ends observed during random walks as a function of resulting word length.

References

[1] Matthew G. Brin and Craig C. Squier. Groups of piecewise linear homeomorphisms of the real line. *Inventiones Mathematicae*, 44:485–498, 1985.

[2] José Burillo. Growth of positive words in Thompson's group F. *Comm. Algebra*, 32(8):3087–3094, 2004.

[3] J.W. Cannon, W.J. Floyd, and W.R. Parry. Introductory notes on Richard Thompson's groups. *Enseign. Math.* (2), 42(3-4):215–256, 1996.

[4] Sean Cleary. Distortion of wreath products in some finitely-presented groups. Preprint arXiv:math.GR/0505288.

[5] Sean Cleary, Murray Elder, and Jennifer Taback. Geodesic languages for lamplighter groups and Thompson's group F. *Quarterly Journal of Mathematics*, to appear.

[6] Sean Cleary and Jennifer Taback. Combinatorial properties of Thompson's group F. *Trans. Amer. Math. Soc.*, 356(7):2825–2849 (electronic), 2004.

[7] Patrick Dehornoy. Geometric presentations for Thompson's groups. Preprint arXiv:math.GT/0407097.

[8] Erling Følner. On groups with full Banach mean value. *Math. Scand.*, 3:243–254, 1955.

[9] S. Blake Fordham. *Minimal Length Elements of Thompson's group F*. PhD thesis, Brigham Young Univ, 1995.

[10] S. Blake Fordham. Minimal length elements of Thompson's group F. *Geom. Dedicata*, 99:179–220, 2003.

[11] R.I. Grigorchuk. An example of a finitely presented amenable group that does not belong to the class EG. *Mat. Sb.*, 189(1):79–100, 1998.

[12] V.S. Guba. On the properties of the Cayley graph of Richard Thompson's group F. *Internat. J. Algebra Comput.*, 14(5-6):677–702, 2004. International Conference on Semigroups and Groups in honor of the 65th birthday of Prof. John Rhodes.

[13] V.S. Guba and M.V. Sapir. The Dehn function and a regular set of normal forms for R. Thompson's group F. *J. Austral. Math. Soc. Ser. A*, 62(3):315–328, 1997.

[14] Victor Guba and Mark Sapir. Diagram groups. *Mem. Amer. Math. Soc.*, 130 (620):viii+117, 1997.

[15] Harry Kesten. Full Banach mean values on countable groups. *Math. Scand.*, 7:146–156, 1959.

[16] Harry Kesten. Symmetric random walks on groups. *Trans. Amer. Math. Soc.*, 92:336–354, 1959.

[17] Alexander Yu. Ol'shanskii and Mark V. Sapir. Non-amenable finitely presented torsion-by-cyclic groups. *Publ. Math. Inst. Hautes Études Sci.*, (96):43–169 (2003), 2002.

[18] Stan Wagon. *The Banach-Tarski paradox*, volume 24 of *Encyclopedia of Mathematics and its Applications*. Cambridge University Press, Cambridge, 1985. With a foreword by Jan Mycielski.

José Burillo
Escola Politècnica Superior
de Castelldefels UPC
Avda del Canal Olímpic s/n
E-08860 Castelldefels, Barcelona, Spain
e-mail: `burillo@mat.upc.es`

Sean Cleary
Department of Mathematics R8133
The City College of New York, Convent Ave & 138th
New York, NY 10031, USA
e-mail: `cleary@sci.ccny.cuny.edu`

Bert Wiest
IRMAR (UMR 6625 du CNRS)
Université de Rennes 1
Campus de Beaulieu
F-35042 Rennes Cedex, France
e-mail: `bertold.wiest@math.univ-rennes1.fr`

Geometric Group Theory

Trends in Mathematics, 37–44

On the Surjunctivity of Artinian Linear Cellular Automata over Residually Finite Groups

Tullio Ceccherini-Silberstein and Michel Coornaert

Abstract. Let M be an Artinian left module over a ring R and let G be a residually finite group. We prove that every injective R-linear cellular automaton $\tau\colon M^G \to M^G$ is surjective.

Mathematics Subject Classification (2000). 68Q80, 37B15, 20F65, 13E10, 16P20.

Keywords. Linear cellular automaton, surjunctive map, residually finite group, Artinian module, Artinian ring.

1. Introduction

Let X be an object in a category $\mathcal{C}$. An endomorphism $f\colon X \to X$ is said to be *surjunctive* if it is surjective or not injective. Thus, saying that an endomorphism is surjunctive is the same as saying that

$$f \text{ injective } \Rightarrow f \text{ surjective.}$$

One says that X is *incompressible* if every endomorphism of X is surjunctive. There are many familiar classes of incompressible objects such as finite sets, finite-dimensional vector spaces, co-hopfian groups, closed topological manifolds, complex algebraic varieties [Ax, Corollary 1], etc. Note that $\mathbb{Z}$ is incompressible in the category of rings since the only ring endomorphism of $\mathbb{Z}$ is the identity map, but not in the category of groups (or $\mathbb{Z}$-modules) since the group endomorphism of $\mathbb{Z}$ given by $x \mapsto 2x$ is injective but not surjective.

Let G be a group. Given a set A, called the *alphabet*, we denote by A^G the set of all maps $x\colon G \to A$, equipped with the right action of G defined by $(x, g) \mapsto x^g$, where $x^g(g') = x(gg')$ for all $g' \in G$.

This article originates from the Geneva conference.

Given two alphabet sets A and B, a map $\tau\colon A^G \to B^G$ is called a *cellular automaton* over G if there exists a finite subset $S \subset G$ and a map $\mu\colon A^S \to B$ such that

$$\tau(x)(g) = \mu(x^g|_S) \quad \text{for all } x \in A^G, g \in G, \tag{1.1}$$

where $x^g|_S$ denotes the restriction of x^g to S. Such a set S is called a *memory set* and μ is called a *local defining map* for τ. Note that every cellular automaton $\tau\colon A^G \to B^G$ is *G-equivariant*, i.e., it satisfies $\tau(x^g) = \tau(x)^g$ for all $g \in G$ and $x \in A^G$.

In symbolic dynamics, one studies the category $\mathcal{C}_G$ whose objects are all G-sets A^G and whose morphisms are all cellular automata $A^G \to B^G$. Lawton (see [Got]) proved that, when G is residually finite, A^G is incompressible in $\mathcal{C}_G$ for any finite set A. It is an open question whether this property holds for all groups. Note however that Lawton's result has been extended recently to the class of sofic groups by Gromov [Gro] and Weiss [Wei] and that no example of a non-sofic group is known up to now.

Consider a ring R. Given two left R-modules M and N, the products M^G and N^G have natural structures of left R-modules. It is easy to see (cf. Lemma 2.3 in [CeC2]) that a map $\tau\colon M^G \to N^G$ is an R-linear cellular automaton if and only if there exist a finite set $S \subset G$ and a family of R-linear maps $u_s\colon M^G \to N^G$, $s \in S$, such that

$$\tau(x)(g) = \sum_{s \in S} u_s(x(gs)) \quad \text{for all } x \in M^G, g \in G.$$

We can form a new category $\mathcal{LC}_{R,G}$ whose objects are the product modules M^G, where M ranges over all left R-modules, and whose morphisms are R-linear cellular automata. The goal of this note is to give a linear analogue of Lawton's surjunctivity theorem in this category. Recall that a module is called *Artinian* if its submodules satisfy the descending chain condition (see Section 2 for more on Artinian modules and Artinian rings). We shall prove in Section 3 the following result which reduces to Theorem 1.3 in [CeC1] when R is a field.

Theorem 1.1. *Let G be a residually finite group and let M be an Artinian left module over a ring R. Then every injective R-linear cellular automaton $\tau\colon M^G \to M^G$ is surjective.*

Since finitely generated left modules over left Artinian rings are Artinian (see Proposition 2.5 below), we deduce from Theorem 1.1 the following:

Corollary 1.2. *Let G be a residually finite group and let M be a finitely generated left module over a left Artinian ring R. Then every injective R-linear cellular automaton $\tau\colon M^G \to M^G$ is surjective.* $\qquad\square$

2. Artinian modules

In this section we collect some basic facts on Artinian modules and Artinian rings (see for example [Hun]).

Let R be a ring.

A left R-module M is said to be *Artinian* if its submodules satisfy the *descending chain condition*, i.e., every decreasing sequence

$$N_1 \supset N_2 \supset \cdots$$

of submodules of M stabilizes (there is an integer $n_0 \geq 1$ such that $N_i = N_j$ for all $i, j \geq n_0$).

Example. Let p be a fixed prime and consider the subgroup M of $\mathbb{Q}/\mathbb{Z}$ consisting of all elements whose order is a power of p. The only proper subgroups of M are the N_i, $i = 1, 2, \ldots$, where N_i is the subgroup of M generated by p^{-i}. We have $N_1 \subsetneq N_2 \subsetneq \cdots$. Thus the $\mathbb{Z}$-module M is Artinian. Note that the submodules of M do not satisfy the ascending chain condition (M is not *Noetherian*) and that in fact M is not even finitely generated.

In the category of left R-modules, Artinian modules are incompressible. In other words, we have:

Proposition 2.1. *Every injective endomorphism of an Artinian left module is surjective.*

Proof. Let $f \colon M \to M$ be an endomorphism of an Artinian left R-module M. Since M is Artinian, the sequence of submodules

$$\mathrm{Im}(f) \supset \mathrm{Im}(f^2) \supset \cdots$$

stabilizes. Thus, there is an integer $n \geq 1$ such that for all $x \in M$ there is $y \in M$ satisfying $f^n(x) = f^{n+1}(y)$. If f is injective, this implies $x = f(y)$. $\square$

Let M be a left R-module. A subset $K \subset M$ is called an *affine subspace* of M if it is the empty set or if there exist a point $x \in M$ and a submodule $N \subset M$ such that $K = x + N$. This is equivalent to say that there exist an endomorphism of left R-modules $f \colon M \to M'$ and a point $x' \in M'$ such that $K = f^{-1}(x')$.

Proposition 2.2. *Let M be an Artinian left R-module. Then the affine subspaces of M satisfy the descending chain condition, i.e., if*

$$K_1 \supset K_2 \supset \cdots$$

is a decreasing sequence of affine subspaces of M, then there is an integer $n_0 \geq 1$ such that $K_i = K_j$ for all $i, j \geq n_0$.

Proof. We can assume that $K_i = x_i + N_i$, where $x_i \in M$ and N_i is a submodule of M for all i. Observe that $x_{i+1} \in x_i + N_i$ and hence $N_{i+1} \subset x_i - x_{i+1} + N_i = N_i$. Since M is Artinian, there exists n_0 such that $N_{i+1} = N_i$ for all $i \geq n_0$. It follows that $K_{i+1} = x_{i+1} + N_i = x_i + N_i = K_i$ for all $i \geq n_0$. $\square$

Remark. By the preceding proposition, every decreasing sequence of non-empty affine subspaces of an Artinian module has a non-empty intersection. This property does not hold even in $\mathbb{Z}$ considered as a $\mathbb{Z}$-module over itself. Indeed, given an integer $a \geq 3$, the subsets

$$K_i = 1 + a + a^2 + \cdots + a^{i-1} + a^i \mathbb{Z}, \quad i = 1, 2, \ldots,$$

form a strictly decreasing sequence of affine subspaces of $\mathbb{Z}$ such that $\bigcap_{i \geq 1} K_i = \varnothing$.

Proposition 2.3. *Let N be a submodule of a left R-module M and let $Q = M/N$. Then M is Artinian if and only if N and Q are both Artinian.*

Proof. Let $p \colon M \to Q$ denote the quotient map. The submodules of N are the submodules of M contained in N and those of Q are in bijection with the submodules of M containing N via p. Thus N and Q are Artinian if M is Artinian.

Conversely, suppose that N and Q are Artinian. Let $A_1 \supset A_2 \supset \cdots$ be a decreasing sequence of submodules of M. Then the sequences $B_i = A_i \cap N \subset N$ and $C_i = p(A_i) \subset Q$ stabilize. Therefore there is an integer $n_0 \geq 1$ such that $B_{i+1} = B_i$ and $C_{i+1} = C_i$ for all $i \geq n_0$. We claim that $A_{i+1} = A_i$ for all $i \geq n_0$. Since $A_{i+1} \subset A_i$, we only need to show that $A_i \subset A_{i+1}$. Thus, let $a_i \in A_i$. Then there exists $a_{i+1} \in A_{i+1}$ such that $p(a_i) = p(a_{i+1})$ and the element $n = a_i - a_{i+1}$ belongs both to $N = \mathrm{Ker}(p)$ and to A_i. Thus $n \in B_i = B_{i+1} \subset A_{i+1}$ and hence $a_i = a_{i+1} + n \in A_{i+1}$. Therefore the sequence A_i also stabilizes. This shows that M is Artinian. $\square$

From the preceding proposition, we deduce that if M_1 and M_2 are Artinian modules then $M_1 \times M_2$ is Artinian. By induction, we get:

Corollary 2.4. *Let $M_1, M_2, \ldots, M_n$ be Artinian left R-modules. Then the left R-module $M = M_1 \times M_2 \times \cdots \times M_n$ is Artinian.* $\square$

A ring R is said to be *left Artinian* if it is Artinian as a left module over itself. In other words, a ring R is left Artinian if and only if every decreasing sequence of left ideals of R stabilizes. *Right* Artinian rings are defined similarly. A ring is said to be *Artinian* if it is both left and right Artinian.

Examples.
1) Every finite ring is Artinian.
2) Every division ring is Artinian. Indeed, if R is a division ring, then its only left (resp. right) ideals are 0 and R.
3) In a left Artinian ring, every element which is not a right zero divisor is left invertible. Indeed, if R is a left Artinian ring and $x \in R$, then the map $f \colon R \to R$ given by right multiplication by x, namely, $f(r) = rx$ for all $r \in R$, is an endomorphism of R viewed as a left module over itself. If x is not a right zero divisor then f is injective and, by Proposition 2.1, it is surjective. Thus, there exists $x' \in R$ such that $1 = f(x') = x'x$. We deduce that an integral domain (i.e., a non-trivial commutative ring without zero divisors) is Artinian if and only if it is a field.

4) It is not difficult to check that the ring of all 2×2 matrices $\begin{pmatrix} a & 0 \\ b & c \end{pmatrix}$ such that $a \in \mathbb{Q}$ and $b, c \in \mathbb{R}$ is left Artinian but not right Artinian.

Proposition 2.5. *Let R be a left Artinian ring. Then every finitely generated left R-module is Artinian.*

Proof. Let M be a finitely generated left R-module. Then M is isomorphic to a quotient of R^n for some $n \in \mathbb{N}$. Since R is left Artinian, the left R-module R^n is Artinian by Corollary 2.4. Therefore M is Artinian by Proposition 2.3. $\qquad\square$

Corollary 2.6. *Let R be a subring of a ring R'. Suppose that R is a left Artinian ring and that R' is finitely generated as a left R-module. Then R' is a left Artinian ring.*

Proof. Every left ideal of R' is a submodule of the left R module R'. $\qquad\square$

Examples.
1) Let R be a left Artinian ring. Then the ring $R' = \mathrm{Mat}_n(R)$ of $n \times n$ matrices with coefficients in R is left Artinian. In particular, the ring $\mathrm{Mat}_n(D)$ is left Artinian for any division ring D. If D is a division ring then the ring $\mathrm{Mat}_n(D)$ is *simple*, that is, its only two-sided ideals are 0 and itself. Conversely, the Wedderburn-Artin theorem asserts that every left Artinian simple ring is isomorphic to $\mathrm{Mat}_n(D)$ for some division ring D and $n \geq 1$ (see for example [Hun, Th.1.14 Ch. IX]).
2) Let R be a left Artinian ring and let G be a group. Let $R[G]$ denote the group ring of G with coefficients in R. By the preceding proposition, $R[G]$ is left Artinian if G is finite. The converse also holds by a theorem of Connell [Con] (see also [Pas]).
3) Let R be a left Artinian ring and let I denote the two-sided ideal of $R[x]$ generated by a monic polynomial $p(x) = x^n + a_{n-1}x^{n-1} + \cdots + a_0 \in R[x]$. Then the quotient ring $R' = R[x]/I$ is left Artinian.

Remark. It follows from Propositions 2.1 and 2.5 that every finitely generated left module over a left Artinian ring is incompressible. Vasconcelos has shown in [Vas] that a commutative ring R is such that every finitely generated R-module is incompressible if and only if the Krull dimension of R is 0, i.e., all prime ideals in R are maximal. Non-commutative rings R such that all finitely generated left modules over R are incompressible have been characterized in [AFS].

3. Proof of the main result

In this section we prove Theorem 1.1 by adapting the proof of Theorem 1.3 in [CeC1].

Let A be a set, G a group, and $\tau\colon A^G \to A^G$ a cellular automaton over G with memory set S and local defining map $\mu\colon A^S \to A$. Let H be a subgroup of G containing S and consider the map $\tau_H\colon A^H \to A^H$ defined by $\tau_H(x)(h) = \mu(x^h|_S)$ for all $x \in A^H$ and $h \in H$. Then τ_H is a cellular automaton over H with memory set S and local defining map μ. The cellular automaton τ_H is called the *restriction* of τ to H ([CeC1]). We shall use the following result whose proof is given in [CeC2].

Proposition 3.1. *The cellular automaton τ is injective (resp. surjective) if and only if its restriction τ_H is injective (resp. surjective). Moreover, if A is a left module over a ring R, then τ is R-linear if and only if τ_H is R-linear.* $\square$

We shall also use the following.

Lemma 3.2 (Closure Lemma). *Let G be a countable group and let M be an Artinian left module over a ring R. Let $\tau\colon M^G \to M^G$ be an R-linear cellular automaton. Suppose that y is an element in M^G such that, for each finite subset $\Omega \subset G$, there exists $x \in M^G$ such that y and $\tau(x)$ coincide on Ω. Then $y \in \tau(M^G)$. (In other words, $\tau(M^G)$ is closed in M^G for the product topology on M^G induced by the discrete topology on M.)*

Proof. Let S be a memory set for τ. Consider a sequence $(A_n)_{n\in\mathbb{N}}$ of finite subsets of G such that $G = \bigcup_{n\in\mathbb{N}} A_n$, $S \subset A_0$ and $A_n \subset A_{n+1}$ for all $n \in \mathbb{N}$. Let B_n denote the S-*interior* of A_n, that is, $B_n = \{g \in G : gS \subset A_n\}$. Note that $G = \bigcup_{n\in\mathbb{N}} B_n$ and $B_n \subset B_{n+1}$ for all n. Consider, for each $n \in \mathbb{N}$, the affine subspace $L_n \subset M^{A_n}$ defined by $L_n = \tau_n^{-1}(y|_{B_n})$, where $\tau_n\colon M^{A_n} \to M^{B_n}$ denotes the R-linear map induced by τ. We have $L_n \neq \varnothing$ for all n by our hypothesis. For $n \leq m$, the restriction map $M^{A_m} \to M^{A_n}$ is R-linear and induces a map $\pi_{n,m}\colon L_m \to L_n$. Consider, for all $n \leq m$, the affine subspace $K_{n,m} \subset L_n$ defined by $K_{n,m} = \pi_{n,m}(L_m)$. We have $K_{n,m'} \subset K_{n,m}$ for all $n \leq m \leq m'$ since $\pi_{n,m'} = \pi_{n,m} \circ \pi_{m,m'}$. Fix $n \in \mathbb{N}$. By Corollary 2.4, the left R-module M^{A_n} is Artinian. As the sequence $K_{n,m}$ ($m = n, n+1, \ldots$) is a decreasing sequence of affine subspaces of M^{A_n}, it follows from Proposition 2.2 that this sequence stabilizes, i.e., for all $n \in \mathbb{N}$ there exist a non-empty affine subspace $J_n \subset L_n$ and an integer $k_n \geq n$ such that $K_{n,m} = J_n$ for all $m \geq k_n$. For $n \leq n' \leq m$, we have $\pi_{n,n'}(K_{n',m}) \subset K_{n,m}$ since $\pi_{n,n'} \circ \pi_{n',m} = \pi_{n,m}$. Therefore, $\pi_{n,n'}$ induces by restriction a map $\rho_{n,n'}\colon J_{n'} \to J_n$ for all $n \leq n'$. We claim that $\rho_{n,n'}$ is surjective. To see this, let $u \in J_n$. Let us choose m large enough so that $J_n = K_{n,m}$ and $J_{n'} = K_{n',m}$. Then we can find $v \in L_m$ such that $u = \pi_{n,m}(v)$. We have $u = \rho_{n,n'}(w)$, where $w = \pi_{n',m}(v) \in K_{n',m} = J_{n'}$. This proves the claim. Now, using the surjectivity of $\rho_{n,n+1}$ for all n, we construct by induction a sequence of elements $x_n \in J_n$, $n \in \mathbb{N}$, as follows. We start by choosing an arbitrary element $x_0 \in J_0$. Then, assuming x_n has been constructed, we take as x_{n+1} an arbitrary element in $\rho_{n,n+1}^{-1}(x_n)$. Since x_{n+1} coincides with x_n

on A_n, there exists $x \in M^G$ such that $x|_{A_n} = x_n$ for all n. We have $\tau(x) = y$ since $\tau_n(x_n) = y|_{B_n}$ for all n by construction. $\qquad\square$

We recall that a group is said to be *residually finite* if the intersection of its subgroups of finite index is reduced to the identity element. The class of residually finite groups is quite large. For instance, every finitely generated linear group is residually finite.

Proof of Theorem 1.1. Let S be a memory set for τ. By Proposition 3.1, we can assume that G is generated by S and hence countable. Let $y \in M^G$. Consider a finite subset $\Omega \subset G$. Since G is residually finite, we can find a normal subgroup of finite index $H \lhd G$ such that the restriction to Ω of the canonical epimorphism $\rho\colon G \to G/H$ is injective. Observe that the R−linear map $\rho^*\colon M^{G/H} \to M^G$, defined by $z \mapsto z \circ \rho$, is injective and that the image of ρ^* coincides with the submodule $\mathrm{Fix}(H) \subset M^G$ consisting of all elements in M^G fixed by H. Since τ is G-equivariant, it induces by restriction an endomorphism $\widetilde{\tau}$ of $\mathrm{Fix}(H)$. The map $\widetilde{\tau}$ is injective since τ is injective by hypothesis. On the other hand, the left R-module $\mathrm{Fix}(H) \cong M^{G/H}$ is Artinian by Corollary 2.4. therefore $\widetilde{\tau}$ is surjective by Proposition 2.1. It follows that we can find $x \in \mathrm{Fix}(H)$ such that $\tau(x)|_\Omega = \widetilde{\tau}(x)|_\Omega = y|_\Omega$. Using Lemma 3.2, we conclude that $y \in \tau(M^G)$. This shows that τ is surjective. $\qquad\square$

Acknowledgments

We express our gratitude to the anonymous referee of our paper [CeC1] for her/his comments and suggestions which gave birth to the present note.

References

[AFS] E.P. Armendariz, J.W. Fisher, R.L. Snider, *On injective and surjective endomorphisms of finitely generated modules*, Comm. Algebra **6** (1978), 159–172.

[Ax] J. Ax, *Injective endomorphisms of varieties and schemes*, Pacific J. Math. **31** (1969), 1–7.

[CeC1] T. Ceccherini-Silberstein, M. Coornaert, *The Garden of Eden theorem for linear cellular automata*, Ergod. Th. & Dynam. Sys. **26** (2006), 53–68.

[CeC2] T. Ceccherini-Silberstein and M. Coornaert, *Injective linear cellular automata and sofic groups*, Israel J. Math. (to appear).

[Con] I.G. Connell, *On the group ring*, Canad. J. Math. **15** (1963), 650–685.

[Got] W. Gottschalk, *Some general dynamical systems*, Recent advances in topological dynamics, Lecture Notes in Mathematics **318**, Springer, Berlin, 120–125.

[Gro] M. Gromov, *Endomorphisms of symbolic algebraic varieties*, J. Eur. Math. Soc. (JEMS) **1** (1999), 109–197.

[Hun] T.W. Hungerford, Algebra, Graduate Texts in Mathematics, Springer-Verlag, New York, 1987.

[Pas] D.S. Passman, The algebraic structure of group rings. Reprint of the 1977 original. Robert E. Krieger Publishing Co., Inc., Melbourne, FL, 1985.

[Vas]　W.V. Vasconcelos, *Injective endomorphisms of finitely generated modules*, Proc. Amer. Math. Soc. **25** (1970), 900–901.

[Wei]　B. Weiss, *Sofic groups and dynamical systems*, (Ergodic theory and harmonic analysis, Mumbai, 1999) Sankhya Ser. A. **62** (2000), 350–359.

Tullio Ceccherini-Silberstein
Dipartimento di Ingegneria
Università del Sannio
C.so Garibaldi 107
I-82100 Benevento, Italy
e-mail: `tceccher@mat.uniroma1.it`

Michel Coornaert
Institut de Recherche Mathématique Avancée
Université Louis Pasteur et CNRS
7 rue René Descartes
F-67084 Strasbourg Cedex, France
e-mail: `coornaert@math.u-strasbg.fr`

Geometric Group Theory

Trends in Mathematics, 45–50

© 2007 Birkhäuser Verlag Basel/Switzerland

Some Residually Finite Groups Satisfying Laws

Yves de Cornulier and Avinoam Mann

Abstract. We give an example of a residually-p finitely generated group, that satisfies a non-trivial group law, but is not virtually solvable.

Denote by F_n the free group on n generators. Recall that, given a group word $m(x_1, \ldots, x_n) \in F_n$, a group G satisfies the law $m = 1$ if for every $u_1, \ldots, u_n \in G$, $m(u_1, \ldots, u_n) = 1$. Given a set $\mathscr{S}$ of group laws, the n-generator free group in the *variety generated by* $\mathscr{S}$ is the quotient of F_n by the intersection of all kernels of morphisms of F_n to a group satisfying all the group laws in $\mathscr{S}$. Taking the quotient by the intersection of all finite index subgroup (resp. of p-power index), we obtain the *restricted* (resp. *p-restricted*) n-generator free group in the variety generated by $\mathscr{S}$.

The celebrated Tits Alternative states that if G is a finitely generated linear group over any field, then either G contains a non-abelian free subgroup, or it is virtually solvable (i.e., contains a solvable subgroup of finite index). It follows that if such a group G satisfies a nontrivial group law, it is virtually solvable. It is natural to ask to what extent the assumption of linearity can be relaxed. Can we, for instance, replace linearity by residual finiteness? Here we show that this is not possible, even under the assumption that G is residually-p (i.e., residually a finite p-group). We provide several constructions. The results we obtain are probably known to the specialists; however, to the best of our knowledge, they do not seem to appear in the literature.

For any $q \in \mathbf{N}$, let G_q be the restricted free 2-generator group in the variety generated by the group law $[x, y]^q = 1$.

We begin by the following elementary result:

Theorem 1. *For $q = 30$, G_q is a 2-generator, residually finite group that satisfies a nontrivial group law, but is not virtually solvable.*

Proof. The only nontrivial verification is that G_{30} is not virtually solvable. To show this, it suffices to show that, for every n, G_{30} has a finite quotient having no solvable subgroup of index $\leq n$.

This article originates from the Geneva conference.

Start with $I = \mathrm{Alt}_5$. Then $|I| = 60$ and any solvable subgroup of I has order ≤ 12. Therefore, for every m, any solvable subgroup of I^m has order $\leq 12^m$, therefore index $\geq 5^m$.

Now, for all $k \geq 2$, the wreath product $I \wr C_k = I^k \rtimes C_k$ is generated by 2 elements: the element $(1, z)$, where z is a generator of C_k (we now identify z and $(1, z)$), and the element $\sigma = ((s, t, 1, \ldots, 1), 1)$, where s has order 2, t has order 3, and s and t generate I. Indeed, $\sigma^3 = (s, 1, \ldots, 1)$, $z^{-1}\sigma^4 z = (t, 1, \ldots, 1)$, so that σ and z generate $I \wr C_k$. Now $I \wr C_k$ has derived subgroup of exponent 30, hence it is a quotient of G_{30}. $\qquad\square$

Remark 2. The wreath product $\mathrm{Alt}_5 \wr \mathbf{Z}$ is not residually finite; actually it has no residually finite quotient bigger than $\mathbf{Z}$ [Gr57].

By taking the profinite completion, we obtain:

Corollary 3. *There exists a profinite group, topologically finitely generated, that satisfies a non-trivial group law, and is not virtually prosolvable.* $\qquad\square$

Theorem 1 can be strengthened by demanding the group to be residually-p. For this purpose, we need to appeal to a deep result of Y.P. Razmyslov (see [VL93]).

Theorem 4 (Razmyslov). *For every prime power $q \geq 4$, there exist finite groups of exponent q and arbitrarily large solvability length.*

Given $n \in \mathbf{N}$ and a set X, denote by $B(X, n)$ the free group of exponent n on the generators $(u_x)_{x \in X}$. Let $R(X, n)$ be the restricted free group of exponent n on the generators $(u_x)_{x \in X}$; namely, the quotient of $B(X, n)$ by the intersection of all its finite index subgroups.

Remark 5. There is a canonical isomorphism between $R(X, n)$ and the direct limit $\varinjlim_F R(F, n)$, where F ranges over all finite subsets of X.

Indeed, for finite $F \subset X$, there is a natural split morphism $B(F, n) \to B(X, n)$, inducing a split (in particular, injective) morphism $i_F : R(F, n) \to R(X, n)$. This induces a morphism of the direct limit $\varinjlim_F R(F, n) \to R(X, n)$. As a direct limit of injective morphisms, it is injective; it is trivially surjective since all marked generators are in the image.

In particular, for all n such that groups of exponent n are locally finite (this is known for $n \leq 4$ and $n = 6$), $R(X, n) = B(X, n)$.

If G is any group, then G acts on $R(G, n)$ by shifting the generators.

Proposition 6. *Suppose that G is residually finite (resp. residually-p). Then so is $G \ltimes R(G, n)$ for all $n \in \mathbf{N}$ (resp. for every $n = p^k$ for some $k \in \mathbf{N}$).*

Proof. Let (g, x) belong to $G \ltimes R(G, n)$ let us show that there exists a residually finite quotient of G in which (g, x) has a nontrivial image. If $g \neq 1$, G is such a quotient. Suppose that $g = 1$. Writing x as a word in the generators (u_g), $g \in G$, involves only a finite subset B of G. By residual finiteness of G, there exists a finite quotient G/N of G such that the quotient morphism is injective in restriction to B. It extends to a morphism of $G \ltimes R(G, n)$ onto $G/N \ltimes R(G/N, n)$, whose restriction to $R(B, n)$ is injective (this follows from Remark 5). It follows that the image of x in $G/N \ltimes R(G/N, n)$ is nontrivial. Finally, $G/N \ltimes R(G/N, n)$ is residually finite[1] since it contains $R(G/N, n)$ as a subgroup of finite index.

The proof of the statement for residually-p groups is similar. $\qquad\square$

Remark 7. A similar result holds if we replace the restricted free groups of exponent n by the restricted (or p-restricted) free groups on any variety.

Theorem 8. *For all n, $\mathbf{Z} \ltimes R(\mathbf{Z}, n)$ is 2-generated, residually finite, residually-p if n is a power of p, and satisfies the group law $[x, y]^n = 1$. If $n = 4, 5$ or $n \geq 7$, it is not virtually solvable.*

Proof. It is clearly generated by $(1, 1)$ and $(0, u_0)$. The statement on residual finiteness follows from Proposition 6. The group law $[x, y]^n = 1$ is trivially satisfied since $\mathbf{Z}$ is abelian. The last statement follows from Theorem 4, which is equivalent to the following statement: for every prime-power $q \geq 4$, $R(\mathbf{Z}, q)$ is not virtually solvable. Indeed, if it were virtually solvable, there would be a bound on the length of solvability of finite groups of exponent q. It immediately generalizes to any non-necessarily prime-power $q \geq 7$: indeed, such a number has a divisor that is either 4, 9, or a prime ≥ 5. $\qquad\square$

Remark 9. In the remaining cases, namely when $n \leq 3$ or $n = 6$, the free Burnside group $B(\mathbf{Z}, n)$ is solvable and locally finite; for $n \leq 2$ it is clearly abelian; for $n = 3$ it is metabelian and for $n = 6$ it is 5-solvable (more precisely, it is (exponent 3)-by-(exponent 2)-by-(exponent 3)), by a result of M. Hall [MH58].

As an application, by taking the pro-p completion, we answer a question asked by E. Breuillard and T. Gelander in an early version of [BG04].

Corollary 10. *There exists a pro-p-group, topologically finitely generated, that satisfies a non-trivial group law, and is not virtually solvable.* $\qquad\square$

Let us provide a second construction.

Let F be a free group of rank 2, and fix a prime power $q = p^a \geq 4$. By Theorem 4, for every n, there exists k and a finite k-generated group of exponent q that is not n-solvable. Choose a normal subgroup N_n of F such that F/N_n is abelian of order r, where $r \geq k - 1$ is a power of p. Let K_n be the smallest normal subgroup of N_n with a finite factor group of exponent q; note that K_n is characteristic in N_n and is therefore normal in F. Besides, since N is free of rank $r \mid 1$, by the assumption on k, N_n/K_n is not n-solvable. Now setting $G = F/\bigcap K_n$, G is 2-generated and residually-p; its derived subgroup has exponent q but is not virtually solvable. Note that this provides another proof that the group G_q

---

[1] By the solution to the restricted Burnside problem, it is even finite. But we do not need this deep result due to Zelmanov (see [VL93]), and our argument is preferable in view of Remark 7.

introduced above is not virtually solvable for all prime power $q \geq 4$ (and therefore for every integer $q \geq 4$, $q \neq 6$).

We now give a third construction, of independent interest, relying on the following theorem from [NN59]; we do not quote it in the utmost generality.

Theorem 11 (B.H. Neumann and H. Neumann, 1959). *Let G be a countable group. Then there exist cyclic groups B, C, and an embedding i of G into the unrestricted wreath product $Q = (G \wr C) \wr B$ so that $i(G)$ is contained in the second derived subgroup Γ'' of a two-generator subgroup Γ of Q (in particular, every group law satisfied by G is satisfied by Γ'').*
Moreover,

   (1) *if G is finitely generated, we can choose for B any cyclic group of sufficiently large order $k \in \mathbf{N} \cup \{\infty\}$ (say[2], $k \geq 4m - 1$, if G is generated by m elements), and*

   (2) *if G is generated by elements whose orders divide n, we can choose $C = \mathbf{Z}/n\mathbf{Z}$ (in particular, we always can choose $C = \mathbf{Z}$).*   □

Here is our third construction.

Let H_q be the free 2-generator group in the variety generated by the group law $[[x, y], [z, t]]^q = 1$. Let $H := (H_q)_{rp}$ be the quotient of G by the intersection of all its normal subgroups of p-power index; it is trivially residually-p. We claim that, for $q \geq 4$, a power of p, it is not virtually solvable.

For every n, by Theorem 4, there exists a finite group F of exponent q, and of solvability length $\geq n$. By Theorem 11, F embeds in the second derived subgroup of a 2-generator subgroup P of $(F \wr C_{p^k}) \wr C_{p^k}$ for sufficiently large k. Then P is a p-group and it is a quotient of H_q, therefore also of H. Suppose that H has a normal solvable subgroup of finite index r and solvability length h. Then P has a normal subgroup Q of index at most r and solvability length at most h. Then P itself has solvability length at most $h + r$, which is impossible if n is large enough.

Question 12. Does there exist a residually finite, finitely presented group, that satisfies a nontrivial group law but is not virtually solvable?

Remarks 13.
   – This question is not trivial at all even without the residual finiteness assumption. A.Olshanskii and M.Sapir [OS02] have constructed, for large n, a finitely presented, non-virtually solvable group (actually non-amenable), whose derived subgroup has exponent n. By the solution to the restricted Burnside problem, their group is not residually finite, since it contains an infinite, finitely generated, finite exponent group.
   – If we also drop the finite presentation assumption, there are many examples. One of the simplest (but not the best known) is given by the standard wreath product $F \wr \mathbf{Z}$, where F is any finite, non-solvable group.

---

[2]Indeed, if $k \geq 4m - 1$, then the elements $b_1 = 1, b_2 = 3, \ldots, b_m = 2m - 1$ of $\mathbf{Z}/k\mathbf{Z}$ satisfy (4.3) of [NN59].

– Again using the solution to the restricted Burnside problem, the residually finite groups whose derived subgroup has finite exponent are (locally finite)-by-abelian.

If q is a prime power, then the free restricted group in the variety generated by the group law $[x, y]^q$ is also (locally finite)-by-abelian, by [Sh99] (a related result appears in [Sh02]). In particular, all these groups are elementary amenable. This motivates the following question:

Question 14. Does there exist a finitely generated, residually finite group, that is not (elementary) amenable and satisfies a nontrivial group law? What about the free restricted group in the variety generated by, for suitable n, the group law $[x, y]^n$? $[[x, y], [z, t]]^n$?

We can ask a similar question concerning semigroup laws

Question 15. Does there exist a topologically finitely generated profinite group (resp. pro-p-group) that does not contain any nonabelian free semigroup, but is not virtually nilpotent?

Remark 16. A deep result of J. Semple and A. Shalev [SS93] states that a residually finite, finitely generated group satisfying a nontrivial semigroup law is virtually nilpotent.

Remark 17. The first Grigorchuk group is a finitely generated, residually-2 torsion group, hence does not contain any free semigroup, and is not virtually nilpotent. However, its pro-2 completion contains nonabelian free subgroups.

In particular, The Grigorchuk group does not satisfy any non-trivial group law. This raises our final

Question 18. Can a finitely generated group satisfying a non-trivial law have intermediate growth?

References

[BG04] E. Breuillard, T. Gelander. *A topological Tits alternative*, Preprint 2004, to appear in Ann. of Math.

[Gr57] K.W. Gruenberg. *Residual properties of infinite soluble groups*, Proc. London Math. Soc. (3) **7**, 1957, 29–62.

[MH58] M. Hall. *Solution of the Burnside problem for exponent six*. Illinois J. Math. **2** (1958), 764–786.

[NN59] B.H. Neumann, H. Neumann. *Embedding Theorems for Groups*, J. London Math. Soc. **34** (1959), 465–479.

[OS02] A. Olshanskii, M. Sapir. *Non-amenable finitely presented torsion-by-cyclic groups*. Publ. Math. Inst. Hautes Études Sci. **96** (2002), 43–169.

[Sh99] P. Shumyatsky. *On groups with commutators of bounded order*, Proc. Amer. Math. Soc. **127** (1999), 2583–2586.

[Sh02]   P. SHUMYATSKY. *Commutators in residually finite groups*, Monatsh. Math. **137** (2002), 157–165.

[SS93]   J. SEMPLE, A. SHALEV. *Combinatorial Conditions in Residually Finite Groups, I*, J. Algebra **157**, 1 (1993), 43–50.

[VL93]   M.R. VAUGHAN-LEE. *The restricted Burnside Problem*, 2nd ed., Oxford Univ. Press, 1993.

Yves de Cornulier
IRMAR
Campus de Beaulieu
F-35042 Rennes Cedex, France
e-mail: `yves.decornulier@univ-rennes1.fr`

Avinoam Mann
Einstein Institute of Mathematics
Hebrew University
Jerusalem 91904, Israel
e-mail: `mann@math.huji.ac.il`

Geometric Group Theory

Trends in Mathematics, 51–64

Classifying Spaces for Wallpaper Groups

Ramón J. Flores

Abstract. In this paper we use the homotopy structure of the classifying space for proper bundles of symmetrics group of the plane to describe the $B\mathbb{Z}/p$ nullification, in the sense of Dror-Farjoun, of the classifying spaces of these groups.

Introduction

The wallpaper groups are the symmetry groups of the plane. More precisely, a group G that acts in the plane $\mathbb{R}^2$ is called *wallpaper* if there exists a compact pattern and two linearly independent translations such that the whole plane is tesselated by the images of the pattern when acted on by the elements of the group. The interest in these groups from the artistic point of view is really old, and representations of them can be found, for example, in the walls of "la Alhambra" in Granada, or in traditional Japanese clothes. More recently, the famous Dutch artist M. Escher drew a number of famous pictures based on these patterns.

The first attempts to classify the patterns on the plane and in space date back to the 19th century, and in this context Jordan identified almost all of the wallpaper groups. Twenty years later, the crystallographer Fedorov, and independently M. Schoenflies, completed the classification of the 17 planar and 320 spatial symmetry groups, and in the seventies, using computers, the classification was finished for the 4-dimensional patterns. In higher dimensions, the problem is still open.

From the (geometric) group-theoretic point of view, the wallpaper groups have been very well studied. They are all extensions of $\mathbb{Z} \oplus \mathbb{Z}$ by a finite group (and then residually finite), can have only torsion in the primes 2 and 3, and very concise presentations for them are available. Models for their action on the plane and the associated tessellations can be found, for example, in [Sc78], [Lee05], and [Joy05], as well as descriptions of the quotient spaces, that will be very useful for us. Moreover, the subgroup growth of these groups has been described by means of

This article originates from the Barcelona conference.

the zeta function ([DMG99]) and they have also been investigated from the point of view of the harmonic analysis ([FL02]).

The starting point of our research is the classifying space for proper G-bundles $\underline{B}G$ (see the definition in 1.2 below), which has deserved a lot of attention in the last decade, because of its relationship with the Baum-Connes conjecture ([BCH94]). For example, it has been proved that every space is, up to homotopy, the classifying space for proper G-bundles of some G ([LN01], Theorem 1); Lück gives in ([Lüc05], Section 6) some concrete models for $\underline{B}G$; Mislin ([Mis03], Section 3) studies the relationship between the Bredon homology of G and the ordinary homology of $\underline{B}G$; and a description of the fundamental group is also available ([LN01], Proposition 3).

In our previous paper ([Flo05], 3.2), we proved that a great part of the homotopy structure of $\underline{B}G$ can be retrieved from the classical classifying space BG (and then from the group-theoretic information of G), at least when $\underline{B}G$ is finite dimensional. For instance, we could describe the behavior of the functor $\underline{B}$ with regard to colimits and fibrations, and we were even able to determine, in some cases, the universal cover of $\underline{B}G$.

The link between the classical and proper classifying spaces was given by the $B\mathbb{Z}/p$-*nullification*, for p prime. Recall that given a space X, the A-nullification is, roughly speaking, a functorial way of isolating the "part" of the space X that is not visible through the mapping space $\mathrm{map}(A, X)$ (see the precise definitions below). The functor $\mathbf{P}_A$ was introduced in [Bou94], where it was also applied first time in the context of p-local homotopy theory, by means of some concrete computations in the case $A = M(\mathbb{Z}/p, 1)$. We are concerned with its infinite dimensional generalization $\mathbf{P}_{B\mathbb{Z}/p}$, which has showed its usefulness since then. For example, we can quote work of Castellana-Crespo-Scherer ([CCS05]), where the authors used it to classify H-spaces with finiteness conditions, work of Dwyer relating the $B\mathbb{Z}/p$-nullification of the classifying space of a simply-connected compact Lie group with its $\mathbb{Z}[1/p]$-localization ([Dwy96]), or the corresponding results for finite groups ([Flo04]). It should be emphasized that $\mathbf{P}_{B\mathbb{Z}/p}$ is a localization in the sense of Dror-Farjoun, see Section 2 and ([Far96], Chapter 1).

The main goal of our work is to show that if there exists a finite dimensional model for $\underline{B}G$, a precise description of its structure can be a good source for getting information about the p-local homotopy of BG. More concretely, the idea is to look at the orbit spaces of the action of every wallpaper group G over $\mathbb{R}^2$ as models for the classifying space for proper G-bundles, and then to use Theorem 3.2 of [Flo05] and the properties of the nullification functors with regard to fibrations to obtain a description of $\mathbf{P}_{B\mathbb{Z}/p}BG$ in every prime. In the background section we provide the needed definitions of the theory of proper actions.

From the point of view of p-local homotopy theory, it is a remarkable fact that, despite of its interest, not much is known in general about the $B\mathbb{Z}/p$-nullification (or, more generally, p-localization or p-completion, see [BK72]) of classifying spaces of infinite discrete groups. A reason for this is that in general, the classifying spaces of groups that are not nilpotent nor perfect are not known to be p-good (i.e., the

p-completion map is a mod p homology equivalence), and moreover, as they are not nilpotent as spaces, they do not admit in general Sullivan's arithmetic square. We prove in this work that for wallpaper groups our methods from the theory of proper actions can avoid these kind of difficulties and are able to describe in this way the p-local structure of the classifying spaces. Observe that, in particular, most of these groups are not nilpotent.

We expect that our tools will be useful in the future to compute localizations of classifying spaces of other classes of crystallographic groups, and that this gives a clue about the mysterious p-local homotopy theory of these spaces.

Acknowledgements. I thank Juan Alfonso Crespo for his careful reading and interesting suggestions, and the referee for the corrections and his/her improvement of the analysis of the trigyro and tryscope groups. The author is supported by MEC grant MTM2004-06686.

1. Background

We give in this section the basic definitions about $B\mathbb{Z}/p$-nullification and proper actions that are needed in the remain of the note.

1.1. A-nullification

The main references on this topic are the book of E. Dror-Farjoun ([Far96]) and the paper of W. Chachólski ([Cha96]).

Definition 1.1. *Let A and X be spaces. X is said to be A-null if the map $X \to \mathrm{map}(A, X)$ induced by inclusion of points as constant maps is a weak equivalence.*

Note that if A and X are pointed and connected, this is equivalent to say that the pointed mapping space $\mathrm{map}_*(A, X)$ is weakly contractible.

Proposition 1.2. *For every space A, there exists a coaugmented idempotent functor $\mathbf{P}_A$ from the category of spaces to itself, such that for every X, the space $\mathbf{P}_A X$ is $B\mathbb{Z}/p$-null, and moreover the coaugmentation induces a weak homotopy equivalence $\mathrm{map}(\mathbf{P}_A X, Y) \simeq \mathrm{map}(X, Y)$ for every A-null space Y.*

As before, a similar statement can be formulated in the pointed case.

Definition 1.3. *A space X is called A-acyclic if its A-nullification is contractible.*

The first construction of $\mathbf{P}_A$ can be found in the paper [Bou94] of Bousfield, which called it "periodization" because of its relationship with the periodic homotopy groups of a space, described in [MT89]. Later, Dror-Farjoun regards the A-nullification as the localization L_c with respect to the constant map $c : A \to *$; this is the simplest case of his concept of "localization with respect to a map", see ([Far96], Chapter 1). In particular, the information available about localizations L_f turns out to be valid for the A-nullification.

Dror-Farjoun gives a functorial construction for L_f (and hence for $\mathbf{P}_A$), and a different approach to the concrete case of the A-nullification can be found in ([Cha96], 17.1). These authors also prove a number of important properties of

$\mathbf{P}_A$, as for example its behavior with regard to products, homotopy colimits, loops and suspensions, connectedness, etc. In particular, a great part of the research developed around $\mathbf{P}_A$ (and L_f) has aimed to describe to what extent these functors preserve fibration sequences. In this sense, the next proposition will be crucial in the remaining of this note.

Proposition 1.4 ([Far96], 1.H.1 and 3.D.3). *Let* $F \longrightarrow E \longrightarrow B$ *be a fiber sequence. Then*

1. *If* $\mathbf{P}_A(F)$ *is a contractible space, then the induced map* $\mathbf{P}_A(E) \longrightarrow \mathbf{P}_A(B)$ *is a homotopy equivalence.*
2. *If* B *is* A*-null then* $\mathbf{P}_A(F)$ *is the homotopy fiber of the induced map* $\mathbf{P}_A(E) \to B$.

See ([Bou94] 8.1, [Far96] 1.F.1, or [BF03], 0.1) for more results of this kind.

1.2. Classifying spaces for families

As we have said in the introduction, the main tool that we are going to use as input for studying the $B\mathbb{Z}/p$-nullification of classifying spaces of wallpaper groups is the classifying space for proper G-bundles, so we will recall its definition. First, we need a previous concept:

Definition 1.5. *Let* G *be a discrete group, and* X *a* G-*CW*-*complex where the isotropy groups are finite (i.e., a* proper *action). The space* X *is a model for the classifying space for proper* G-*actions* if for every finite subgroup $H < G$ the fixed-point set X^H is contractible.

It can be proved (see for example [Die87] Section 1.6) that the classifying space for proper G-actions is unique up to G-homotopy equivalence. It is usually denoted by $\underline{E}G$, by analogy with the universal space for principal bundles.

This space was first defined by Serre ([Ser71]), and gained great importance as the main ingredient of the Baum-Connes conjecture, in its nearly definitive shape of [BCH94]. In this last paper also appears first time in an explicit way the orbit space $\underline{E}G/G$, in the following context: it is defined a concept of proper G-bundle, and it is proved that the orbit space classifies them up to homotopy of bundles. For this reason, it is called *classifying space for proper G-bundles* and denoted $\underline{B}G$. Further information about these spaces from the point of view of G-actions can be found in [Die87] or [Lüc89].

The assumption that the action of G is proper can be slightly changed, giving rise to a more general concept.

Definition 1.6. *Let* G *be a discrete group, and let* $\mathcal{F}$ *be a family of subgroups of* G *which is subgroup-closed and conjugation-closed. A* G-*space* X *is a* classifying space for the family $\mathcal{F}$ *if the isotropy groups of the action belong to* $\mathcal{F}$*, and moreover for any* $H \in \mathcal{F}$*, the fixed-point set* X^H *is contractible.*

If the family $\mathcal{F}$ is defined, its classifying space is denoted by $E_\mathcal{F}G$ and its quotient by $B_\mathcal{F}G$. It is clear that if $\mathcal{F}$ is the family of finite subgroups of G we are in the previous case, while if $\mathcal{F} = \{1\}$ we have $\mathrm{E}_{\mathcal{F}}G = \mathrm{E}G$. In the sequel, we will denote by $\mathcal{F}_p$ the family of finite p-groups of G. More information about classifying spaces for families can be found in [Lüc00].

Now we have all the ingredients we needed, and we can state the two key results that will allow us to compute the nullifications of the classifying spaces of wallpaper groups. The first one gives a model for the classifying space for proper actions of each wallpaper group, and can be found for example in ([LS00], Section 4).

Proposition 1.7. *Let G a wallpaper group. Then the plane, with the natural action of G, is a model for $\underline{\mathrm{E}}G$.*

Moreover, as we said before, very well-known standard models are known for the orbit spaces $\mathbb{R}^2/G$, which in fact are always 2-dimensional orbifolds. We refer the reader to [Sc78] and [Lee05] for the models that are not explicitly described in next section.

Once we know the shape of $\underline{\mathrm{B}}G$ for the wallpaper groups, we need to extract from it the desired information about the nullification of $\mathrm{B}G$. For this, we use our second key result, which in fact is the main achievement of [Flo05]:

Proposition 1.8 ([Flo05] 3.2). *Let G be a discrete group, W the wedge of the classifying spaces of all the p-groups of prime order, and suppose that there exists a finite dimensional model for $\underline{\mathrm{B}}G$. Then the W-nullification of $\mathrm{B}G$ has the homotopy type of $\underline{\mathrm{B}}G$.*

According to the previous result 1.7, this result can be applied to classifying spaces of wallpaper groups. On the other hand, if a group G only have torsion in the primes $\{p_1 \ldots p_n\}$, it is easy to deduce from the proof of the theorem that $\mathbf{P}_W \mathrm{B}G \simeq \mathbf{P}_{\mathrm{B}p_1 \vee \cdots \vee \mathrm{B}p_n} \mathrm{B}G \simeq \underline{\mathrm{B}}G$.

Now we have described all the ingredients we needed.

2. Classifying spaces of wallpaper groups

In this section we compute the $\mathrm{B}\mathbb{Z}/p$-nullification of the classifying spaces of the seventeen wallpaper groups. Although they are really well known, no standard notation is available for them, and various of them are widely used, as for example Conway's, Speiser's or Fejer-Toth's one; we have chosen the one adopted by the International Union of Crystallography in 1952.

We do not intend to give an exhaustive description of the geometrical features of these groups, that the interested reader can find, for example, in [CM65], [Sc78], or [Joy05]; we strongly recommend the reader to be careful with the notation when he consults other sources, because sometimes even the names of some groups are interchanged. For every group, we describe the geometrical meaning of the generators and the distinguished isometries, and we give a model for the orbit space $\mathbb{R}^2/G$. The presentations we use here are essentially taken from the paper [DMG99], because in it the generators that represent the translations that define the pattern (usually x and y, see below) are clearly distinguished from ones that do not. We will usually refer to the remaining generators as the *exotic* generators.

In our study, just a couple of cases remain unsolved, because of reasons that will be explained in its place. We do not include the results of the computations for the groups **pmm**, **p3** and **p3m1**, that were undertaken in ([Flo05], Section 6); we only put here the descriptions of the nullifications, for the sake of completeness.

So let us go with the computations:

1. **The monotrope group p3** $= \{x, y; xyx^{-1}y^{-1} = 1\}$.

As we have said, this group was studied in [Flo05], where we prove that for every prime p, $\mathbf{P}_{\mathrm{B}\mathbb{Z}/p}\mathbf{Bp3}$ has the homotopy type of a torus.

2. **The ditrope group p2** $= \{x, y, z; xyx^{-1}y^{-1} = 1, z^2 = 1, zxzx = 1, zyzy = 1\}$.

A fundamental region for this group is a triangle which is exactly the half of the parallelogram that the two generating translations determine. The generator z represents a rotation of angle π around the center of that parallelogram, and no other distinguished isometries appear; hence, it only appears 2-torsion.

The orbit space of the plane under the action of the the ditrope group has the shape of a non-slit turnover, which is homotopy equivalent to a sphere. Then, by 1.8, 1.7 and ([Mil84] 9.9), $\mathbf{P}_{\mathrm{B}\mathbb{Z}/2}\mathbf{Bp2} \simeq \mathrm{B}_{\mathcal{F}_2}\mathbf{p2} \simeq S^2$. This line of reason will be repeated in the sequel with no express mention.

3. **The monoscopic group**
   $$\mathbf{pm} = \{x, y, z; xyx^{-1}y^{-1} = 1, z^2 = 1, zxzx^{-1} = 1, zyzy = 1\}.$$

This group is quite similar to the previous one, because it only has an "exotic" generator of order 2, and the fundamental region is again the half of the rectangle determined by the translations x and y. However, in this case z represents a reflection whose axis passes through the middle points of two opposite sides of the rectangle, and then a model for the orbit space is given by a cylinder, corresponding the circle of one base with the mentioned axis. As the monoscopic group only has 2-torsion (for the same reason as the previous example) and the cylinder retracts over the circle, we conclude that the $\mathrm{B}\mathbb{Z}/2$-nullification of $\mathbf{Bpm}$ has the homotopy type of S^1.

4. **The monoglide group pg** $= \{x, y, z; xyx^{-1}y^{-1} = 1, z^2x^{-1} = 1, zyz^{-1}y = 1\}$.

The group **pg** is generated by the two usual translations and one glide-reflection, where the glide-reflection vector goes from the middle point of one of the sides of the rectangle generated by the translations to the center of the rectangle. The fundamental region is, just like in the previous example, a rectangle which is the half of the previous one, but in this case the glide-reflection axis cuts the small rectangle in two equal halves. Hence, the orbit space is different, and in fact it is homotopy equivalent to the Klein bottle.

As the monoglide group has no reflections nor rotations, it is torsion-free (in fact, it is the fundamental group of the Klein bottle), being the second and last of the wallpaper groups for which this happens. Hence, for every prime p,

$\mathbf{P}_{\mathrm{B}\mathbb{Z}/p}\mathbf{Bpg} \simeq \underline{\mathbf{Bpg}} \simeq K$. The statement could also have been proved by checking directly that $\mathbf{pg}$ is the fundamental group of the Klein bottle.

5. The monorrhombic group
$$\mathbf{cm} = \{x, y, z; xyx^{-1}y^{-1} = 1, z^2 = 1, zyzy = 1, zxzy^{-1}x^{-1} = 1\}.$$

The name of this group comes from the fact that the generating translations determine a rhombus. The generator z represents in this case a reflection with respect to one of the diagonals of the rhombus, so the fundamental region is one of the two triangles in which the rhombus is divided. It is not hard to see that the orbit space of the plane under the action of $\mathbf{cm}$ is a Möbius band, which is known to retract to a circle.

On the other hand, as there are no rotations, the group has only torsion in the prime 2. Hence, the $\mathrm{B}\mathbb{Z}/2$ nullification of the monorrhombic group has the homotopy type of S^1.

6. The discope group
$$\mathbf{pmm} = \{x, y, z, t; xyx^{-1}y^{-1} = 1, z^2 = t^2 = 1, ztz^{-1}t^{-1} = 1,$$
$$zxzx^{-1} = 1, txtx = 1, zyzy = 1, tyty^{-1} = 1\}.$$

This is the first example where two classes of different distinguished isometries appear in the group: the generators z and t represent reflections, while a glide-reflection is given by the composition of a generating reflection with the corresponding generating translation. The translations are supposed here to determine a rectangle, and the reflection axes join opposite middle points of the sides of the rectangle. Hence, a fundamental region is given by a little rectangle which is a quarter of the original one. As the identification do not affect now the sides of the fundamental region, it is a model for the classifying space for proper $\mathbf{pmm}$-bundles. Thus, in this case $\underline{\mathbf{Bpmm}}$ is contractible.

On the other hand, as there are no rotations, there is only torsion at the prime 2, and hence we have that $\mathbf{P}_{\mathrm{B}\mathbb{Z}/2}\mathbf{Bpmm}$ has the homotopy type of a point. Note in particular that according to ([LN01], proposition 3), the group is generated by torsion, as one can directly see by changing the family of given generators to $\{z, t, tx, zy\}$.

7. The digyro group
$$\mathbf{pmg} = \{x, y, z, t; xyx^{-1}y^{-1} = 1, t^2 = 1, z^2y^{-1} = 1,$$
$$zxz^{-1}x = 1, txtx^{-1} = 1, tztz = 1, tyty = 1\}.$$

This group contains 2-rotations and reflections as distinguished isometries, and the generating translations are supposed to span a rectangle. The generator t represents a rotation of angle π around the center c of the rectangle, and z is the composition of t with a reflection with respect to a line that passes through the middle point of the segment determined by c and the middle point of the upper side, and is parallel to this side. There is another reflection axis, which is symmetric to this one, and 2-rotation centers in the vertices and the middle points of the sides of the rectangle. Of course, there is again only torsion in the prime 2.

The fundamental region is again a rectangle, which is the smallest among the two determined by one of the reflection axes previously described. The identifications in the quotient do not affect the sides of the fundamental region, and hence this is a contractible model for $\underline{\mathrm{B}}\mathbf{pmg}$. So, we can conclude that the $\mathrm{B}\mathbb{Z}/2$-nullification of the classifying space of the digyro group is trivial.

8. The diglide group

$$\mathbf{pgg} = \{x, y, z, t; xyx^{-1}y^{-1} = 1, t^2y^{-1} = 1, z^2x^{-1} = 1,$$
$$txt^{-1}x = 1, ztzt = 1, zyz^{-1}y = 1\}.$$

The name of the group comes from the fact that the generators z and t represent two glide-reflections. It is also supposed here that the generating translations x and y give rise to a rectangle, and then the glide-reflection vectors are perpendicular, two-by-two parallel to the sides of the rectangle, and they quarter precisely a quarter of the rectangle. Hence, this is the fundamental region for $\mathbf{pmg}$. The orbit space $\mathbb{R}^2/\mathbf{pgg}$ can be seen to have the shape of a "non-orientable football", which is in fact homotopy equivalent to the projective plane.

The group contains also rotations of angle π, as for example the composition of z and t, and no other distinguished isometries. Hence, there appears only torsion in the prime 2, and then the $\mathrm{B}\mathbb{Z}/2$-nullification of $\mathbf{Bpgg}$ is homotopy equivalent to $\mathbb{R}P^2$. This is the first case where there is some torsion in the fundamental group of the nullification.

9. The dirrhombic group

$$\mathbf{cmm} = \{x, y, z, t; xyx^{-1}y^{-1} = 1, z^2 = t^2 = 1,$$
$$zyzy = 1, zxzy^{-1}x^{-1} = 1, ztzt = 1, tyty = 1, txtx = 1\}.$$

As in the previous example $\mathbf{cm}$, the generating translations determine a rhombus. The other two generators represent reflections whose axes are precisely the diagonals of the rhombus. There are rotations of angle π, as for example the composition of z and t, whose center is the center of the rhombus. Moreover there appear glide-reflections, whose axes join middle points of adjacent sides, but there are no other rotations. So, we only have again to deal with 2-torsion.

It is clear from the structure of the generators that a fundamental region is a quarter of the rhombus, i.e., a rectangle triangle whose sides are the halves of the reflection axis and a side of the original rhombus. The quotient of the plane by the action of the dirrhombic induces an identification on this last side, and the classifying space for proper $\mathbf{cmm}$-bundles has the shape of a slit turnover, with a corner and two mirror points, which is contractible. Then, $\mathrm{B}_{\mathcal{F}_2}\mathbf{cmm} \simeq *$ is also homotopy equivalent to a point, and the classifying space of the group is $\mathrm{B}\mathbb{Z}/2$-acyclic.

10. The tetratrope group

$$\mathbf{p4} = \{x, y, z; xyx^{-1}y^{-1} = 1, z^4 = 1, zyz^{-1}x = 1, zxz^{-1}y^{-1} = 1\}.$$

In this group the translations x and y give rise to a square, and the generator z is simply a rotation of angle $\pi/2$ around the center of the square. The group contains also some 2-rotations, but no other distinguished isometries. So, there is only torsion in the prime 2, with some elements of order 4. As the generator z interchanges the four quarters of the square, we can take one to be the fundamental region. In the orbit space $\mathbb{R}^2/\mathbf{p4}$ the rotation z has identified the two sides of the quarter that are not over the big square. Then, the classifying space for the family of finite 2-groups has again the shape of a turnover, but unlike the previous example, it is *non*-slit. This changes radically its homotopy type, and in fact it is not hard to see that $\underline{\mathrm{B}}\mathbf{p4}$ is homotopy equivalent to a sphere. Hence, $\mathbf{P}_{\mathrm{B}\mathbb{Z}/2}\mathrm{B}\mathbf{p4} \simeq \mathrm{B}_{\mathcal{F}_2}\mathbf{p4} \simeq S^2$.

11. The tetrascope group

$$\mathbf{p4} = \{x, y, z, t; xyx^{-1}y^{-1} = 1, z^4 = t^2 = 1,$$
$$zyz^{-1}x = 1, zxz^{-1}y^{-1} = 1, txty^{-1} = 1, tztz = 1\}.$$

As in the previous case, the translations x and y generate a square. The generator z represents again a rotation of angle $\pi/2$ around the center of the square, and t is a reflection with respect to an axis that joins two opposite middle points of sides of the square. Moreover, the perpendicular line to this one is another reflection axis, and also the diagonals of the square. There are 2-rotations centered in the middle points of the sides, and glide-reflections whose axes join adjacent middle points of the sides. As there are no other rotations, there is again only 2-torsion.

Recall that in the previous group the fundamental region was a quarter of the original square. Here, as the reflection axis of the new generator divide the square in two halves, the fundamental region is a little triangle. In this case, the identifications do not affect the sides of the triangle, so this is a model for $\mathrm{B}_{\mathcal{F}_2}\mathbf{p4}$, and as it is contractible, $\mathrm{B}\mathbf{p4}$ is $\mathrm{B}\mathbb{Z}/2$-acyclic. Observe that the vertices on the diagonal are 4-mirror points, that correspond to centers of 4-rotations, and one 2-mirror point, that correspond to a center of 2-rotation.

12. The tetragyro group

$$\mathbf{p4g} = \{x, y, z, t; xyx^{-1}y^{-1} = 1, z^4 = t^2 = 1,$$
$$zyz^{-1}x = 1, zxz^{-1}y^{-1} = 1, txty^{-1} = 1, tztxz = 1\}.$$

This is the last of the wallpaper groups for which the generating translations give rise to a square. Although the relations seem almost identical to those of $\mathbf{p4}$, they are quite different. The generator z represents again a 4-rotation around the center of the square, but this t stands for a 2-reflection through an axis that joins middle points of *adjacent* sides of the square. There are again 2-rotations, whose centers are the middle points of the sides of the square, and glide-reflections whose axes are this time parallel to the sides. This is the last wallpaper group where the torsion is concentrated in the prime 2.

The generator t interchanges again the two halves of the quarter of the square (although perpendicularly to the one of the previous example) and then the generating region is again a triangle. However, this time there are non-trivial identifications in the orbit space, that has the shape of a slit turnover with a corner and a mirror point. This space is known to be contractible, and then the $\mathrm{B}\mathbb{Z}/2$-nullification of the classifying space of the tetragyro group is trivial.

13. The tritrope group
$$\mathbf{p3} = \{x, y, z; xyx^{-1}y^{-1} = 1, z^3 = 1, zxz^{-1}y^{-1}x = 1, zyz^{-1}x = 1\}.$$

This example was already studied in [Flo05], where it is shown that there only appears 3-torsion, and the $B\mathbb{Z}/3$-nullification of $\mathbf{Bp3}$ has the homotopy type of the sphere.

14. The trigyro group
$$\mathbf{p31m} = \{x, y, z, t; xyx^{-1}y^{-1} = 1, z^3 = t^2 = 1, tztz = 1,$$
$$tzxtzx^{-1} = 1, tyty^{-1} = 1, txty^{-1}x = 1, tzytzyx = 1\}.$$

In this example, the generating translations span a rhombus. The generator t represents a reflection with respect to the short diagonal of the rhombus, while z is a 3-rotation around one of the two equilateral triangles that determine t. There are more reflections, whose axes stay over all the sides of the rhombus, and glide-reflections, whose vectors join the middle points of opposite sides of the rhombus. Actually, this group is very similar to the previous one, except for the fact that here the rotation centers are *not* in the reflection axes. There appears torsion at the primes 2 and 3.

As the angle of the rotation z is $2\pi/3$, a fundamental region for the trigyro group is an obtuse triangle that is exactly a third of the equilateral triangle, and such that its obtuse angle is in the center of rotation of z and measures $2\pi/3$. As z identifies two sides of this little triangle in the quotient space, it has the shape of a slit turnover, which is contractible, as we already know. Hence, we have that $\mathbf{P}_{B\mathbb{Z}/2 \vee B\mathbb{Z}/3} \mathbf{Bp31m} \simeq \underline{\mathbf{Bp31m}} \simeq *$.

Now we concentrate in the prime 3. It is known (see [CM65]) that the tritrope group $\mathbf{p3}$ is a normal subgroup of index 2 of the trigyro group. Then we have a fibration of classifying spaces:

$$\mathbf{Bp3} \longrightarrow \mathbf{Bp31m} \longrightarrow B\mathbb{Z}/2.$$

Now, as the base is $B\mathbb{Z}/3$-null, the fibration is preserved by the $B\mathbb{Z}/3$-nullification functor, and hence $\mathbf{P}_{B\mathbb{Z}/3}\mathbf{Bp31m}$ fits in a covering fibration where the base is the infinite projective space and the universal cover is the sphere.

Here, the action of the fundamental group of $\mathbf{P}_{B\mathbb{Z}/3}\mathbf{Bp31m}$ on the fiber is induced by the one of $\mathbb{Z}/2$ over $\mathbf{p3}$. If we identify this fiber with $\underline{\mathbf{Bp3}}$, this action is by reflection in the equator.

Now, if we consider $S^2 \times_{\mathbb{Z}/2} E\mathbb{Z}/2$, the Borel construction of the 2-sphere with respect to this action, it is a consequence of ([Las56], 5.6) that the $B\mathbb{Z}/3$-nullification of $\mathbf{Bp31m}$ has the homotopy type of $S^2 \times_{\mathbb{Z}/2} E\mathbb{Z}/2$. In particular, it is apparent from this description that $\mathbf{P}_{B\mathbb{Z}/3}\mathbf{Bp31m}$ cannot be either the product bundle nor the projective plane.

We could not use these tools to obtain a model for the $B\mathbb{Z}/2$-nullification of $\mathbf{Bp31m}$, because this group has no normal subgroup whose quotient is a $B\mathbb{Z}/2$-null space.

15. The tryscope group

$$\mathbf{p3m1} = \{x, y, z; xyx^{-1}y^{-1} = 1, z^3 = t^2 = 1,$$
$$tztz = 1, zxz^{-1}y^{-1}x = 1, zyz^{-1}x = 1, txtx = 1, tyty^{-1}x = 1\}.$$

This was the last wallpaper group whose nullification was described in [Flo05]. It is proved there that $\mathbf{Bp3m1}$ is $B\mathbb{Z}/2 \vee B\mathbb{Z}/3$-acyclic, and that its nullification fits in a covering fibration:

$$S^2 \longrightarrow \mathbf{P}_{B\mathbb{Z}/3}\mathbf{Bp3m1} \longrightarrow B\mathbb{Z}/2.$$

As in the trigyro group, the total space of this fibration has the homotopy type of the Borel construction $S^2 \times_{\mathbb{Z}/2} E\mathbb{Z}/2$, with respect to the action of $\mathbb{Z}/2$ on the fibre.

We cannot compute the $B\mathbb{Z}/2$-nullification because of similar reasons to those of the previous group. It is worth to remark that a model for $\mathbf{P}_{B\mathbb{Z}/2}\mathbf{Bp31m}$ would immediately give another one for $\mathbf{P}_{B\mathbb{Z}/2}\mathbf{Bp3m1}$, because $\mathbf{p31m}$ is an index 3 normal subgroup of $\mathbf{p3m1}$.

16. The hexatropic group

$$\mathbf{p6} = \{x, y, z; xyx^{-1}y^{-1} = 1, z^6 = 1, zxz^{-1}y^{-1} = 1, zyzy^{-1}x = 1\}.$$

In this wallpaper group and the following one, the generating translations will span a rhombus, just like in the trigyro group. As the rest of the $\mathbf{pn}$-groups, the unique distinguished isometries of $\mathbf{p6}$ group are rotations. More concretely, the generator z represents a 6-rotation around the point of application of the direction vectors of the generating translations, and the group also contains 6-rotations around the other vertices of the rhombus, 2-rotations around the middle point of the sides and the center of the rhombus, and 3-rotations around the centers of the two equilateral triangles that the short diagonal determine.

The fundamental region is, like in the trygiro group, a obtuse triangle that is a third of one of the equilateral ones. The quotient by the action of the hexatropic group identify the short sides of the little triangle, and bends over itself the long one, so a model for the classifying space for proper $\mathbf{p6}$-bundles is a 2-sphere. In this group torsion appears for the primes 2 and 3, so $\mathbf{P}_{B\mathbb{Z}/2\vee B\mathbb{Z}/3}\mathbf{Bp6} \simeq S^2$.

As it happens for the two previous groups, the tritrope group is included in $\mathbf{p6}$ as a normal group of index 2, and then we have again a universal covering fibration:

$$S^2 \longrightarrow \mathbf{P}_{B\mathbb{Z}/3}\mathbf{Bp6} \longrightarrow B\mathbb{Z}/2.$$

Fortunately, the group $\mathbf{p6}$ also includes the ditrope group as an index *three* normal subgroup, so we can also characterize the $B\mathbb{Z}/2$-nullification of the hexatropic group by means of a universal covering fibration

$$S^2 \longrightarrow \mathbf{P}_{B\mathbb{Z}/3}\mathbf{Bp6} \longrightarrow B\mathbb{Z}/3.$$

taking account of the fact that $\mathbf{P}_{B\mathbb{Z}/2}\mathbf{p2}$ is homotopy equivalent to the sphere.

17. The hexascopic group

$$\mathbf{p6} = \{x, y, z, t; xyx^{-1}y^{-1} = 1, z^6 = t^2 = 1, zyz^{-1}y^{-1}x = 1,$$
$$zxz^{-1}y^{-1} = 1, txtx = 1, tyty^{-1}x = 1, tzty^{-1}z = 1\}.$$

The last of the wallpaper groups is probably the most complex of all of them. Fourteen of them can be immersed in it, and it is the only one that contains 6-rotations, reflections and glide-reflections. The generators are a rotation z of angle $\pi/3$ around the vertex that determine the direction vectors of the generating translations, and a reflection whose axis is the long diagonal of the rhombus. There are again 6-rotations around the other vertices of the rhombus, 2-rotations around the center of the rhombus and the middle point of the sides, 3-rotations around the centers of the two equilateral triangles determined by the short diagonal, and twelve glide-reflections vectors.

The fundamental region is a rectangle triangle determined by the long diagonal, the direction vector of one of the translations and the perpendicular to the latter that passes by the arrow of the other vector. In the orbit space there are no identifications, so this little triangle is a model for $\underline{\mathrm{B}}\mathbf{p6m}$, and then for the $\mathrm{B}\mathbb{Z}/2 \vee \mathrm{B}\mathbb{Z}/3$-nullification of the classifying space of the hexascopic group.

For the same reasons as above, we are not able to compute the $\mathrm{B}\mathbb{Z}/2$-nullification of the classifying space of the group, so we will finish by computing $\mathbf{P}_{\mathrm{B}\mathbb{Z}/3}\mathrm{B}\mathbf{p6m}$. Again we use as input the tritrope group $\mathbf{p3}$, that is contained in $\mathbf{p6m}$ as an index 4 subgroup. Hence, the desired nullification is defined by an universal covering fibration:

$$S^2 \longrightarrow \mathbf{P}_{\mathrm{B}\mathbb{Z}/3}\mathrm{B}\mathbf{p6m} \longrightarrow \mathrm{B}\mathbb{Z}/2 \times \mathrm{B}\mathbb{Z}/2.$$

References

[BCH94] P. Baum, A. Connes and N. Higson. Classifying space for proper actions and K-Theory of group C^*-Algebras *Contemp. Math*, 167:241–291, 1994.

[BF03] A.J. Berrick and E. Dror-Farjoun Fibrations and nullifications. *Israel J. Math.*, 135:205–220, 2003.

[BK72] A.K. Bousfield and D.M. Kan. *Homotopy limits, completions and localizations.* Springer-Verlag, Berlin, 1972. Lecture Notes in Mathematics, Vol. 304.

[Bou94] A.K. Bousfield. Localization and periodicity in unstable homotopy theory. *J. Amer. Math. Soc.*, 7(4):831–873, 1994.

[CCS05] N. Castellana, J.A. Crespo and J. Scherer. Deconstructing Hopf spaces. Preprint 2004, available at http://xxx.sf.nchc.gov.tw/abs/math.AT/0404031.

[Cha96] W. Chachólski. On the functors CW_A and P_A. *Duke Math. J.*, 84(3):599–631, 1996.

[CM65] H. Coxeter and W. Moser. *Generators and relations for discrete groups*, volume 14 of *Ergebnisse der Mathematik und ihrer Grenzgebiete*. Springer-Verlag, 2nd edition, Berlin, 1965.

[Die87]   T. tom Dieck. *Transformation groups*, volume 8 of *de Gruyter Studies in Mathematics*. de Gruyter and Co., 1987.

[DMG99] J. Dermott, M. du Sautoy and G. Smith. Zeta functions of crystallographic groups and analytic continuation. *Proc. Lond. Math. Soc.*, 79(3):511–534, 1999.

[Dwy96]   W.G. Dwyer. The centralizer decomposition of BG. In *Algebraic topology: new trends in localization and periodicity (Sant Feliu de Guíxols, 1994)*, volume 136 of *Progr. Math.*, pages 167–184. Birkhäuser, Basel, 1996.

[Far96]   E. Dror Farjoun. *Cellular spaces, null spaces and homotopy localization*, volume 1622 of *Lecture Notes in Mathematics*. Springer-Verlag, Berlin, 1996.

[FL02]   F.A. Farris and R. Lanning. Wallpaper functions. *Expo. Math.*, 20(3):193–214, 2002.

[Flo04]   R.J. Flores. Nullification and cellularization of classifying spaces of finite groups. To appear in *Trans. Amer. Math. Soc.*

[Flo05]   R.J. Flores. Nullification functors and the homotopy type of the classifying space for proper actions. *Algebr. Geom. Topol.*, 5(46):1141–1172. 2005.

[Joy05]   D.E. Joyce. http://www.clarku.edu/~djoyce/wallpaper/.

[Las56]   R. Lashof. Classification of fibre bundles by the fibre space of the base. *Ann. of Math.*, 64(3):436–446, 1956.

[LN01]   I.J. Leary and B.E.A. Nucinkis. Every CW-complex is a classifying space for proper bundles. *Topology*, 40(3):539–550, 2001.

[Lee05]   X. Lee. http://www.xahlee.org/wallpaper_dir/c5_17wallpapergroups.html.

[LS00]   W. Lück and R. Stamm. Computations of K- and L-theory of cocompact planar groups. *K-theory*, 21(3):242–292, 2000.

[Lüc89]   W. Lück. *Transformation groups and algebraic K-theory*, volume 1408 of *Lecture Notes in Mathematics*. Springer-Verlag, Berlin, 1989.

[Lüc00]   W. Lück. The type of the classifying space for a family of subgroups. *J. Pure Appl. Algebra*, 149(2):177–203, 2000.

[Lüc05]   W. Lück. Survey on classifying spaces for families of subgroups. Preprint, available at http://www.math.uni-muenster.de/u/lueck/publ/lueck/classi.pdf.

[MT89]   M. Mahowald and R. Thompson. *K-Theory and unstable homotopy groups*, volume 96 of *Contemp. Math.* Amer. Math. Soc., 1989.

[Mil84]   H. Miller. The Sullivan conjecture on maps from classifying spaces. *Ann. of Math.* (2), 120(1):39–87, 1984.

[Mis03]   G. Mislin. Equivariant K-homology of the classifying space for proper actions. In *Proper group actions and the Baum-Connes conjecture (Bellaterra, 2001)*, volume 3 of *Adv. Cours. Math. CRM*, pages 1–78. Birkhäuser, Basel, 2003.

[RS01]   J.L. Rodríguez and J. Scherer. Cellular approximations using Moore spaces. In *Cohomological methods in homotopy theory (Bellaterra, 1998)*, volume 196 of *Progr. Math.*, pages 357–374. Birkhäuser, Basel, 2001.

[Sc78]   D. Schattschneider. The plane symmetry groups: their recognition and notation. *Amer. Math. Monthly*, 85(6):439–450, 1978.

[Ser71]   J.P. Serre. Cohomologie des groupes discrets. In Prospects in Mathematics, *Proc. Sympos. Princeton Univ. (Princeton, N.J.,* 1970), volume 70 of *Ann. of Math. Studies.*, pages 77–169. Princeton University Press, 1971.

Ramón J. Flores
Departamento de Estadística
Universidad Carlos III
E–28270 Colmenarejo, Madrid, Spain
e-mail: `rflores@est-econ.uc3m.es`

Geometric Group Theory
Trends in Mathematics, 65–73
© 2007 Birkhäuser Verlag Basel/Switzerland

A General Construction of JSJ Decompositions

Vincent Guirardel and Gilbert Levitt

This is an extended and updated version of a talk given by Gilbert Levitt in Barcelona on July 2, 2005 about joint work with Vincent Guirardel, currently being written. It is based on notes TEX'd by Peter Kropholler. I thank Peter for making his notes available to me. Of course, he shouldn't be held responsible for the content. Thanks also to the organizers of the Barcelona Conference on Geometric Group Theory, and to the referee for helpful suggestions.

1. Introduction

JSJ theory has its roots in the work of Jaco-Shalen and Johannson on 3-manifolds. In the context of group theory, JSJ decompositions were constructed by Kropholler for duality groups, and in general by Rips-Sela [RS], Dunwoody-Sageev [DS], Fujiwara-Papasoglu [FP], Scott-Swarup [SS]. Roughly speaking, the main purpose of a JSJ decomposition is to describe all splittings of a given group G over a certain class $\mathcal{A}$ of subgroups.

Here are a few ideas that will be developed in this talk:
- Rather than viewing the JSJ decomposition of G over $\mathcal{A}$ as a G-tree (a splitting) satisfying certain properties (which in general don't define it uniquely), one should *view the JSJ decomposition as a well-defined deformation space* $\mathcal{D}_{JSJ}$ satisfying a universal property.

  A deformation space [Fo 1] is a collection of G-trees (and $\mathcal{D}_{JSJ}$ contains the trees constructed in [RS], [DS], [FP]). It is a contractible complex [GL]. A basic example: when $G = F_n$, the JSJ deformation space (over any $\mathcal{A}$) is Culler-Vogtmann's Outer space, consisting of all free F_n-trees (see [Vo]); it has no preferred element.
- *A deformation space $\mathcal{D}$ carries a lot of information* (see [GL]). Trees in a given deformation space have a lot in common, so many invariants may be extracted from $\mathcal{D}$ (in particular, the set of vertex stabilizers not in $\mathcal{A}$). One may also say that two trees in the same deformation space are always related

This article originates from the Barcelona conference.

by certain types of moves (expansions and collapses, sometimes slides), so the JSJ splitting of G over $\mathcal{A}$ is well defined up to these moves.

- *If G is finitely presented, the JSJ deformation space $\mathcal{D}_{JSJ}$ always exists.* No assumption on $\mathcal{A}$ is necessary. Vertex stabilizers of trees in $\mathcal{D}_{JSJ}$ yield a canonical finite set of conjugacy classes of subgroups of G.

- The JSJ deformation space $\mathcal{D}_{JSJ}$ usually contains no preferred splitting. In order to get a canonical tree, we construct (for G finitely presented) a second deformation space, the compatibility JSJ deformation space of G over $\mathcal{A}$. Unlike $\mathcal{D}_{JSJ}$, it always has a preferred element, yielding an $\mathrm{Aut}(G)$-*invariant splitting which we call the compatibility JSJ tree.* For instance, the Bass-Serre tree of the HNN extension defining $BS(m,n) = \langle a, t \mid ta^m t^{-1} = a^n \rangle$ is the cyclic compatibility tree of $BS(m,n)$ when none of m, n divides the other.

2. Assumptions and notations

We fix a finitely generated group G. Let $\mathcal{A}$ be a family of subgroups which is subgroup closed and closed under conjugation ($\mathcal{A}$ is usually defined by restricting the isomorphism type: trivial, finite, cyclic, abelian, slender, small...).

We always consider simplicial G-trees T whose edge stabilizers belong to $\mathcal{A}$. We make the standard assumptions on T: there are no inversions, and T is minimal (it contains no proper G-invariant subtree). The trivial tree ($T = $ point) is permitted (the JSJ decomposition is sometimes trivial...).

An element of G, or a subgroup, is elliptic in T if it fixes a point.

We denote by Γ the quotient graph of groups T/G; this is a marked graph of groups.

All maps between trees will be G-equivariant, but otherwise arbitrary. Of course, any map may be redefined on edges so as to be linear or constant on every edge.

Typical examples of maps f are collapses and folds. *Collapses* will be important for us. In a collapse, an edge e of Γ is collapsed to a single vertex (Figure 1). In T, each edge in a given G-orbit is collapsed to a point.

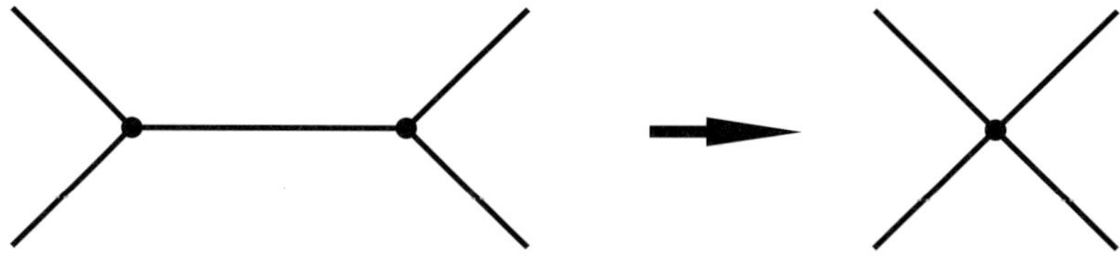

FIGURE 1. Collapsing.

3. Deformation spaces

Deformation spaces were introduced by Forester [Fo 1]. See [GL] for a detailed study.

Definition 1 (domination). T *dominates* T' if there is a map $f : T \to T'$. By abstract nonsense, this is equivalent to saying that, if a subgroup H is elliptic in T, then it is elliptic in T' (recall that all maps are equivariant).

Definition 2 (deformation space). T and T' are in the same *deformation space* if they dominate each other. Equivalently, they have the same elliptic subgroups. (For experts: this is slightly stronger than saying that they have the same elliptic elements.)

In this talk we will think of a deformation space as a set of (non-metric) simplicial trees. It is actually a contractible complex.

Definition 3 (expansion, admissible). An *expansion* is the opposite of a collapse. A collapse or an expansion is *admissible* if it does not change the deformation space. More on admissibility later (Remark 5).

Theorem 4. *Given T, T', the following are equivalent:*

(1) *T, T' are in the same deformation space $\mathcal{D}$.*
(2) *They are equivariantly quasi-isometric.*
(3) *The length functions are bi-Lipschitz equivalent.*
(4) *T, T' are related by a finite sequence of admissible collapses and expansions.*

(4) means that T may be "deformed" to T' *within* $\mathcal{D}$. Most of this theorem, especially the hard part (1) $\Rightarrow$ (4), is due to Forester [Fo 1]. A proof appers in [GL]. (For experts: condition (3) is equivalent to the others only when the trees are irreducible.)

Domination induces a *partial ordering* on the set of deformation spaces of G. The trivial tree is the smallest element. In general, there is no highest element.

Example 1 (free group). G is the free group F_n, and $\mathcal{A}$ consists of the trivial group. There is a highest $\mathcal{D}$, the set of free G-trees. As a complex this is Culler–Vogtmann's Outer space. In this case, the minimality assumption says that the quotient graph has no valence one vertices.

Let F be a finite subgroup of $\mathrm{Aut}(F_n)$, and $G_F = F_n \rtimes F$. The set of free F_n-trees which are F-invariant may be identified with a deformation space of G_F-trees.

Example 2 (surface group). G is the fundamental group of a closed orientable surface, $\mathcal{A}$ is the class of cyclic subgroups. Any tree is dual to a family $\mathcal{C}$ of disjoint, non-parallel, essential simple closed curves. Deformation spaces (viewed as sets) are singletons, one for each $\mathcal{C}$. A deformation space is maximal for domination if and only if the corresponding $\mathcal{C}$ is a pair of pants decomposition. The poset of deformation spaces (ordered by domination) is equivalent to the curve complex of the surface.

Example 3 (GBS groups). Suppose G acts on T with all edge and vertex stabilizers infinite cyclic (we say that G is a generalized Baumslag-Solitar group, or GBS group). If G is not $\mathbb{Z}^2$ or a Klein bottle group, the set of all such trees is a deformation space [Fo 1] (exercise: there are infinitely many spaces if $G = \mathbb{Z}^2$, two if G is a Klein bottle group).

Remark 5. More on admissibility. What does it mean to be admissible in the context of Outer space? It means that you can collapse a forest in the quotient graph but not a loop, since collapsing a loop creates a non-free tree. Theorem 4 is basically the statement that two marked graphs in Outer space are related by forest collapses and expansions.

In a general deformation space, collapsing an edge of Γ is admissible if and only if the edge has distinct end points and the edge group maps onto at least one of the vertex groups: admissible collapses are just consequences of the isomorphism $A *_B B \simeq A$. On the other hand, collapsing an edge carrying a non-trivial amalgam $A *_C B$ creates new elliptic elements (e.g., ab with $a \in A \setminus C$ and $b \in B \setminus C$).

Definition 6 (reduced). T is *reduced* if no collapse is admissible. (For experts: This terminology comes from [Du] and [Fo 1]. Scott-Wall use the name "minimal". Reduced in this sense is stronger than reduced in the sense of Bestvina-Feighn's accessibility paper.)

As reduced trees are the simplest elements in a deformation space, one may try to connect them in a more direct way than by expansions and collapses.

In Outer space, reduced trees are just roses (the quotient graph has only one vertex). Two roses differ by an automorphism of F_n, and any generating set of $\mathrm{Aut}(F_n)$ gives a way to connect roses.

In particular, Nielsen automorphisms correspond to slides. For instance, the automorphism $a \mapsto a$, $b \mapsto ba$ in the free group on a, b corresponds to sliding the edge of Γ labelled b over that labelled a.

Definition 7 (slide). In general, slides are based on the isomorphism $A *_E B *_F C \simeq A *_E C *_F B$, valid if $E \subset F$. A slide looks in Γ like one end of an edge e is sliding along a different edge f satisfying $G_e \subset G_f$ (one or both edges may be loops), see Figure 2. Note that slide moves do not change edge and vertex groups (they are always admissible).

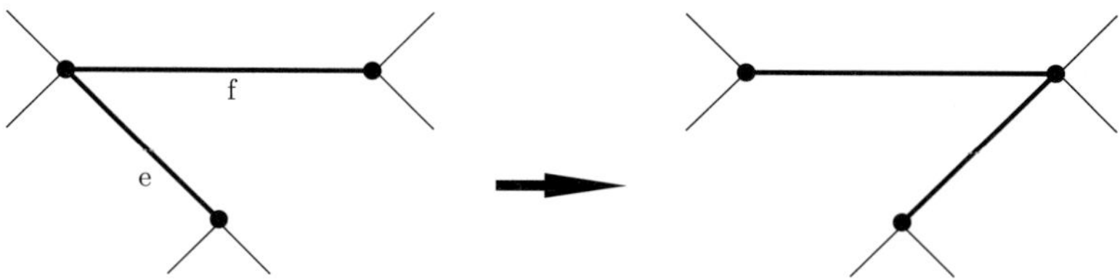

FIGURE 2. Sliding.

In Outer space, two reduced trees are connected by slides (because Nielsen automorphisms (almost) generate $\mathrm{Aut}(F_n)$). But this is not always true.

Here is an example where slide moves are not sufficient to achieve everything.

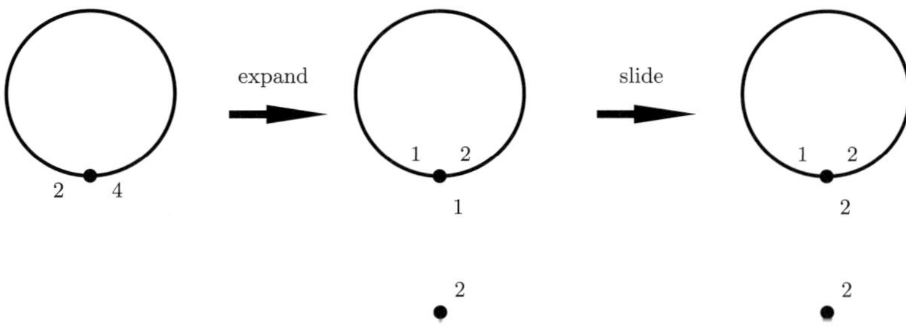

FIGURE 3

Example 4. Start with the standard HNN presentation $\langle a, t \mid ta^2t^{-1} = a^4 \rangle$ of the Baumslag–Solitar group $BS(2,4)$. After an admissible expansion (Figure 3), one gets a non-reduced graph of groups with two edges, corresponding to the presentation $\langle a, t, b \mid tbt^{-1} = b^2, b = a^2 \rangle$. Sliding around the loop gives a reduced graph of groups with two edges, corresponding to $\langle a, t, \tilde{b} \mid t\tilde{b}t^{-1} = \tilde{b}^2, \tilde{b}^2 = a^2 \rangle$ (with $\tilde{b} = t^{-1}bt$). The associated Bass-Serre tree cannot be connected to the original one by slides, as the quotient graphs do not have the same number of edges (other examples in [Fo 2]).

Theorem 8. *Let T, T' be reduced trees in the same deformation space $\mathcal{D}$. If no group in $\mathcal{A}$ properly contains a conjugate of itself, then T, T' are connected by slides.*

This was proved by Forester [Fo 3] for the deformation spaces of Example 3 (GBS groups). His proof works in general. See [GL] for another proof.

Corollary 9. *If $T, T', \mathcal{A}$ are as above, then T, T' have the same edge stabilizers and vertex stabilizers.*

In fact the corollary remains true without assumptions on $\mathcal{A}$, but one has to consider generalized edge stabilizers (groups H such that $G_e \subset H \subset G_{e'}$ for edges e, e') and restrict to "big" vertex stabilizers (those not in $\mathcal{A}$).

Theorem 10. *Two reduced trees belonging to the same deformation space $\mathcal{D}$ have the same generalized edge stabilizers and the same "big" vertex stabilizers.*

4. The JSJ deformation space

The JSJ decomposition of G over $\mathcal{A}$ is supposed to allow a description of all splittings of G over groups in $\mathcal{A}$. In what sense?

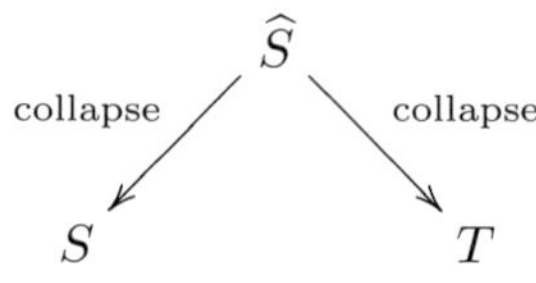

FIGURE 4

First consider an optimal situation. Suppose G is a one-ended hyperbolic group and $\mathcal{A}$ is the set of virtually cyclic subgroups. Then there is a canonical JSJ tree S, constructed purely in terms of the topology of ∂G (Bowditch [Bo]). In this case, the answer to the question asked above is the following [Gu] (Figure 4): every T (with virtually cyclic edge groups) may be obtained from S by first expansions and secondly collapses, but no zig-zag (later on, we will say that S and T are compatible). Note that we are now using non-admissible moves: we want to obtain all trees, in all deformation spaces.

In general, requiring that any tree may be obtained from S as on Figure 4 may force S to be the trivial tree (we will come back to this when we discuss compatibility). To obtain something more interesting, one settles for a bit less (Figure 5): *every T (with edge groups in $\mathcal{A}$) should be obtainable from S by expanding to some $\widehat{S}$ (which depends on T) and mapping $\widehat{S}$ to T (but the map $\widehat{S} \to T$ may fold, rather than just collapse).*

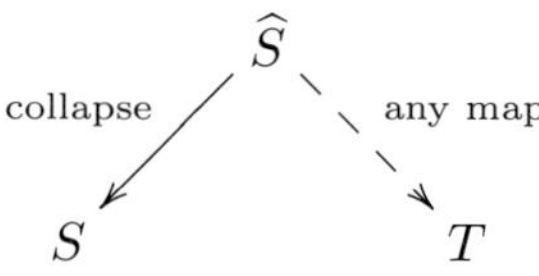

FIGURE 5

Note that, if S has this property, its edge stabilizers are elliptic in every T (because, unlike vertex stabilizers, edge stabilizers of S are always elliptic in $\widehat{S}$). The converse is also true: if edge stabilizers of S are elliptic in T, there is an $\widehat{S}$ as on Figure 5.

Definition 11 (universally elliptic). We say that S is *universally elliptic* if its edge stabilizers are elliptic in every T. We say that $\mathcal{D}$ is *universally elliptic* if some (hence every) reduced tree in $\mathcal{D}$ is universally elliptic.

It follows from this discussion that the JSJ decomposition should be universally elliptic. It should also be as large as possible (note that the trivial tree is universally elliptic).

Theorem 12. *If G is finitely presented, then there is a unique highest universally elliptic deformation space $\mathcal{D}$.*

Definition 13 (JSJ deformation space). We call it the *JSJ deformation space of G over $\mathcal{A}$, denoted $\mathcal{D}_{JSJ}$*. It follows from the above discussion that, given any reduced $S \in \mathcal{D}_{JSJ}$ and any T with edge groups in $\mathcal{A}$, there exists $\widehat{S}$ as in Figure 5.

The proof of the theorem is not hard and uses a form of Dunwoody accessibility: given a sequence of collapses $\cdots \to T_k \to \cdots \to T_1 \to T_0$, there exists T dominating every T_k.

Examples

- If G is free (and $\mathcal{A}$ is arbitrary), then $\mathcal{D}_{JSJ}$ is Outer space.
- If G is a surface group, and $\mathcal{A}$ is the class of cyclic subgroups, then $\mathcal{D}_{JSJ}$ is trivial.
- If G is accessible and $\mathcal{A}$ is the class of finite subgroups, then $T \in \mathcal{D}_{JSJ}$ if and only if vertex groups have at most one end. Note that all reduced trees in $\mathcal{D}_{JSJ}$ have the same edge and vertex groups by Corollary 9.
- If G is a GBS group, and $\mathcal{A}$ is the class of cyclic subgroups, then $\mathcal{D}_{JSJ}$ is the deformation space of Example 3 (trees with cyclic vertex groups).
- The JSJ splittings of [RS], [DS], [FP], [Bo] belong to $\mathcal{D}_{JSJ}$.

5. The compatibility JSJ tree

Now we look at finding preferred (in particular, $\mathrm{Aut}(G)$-invariant) trees, in $\mathcal{D}_{JSJ}$ when possible, or in a lower $\mathcal{D}$. This is similar to the approach of Scott-Swarup [SS], who insist on invariance under automorphisms.

When (as is usually the case) $\mathcal{A}$ is invariant under automorphisms of G, the group $\mathrm{Out}(G)$ acts on $\mathcal{D}_{JSJ}$. Deformation spaces are contractible and this gives information about $\mathrm{Out}(G)$, but having an (interesting) invariant tree is much stronger.

Sometimes you have to go very low to find a tree invariant under automorphisms: with the free group only the trivial tree is invariant. At the other extreme we have the Bowditch tree S of a one-ended hyperbolic group: being constructed in a canonical way from the boundary, S is $\mathrm{Aut}(G)$-invariant. Having this invariant tree makes it possible to determine $\mathrm{Out}(G)$ (Sela, see [Le]).

Recall the expand-then-collapse picture (Figure 4). We shall call S and T compatible, because they have a common expansion $\widehat{S}$ (as in [SS], we also use the term common refinement which seems more natural in this context). For instance, trees dual to disjoint curves on a surface are compatible. But two curves which meet in an essential way give rise to non-compatible splittings.

Definition 14 (compatible). Two trees are *compatible* if they have a common refinement. (For experts: irreducible trees are compatible if and only if the sum of their length functions is a length function.) A tree T is *universally compatible* if it is compatible with every T'. It is *$\mathcal{D}$-compatible* if it is compatible with every T' in $\mathcal{D}$.

Facts

- If $T_1, \ldots, T_k$ are pairwise compatible, then they have a *least common refinement (l.c.r.)*. This was also proved by Scott and Swarup (when edge groups are finitely generated).
- A deformation space $\mathcal{D}$ may contain only finitely many reduced $\mathcal{D}$-compatible trees (this is an easy form of accessibility).

Consequence of these facts: If $\mathcal{D}$ is a deformation space which contains a $\mathcal{D}$-compatible tree, then it has a preferred element $T_{\mathcal{D}}$: the l.c.r. of the reduced $\mathcal{D}$-compatible trees. In general, $T_{\mathcal{D}}$ is not reduced.

Theorem 15. *If G is a finitely presented group, then the set of deformation spaces containing a universally compatible tree has a unique highest element $\mathcal{D}_c$.*

This is quite harder to prove than Theorem 12.

Definition 16 (compatibility JSJ tree). The preferred element T_c of $\mathcal{D}_c$ is called the *compatibility JSJ tree* of G over $\mathcal{A}$. It is invariant under automorphisms if $\mathcal{A}$ is invariant. Any tree T (with edge groups in $\mathcal{A}$) may be obtained from T_c by expanding and collapsing as on Figure 4.

Examples ($\mathcal{A}$ as above)

- Free group, surface group: T_c is trivial.
- One-ended hyperbolic group: T_c is the Bowditch tree.
- $G = BS(2,4)$. Now T_c is trivial. But for $BS(2,3)$ then T_c is the Bass–Serre tree of the HNN extension. Note that the canonical decomposition of [SS] is trivial in this case.
- G a one-ended limit group (fully residually free group), $A = \{\text{cyclic groups}\}$, then T_c (almost) belongs to $\mathcal{D}_{JSJ}$ (almost, because $\mathbb{Z}^2$ subgroups are elliptic in T_c but not necessarily in $\mathcal{D}_{JSJ}$).

In the last example, T_c may be constructed directly as the *tree of cylinders* of any T in $\mathcal{D}_{JSJ}$.

Take an edge e of T, and look at all edges whose stabilizers are commensurable with G_e. The *cylinder* of e is the union of all these edges. Cylinders are subtrees meeting in at most one point. Given a cylinder C, let ∂C be the set of vertices of C also belonging to another cylinder. Then T_c is obtained from T by replacing every C by the cone over ∂C.

This type of construction works for cyclic or abelian splittings of CSA groups (for instance, relatively hyperbolic groups with abelian parabolic subgroups).

References

[Bo]   B. Bowditch, *Cut points and canonical splittings of hyperbolic groups*, Acta Math. **180** (1998), 145–186.

[Du]   M.J. Dunwoody, *Folding sequences*, Geometry and Topology monographs **1** (1998), 139–158.

[DS]   M.J. Dunwoody, M.E. Sageev, *JSJ-splittings for finitely presented groups over slender groups*, Invent. Math. **135** (1999), 25–44.

[Fo 1]   M. Forester, *Deformation and rigidity of simplicial group actions on trees*, Geom. & Topol. **6** (2002), 219–267.

[Fo 2]   M. Forester, *On uniqueness of JSJ decompositions of finitely generated groups*, Comm. Math. Helv. **78** (2003), 740–751.

[Fo 3]   M. Forester, *Splittings of generalized Baumslag-Solitar groups*, Geom. Dedicata **121** (2006), 43–59.

[FP]   K. Fujiwara, P. Papasoglu, *JSJ-decompositions of finitely presented groups and complexes of groups* GAFA **16** (2006), 70–125.

[Gu]   V. Guirardel, *Reading small actions of a one-ended hyperbolic group on **R**-trees from its JSJ splitting*, Am. J. Math. **122** (2000), 667–688.

[GL]   V. Guirardel, G. Levitt, *Deformation spaces of trees*, Groups, Geometry, and Dynamics **1** (2007), 135–181.

[Le]   G. Levitt, *Automorphisms of hyperbolic groups and graphs of groups*, Geom. Dedic. **114** (2005), 49–70.

[RS]   E. Rips, Z. Sela, *Cyclic splittings of finitely presented groups and the canonical JSJ decomposition*, Ann. Math. **146** (1997), 55–109.

[SS]   P. Scott, G.A. Swarup, *Regular neighbourhoods and canonical decompositions for groups*, Astérisque **289** (2003).

[Vo]   K. Vogtmann, *Automorphisms of free groups and Outer space*, Geom. Dedicata **94** (2002), 1–31.

Vincent Guirardel
Laboratoire Émile Picard
umr cnrs 5580, Université Paul Sabatier
F-31062 Toulouse Cedex 4, France.
e-mail: `guirardel@math.ups-tlse.fr`

Gilbert Levitt
LMNO, umr cnrs 6139
BP 5186, Université de Caen
F-14032 Caen Cedex, France
e-mail: `levitt@math.unicaen.fr`

Geometric Group Theory
Trends in Mathematics, 75–102

Décompositions de Groupes par Produit Direct et Groupes de Coxeter

Yves de Cornulier and Pierre de la Harpe

Abstract. We provide examples of groups which are indecomposable by direct product, and more generally which are uniquely decomposable as direct products of indecomposable groups. Examples include Coxeter groups, for which we give an alternative approach to recent results of L. Paris.

For a finitely generated linear group Γ, we establish an upper bound on the number of factors of which Γ can be the direct product. If moreover Γ has a finite centre or a finite abelianization, it follows that Γ is uniquely decomposable as direct product of indecomposable groups.

Résumé. Nous montrons des exemples de groupes indécomposables par produit direct, et plus généralement uniquement décomposables en produits de groupes indécomposables. Les exemples considérés incluent les groupes de Coxeter, pour lesquels nous redémontrons des résultats récents de L. Paris.

Pour un groupe linéaire de type fini Γ, nous établissons une borne supérieure sur le nombre de facteurs dont Γ puisse être produit direct. Si de plus Γ est à centre fini ou à abélianisé fini, il en résulte que Γ est uniquement décomposable en produit de groupes indécomposables.

There is an abridged English version at the end of the present paper.

Mathematics Subject Classification (2000). Primary 20E22, 20E34. Secondary 20D40.

Keywords. Indecomposable groups, direct products, uniquely directly decomposable groups, Coxeter groups, Wedderburn–Remak–Krull–Schmidt theorem.

Mots et phrases clé. Groupes indécomposables, produits directs, groupes uniquement directement décomposables, groupes de Coxeter, théorème de Wedderburn–Remak–Krull–Schmidt.

This article originates from the Geneva conference.

1. Introduction

Il existe plusieurs procédés pour «décomposer» des groupes en constituants qu'on peut imaginer d'étude «plus simple». L'objet du présent article est la décomposition par *produit direct.*

Un groupe est *indécomposable* s'il n'est pas un produit direct de manière non banale. Notre premier but est d'établir l'indécomposabilité de certains groupes apparaissant dans des contextes géométriques; par exemple des sous-groupes «assez grands» ou des réseaux dans certains groupes de Lie ou groupes algébriques.

Il est bien connu que de nombreux groupes possèdent plusieurs décompositions par produits directs qui sont essentiellement différentes. Voir l'exemple $A_{2,2} \times A_{3,3}$ de [Kuro, § 42], un exemple de Scott et les exemples abéliens de [Fuch], tous rappelés ci-dessous (numéros (xiv), (xv) et (xviii) du chapitre 2). Il y a aussi des exemples plus «dramatiques» dus à Baumslag: pour toute paire d'entiers $m, n \geq 2$, il existe des groupes de type fini $A_1, \ldots, A_m, B_1, \ldots, B_n$, nilpotents et sans torsion, indécomposables et non isomorphes deux à deux, tels que les produits directs $A_1 \times \cdots \times A_m$ et $B_1 \times \cdots \times B_n$ sont isomorphes [Baum]. Il existe de nombreux autres cas de «mauvais comportements» du point de vue des produits directs: par exemple des groupes Γ de type fini[1] isomorphes à $\Gamma \times \Gamma$ ([Ty74], [Ty80], [Meie]), ou à $\Gamma \times \Delta$, pour un groupe de présentation finie $\Delta \neq \{1\}$ [Hi86], ou des groupes de présentation finie $\Gamma \neq \{1\}$ qui se surjectent sur $\Gamma \times \Gamma$ [BaMi].

La situation est donc beaucoup moins simple que celle qui prévaut pour les produits libres, puisque le théorème de Grushko assure que tout groupe de type fini est produit libre d'un nombre fini de groupes librement indécomposables, et ceci de manière essentiellement unique (voir par exemple [Stal] ou [ScWa]).

La plupart des complications qui apparaissent dans les décompositions d'un groupe G par produit direct sont liées au centre de G. En effet, si celui-ci est réduit à un élément, une décomposition de G en produit fini de groupes indécomposables est nécessairement unique; ce fait, bien connu, est rappelé ci-dessous comme conséquence de la proposition 2 (voir aussi la proposition 9).

Convenons qu'un groupe est *uniquement directement décomposable*[2] s'il est somme restreinte de groupes indécomposables, ceci de manière unique à isomorphisme près et à l'ordre près des facteurs, et si de plus toute décomposition en somme restreinte admet un raffinement en somme restreinte de groupes indécomposables. (Rappelons que la *somme restreinte* d'une famille $(\Gamma_\iota)_{\iota \in I}$ désigne le sous-groupe du groupe produit des G_ι formé des éléments $(\gamma_\iota)_{\iota \in I}$ tels que $\gamma_\iota = 1$ pour presque tout ι; dans le cas où I est fini, nous suivons Bourbaki en écrivant aussi

---

[1]C'est un problème ouvert bien établi de savoir s'il existe un groupe Γ infini de présentation finie qui soit isomorphe à $\Gamma \times \Gamma$. Voici une question peut-être moins ambitieuse: existe-t-il un groupe Γ infini de présentation finie tel que, pour tout entier $m_0 \geq 1$, il existe des groupes $\Gamma_1, \ldots, \Gamma_m$ ($m \geq m_0$) non réduits à un élément dont le produit direct soit isomorphe à Γ?

[2]D'autres auteurs [Hi90] écrivent «groupe R.K.S.», en référence à Remak, Krull et Schmidt. Pour une notion de décomposabilité apparentée mais distincte, voir ci-dessous le chapitre 5.

«produit direct» au lieu de «somme restreinte».) Remarquons qu'un groupe uniquement directement décomposable qui est de type fini est nécessairement produit direct d'un nombre *fini* de facteurs indécomposables.

Notre second but est d'établir que certaines classes de groupes sont uniquement directement décomposables (propositions 3, 5 et 14). Après des rappels d'exemples standard, nous analysons en particulier la décomposabilité des *sous-groupes Zariski-denses de groupes algébriques*.

Nous illustrons notre méthode par les *groupes de Coxeter* (proposition 8). Dans [Par2], Paris a montré de manière complètement différente que les groupes de Coxeter de type fini sont uniquement directement décomposables. L'origine du présent travail est la recherche d'une autre preuve de ce fait.

Enfin, nous établissons une borne supérieure pour le nombre de facteurs non réduits à un élément dans une décomposition par produit direct d'un groupe linéaire de type fini (théorème 13).

La décomposition des groupes par produit direct intervient dans des résultats de décomposition d'espaces topologiques. Par exemple, soit Γ le groupe fondamental d'une variété riemannienne compacte à courbure négative ou nulle; supposons le centre de Γ réduit à un élément. C'est un cas particulier de résultats classiques de Gromoll–Wolf (1971) et Lawson–Yau (1972) qu'une décomposition du groupe en produit direct $\Gamma = \Gamma_1 \times \Gamma_2$ correspond nécessairement à une décomposition de la variété en produit riemannien. Voici un corollaire de résultats plus récents (voir [BrHa], chapitre II.6, ainsi que [Schr] et le chapitre 10 de [Eber]).

> *Soit Y un espace géodésique compact qui possède la propriété d'extension des géodésiques et qui est à courbure négative ou nulle. Supposons que le groupe fondamental de Y est un produit direct $\Gamma = \Gamma_1 \times \Gamma_2$ et que le centre de Γ est réduit à un élément. Alors Y est un produit d'espaces métriques $Y_1 \times Y_2$ de telle sorte que le groupe fondamental de Y_i est Γ_i ($i = 1, 2$).*

Mentionnons encore que, en théorie ergodique, certains produits directs donnent lieu à des phénomènes de rigidité remarquables [MoSh].

2. Premiers exemples

Un groupe Γ est *indécomposable* si[3] $\Gamma \neq \{1\}$ et si, pour tout isomorphisme de Γ avec un produit direct $\Gamma_1 \times \Gamma_2$, l'un des groupes Γ_1, Γ_2 est réduit à un élément. La notion est classique: voir parmi d'autres [Kuro, § 17], [Rotm, chapitre 6] et [Suzu, § 2.4].

---

[3]Nous adoptons ici une terminologie selon laquelle le groupe $\{1\}$ n'est ni indécomposable, ni décomposable (mais néanmoins le produit de la famille vide de groupes indécomposables). Ceci est cohérent avec la convention (par exemple de Bourbaki) selon laquelle le groupe à un élément n'est pas simple!

(i) Tout groupe simple est indécomposable.

(ii) Il existe de nombreux groupes finis indécomposables qui ne sont pas simples. Par exemple, pour tout $n \geq 2$, le groupe symétrique Sym_n, qui est un groupe de Coxeter de type A_n, est indécomposable (cela résulte de ce que tout sous-groupe normal distinct de $\{1\}$ contient le groupe alterné simple Alt_n si $n \neq 2, 4$, et d'un argument direct si $n = 2$ ou 4); pour tout $n \geq 3$, le produit semi-direct standard $\mathrm{Sym}_n \ltimes (\mathbf{Z}/2\mathbf{Z})^{n-1}$, qui est un groupe de Coxeter de type D_n, est indécomposable (l'argument de la proposition 1 ci-dessous s'applique). En revanche, il existe des groupes de Coxeter irréductibles finis qui sont décomposables; un tel groupe est produit direct de son centre à deux éléments et d'un groupe indécomposable. C'est le cas des groupes diédraux d'ordres $8n + 4$, des groupes de Coxeter de type B_n pour $n \geq 3$ impair (la vérification est laissée au lecteur), ainsi que des groupes de type H_3 et E_7 (voir [BouL], chapitre 6, §4, exercices 3 et 11).

Pour tout entier n pair, il existe un groupe fini métabélien indécomposable d'ordre n; par exemple le produit semi-direct

$$(\mathbf{Z}/2^k\mathbf{Z}) \ltimes_{\pm} (\mathbf{Z}/m\mathbf{Z}) = \langle a, b \mid a^{2^k} = 1, b^m = 1, a^{-1}ba = b^{-1} \rangle,$$

où $k, m \geq 1$ sont tels que $n = 2^k m$ avec m impair. En revanche, les entiers impairs qui sont des ordres de groupes finis indécomposables constituent un ensemble d'entiers de densité nulle [ErPa].

(iii) Les groupes abéliens $\mathbf{Z}$ et $\mathbf{Q}$ sont indécomposables; plus généralement, tout sous-groupe de $\mathbf{Q}$ non réduit à un élément est indécomposable (pour la description de ces groupes, voir par exemple le chapitre 10 de [Rotm]). Soient p un nombre premier et $m \geq 1$ un entier; un groupe cyclique d'ordre p^m est indécomposable, de même que le sous-groupe $\mathbf{Z}(p^\infty)$ de $\mathbf{C}^*$ des racines de l'unité d'ordres des puissances de p. En effet, dans chacun de ces groupes, l'intersection de deux sous-groupes non réduits à un élément n'est jamais réduite à un élément. Le groupe additif $\mathbf{Z}_p$ des entiers p-adiques (vu comme groupe discret) est indécomposable [Kapl, Section 15].

Un groupe abélien divisible indécomposable est isomorphe à l'un de $\mathbf{Q}$, $\mathbf{Z}(p^\infty)$; les groupes abéliens divisibles sont uniquement directement décomposables [Fuch, §23]. Un groupe abélien indécomposable est ou bien de torsion ou bien sans torsion; s'il est de torsion, il est isomorphe à l'un de $\mathbf{Z}/p^m\mathbf{Z}$, $\mathbf{Z}(p^\infty)$; voir [Kapl, Section 9]. Les groupes abéliens sans torsion sont beaucoup moins bien compris, et certainement pas tous uniquement directement décomposables; voir (xviii) ci-dessous.

(iv) Un groupe nilpotent Γ dont le centre est indécomposable est lui-même indécomposable. En effet, soit $\Gamma = \Gamma_1 \times \Gamma_2$ une décomposition en produit direct. Le centre de Γ étant égal au produit des centres de Γ_1 et Γ_2, l'un de ceux-ci est réduit à un élément. L'assertion résulte de ce qu'un groupe nilpotent dont le centre est réduit à un élément est lui-même réduit à un élément.

En particulier, le *groupe de Heisenberg* $\begin{pmatrix} 1 & \mathbf{Z} & \mathbf{Z} \\ 0 & 1 & \mathbf{Z} \\ 0 & 0 & 1 \end{pmatrix}$ est indécomposable.

Notons $\Gamma = C^1(\Gamma) \supset \cdots \supset C^{j+1}(\Gamma) = [\Gamma, C^j(\Gamma)] \supset \cdots$ la série centrale descendante d'un groupe Γ. Soient $k, j \geq 2$ des entiers et F_k le groupe non abélien libre à k générateurs; le *groupe nilpotent libre* de classe j à k générateurs $\Gamma = F_k/C^{j+1}(F_k)$ est indécomposable.

En effet, soit $\Gamma = \Gamma_1 \times \Gamma_2$ une décomposition en produit direct. Soient l, m les rangs des groupes abéliens libres $\Gamma_1/C^2(\Gamma_1), \Gamma_2/C^2(\Gamma_2)$, respectivement; notons que le rang k du groupe abélien libre $\Gamma/C^2(\Gamma) = F_k/C^2(F_k)$ est égal à la somme $l + m$. D'une part, le rang du groupe abélien libre $C^2(\Gamma)/C^3(\Gamma) = C^2(F_k)/C^3(F_k)$ est le coefficient binomial $\binom{k}{2}$. D'autre part, le rang du groupe abélien libre $C^2(\Gamma_1)/C^3(\Gamma_1)$ est majoré par le rang $\binom{l}{2}$ de $C^2(F_l)/C^3(F_l)$, et de même pour Γ_2 et $\binom{m}{2}$. Comme

$$C^2(\Gamma)/C^3(\Gamma) \simeq \left(C^2(\Gamma_1)/C^3(\Gamma_1)\right) \times \left(C^2(\Gamma_2)/C^3(\Gamma_2)\right),$$

il en résulte que $\binom{k}{2} \leq \binom{l}{2} + \binom{m}{2}$, donc que l'un de k, l est zéro, et par suite que l'un de Γ_1, Γ_2 est réduit à un élément.

(v) Un groupe résiduellement résoluble Γ dont l'abélianisé $\Gamma/[\Gamma, \Gamma]$ est indécomposable est lui-même indécomposable. Par exemple, le groupe $\begin{pmatrix} 2^{\mathbf{Z}} & \mathbf{Z}[1/2] \\ 0 & 1 \end{pmatrix}$ est indécomposable. (Voir juste après la proposition 3 un exemple généralisant celui-ci.)

Plus généralement un groupe résiduellement résoluble $\Gamma \neq \{1\}$ est indécomposable dès qu'un des quotients $\Gamma/C^j(\Gamma)$ ou $\Gamma/D^j(\Gamma)$ est indécomposable pour un entier $j \geq 2$, où $\Gamma = D^0(\Gamma) \supset \cdots \supset D^{i+1}(\Gamma) = [D^i(\Gamma), D^i(\Gamma)] \supset \cdots$ désigne la série dérivée.

Détaillons par exemple l'argument pour C^j: soit $\Gamma = \Gamma_1 \times \Gamma_2$ un groupe résiduellement résoluble tel que $\Gamma/C^j(\Gamma)$ est indécomposable; nous pouvons supposer les notations telles que $\Gamma_2/C^j(\Gamma_2)$ est réduit à un élément; *a fortiori*, $\Gamma_2/[\Gamma_2, \Gamma_2]$ est réduit à un élément, ce qui implique que $\Gamma_2 = \{1\}$ puisque Γ_2 est résiduellement résoluble.

(vi) Le groupe fondamental $\Gamma = \langle x, y \mid xyx^{-1} = y^{-1} \rangle$ d'une bouteille de Klein est indécomposable.

En effet, soit $\Gamma = \Gamma_1 \times \Gamma_2$ une décomposition en produit direct. Le groupe dérivé $D(\Gamma) = D(\Gamma_1) \times D(\Gamma_2)$ est engendré par y^2; il est cyclique infini, et donc indécomposable, de sorte qu'on peut supposer $D(\Gamma_2) = \{1\}$, c'est-à-dire Γ_2 abélien. Pour $j = 1, 2$, notons x_j [respectivement y_j] la projection de x [resp. y] sur Γ_j. Alors $D(\Gamma_2) = \{1\}$ implique $y_2^2 = 1$, et même $y_2 = 1$ puisque Γ est sans torsion. Le centre $Z(\Gamma) = Z(\Gamma_1) \times Z(\Gamma_2)$ est engendré par x^2; il est également cyclique infini, donc indécomposable. Si nous avions $Z(\Gamma_1) = \{1\}$, nous aurions aussi $x_1 = 1$ comme ci-dessus, donc $\Gamma = \langle y_1 \rangle \times \langle x_2 \rangle$ serait abélien, ce qui est absurde. C'est donc $Z(\Gamma_2)$ qui est réduit à un élément, ce qui montre enfin que Γ_2 lui-même est réduit à un élément.

(vii) Un produit libre de deux groupes non réduits à un élément est indécomposable.[4] Plus généralement, un sous-groupe d'un produit libre $\Gamma_1 * \Gamma_2$ qui n'est conjugué ni à un sous-groupe de Γ_1 ni à un sous-groupe de Γ_2 est indécomposable. Voir [Kuro], §§ 34 et 35.

(viii) Soit Γ un sous-groupe non réduit à un élément d'un groupe qui est hyperbolique au sens de Gromov et sans torsion. Alors Γ est indécomposable car, pour tout élément $\gamma \neq 1$ dans un tel groupe, le centralisateur de γ est cyclique infini.

(ix) Soit G un groupe de Lie réel connexe semi-simple, à centre réduit à un élément, sans facteur compact, de rang réel au moins deux, et soit Γ un réseau irréductible dans G. C'est une conséquence immédiate du théorème de Margulis concernant les sous-groupes normaux de Γ (voir le chapitre 8 de [Zimm]) que Γ est indécomposable; ceci s'applique par exemple à $\Gamma = \mathrm{PSL}_d(\mathbf{Z})$ pour $d \geq 3$. Avec des formulations plus générales (voir le chapitre VIII de [Marg]), on montre de même que des groupes comme $\mathrm{PSL}_d(\mathbf{Z}[1/p])$, qui est un réseau irréductible dans $\mathrm{PSL}_d(\mathbf{R}) \times \mathrm{PSL}_d(\mathbf{Q}_p)$, sont indécomposables ($p$ est un nombre premier, d un entier, $d \geq 2$).

Le même type de résultat (et d'argument) vaut encore plus généralement pour des réseaux irréductibles dans certains produits de groupes localement compacts [BaSh].

Dans de nombreux cas, ces affirmations peuvent être démontrées par des arguments plus économiques. Par exemple, les groupes $\mathrm{PSL}_d(\mathbf{Z}[1/p])$ sont denses dans le groupe de Lie $\mathrm{PSL}_d(\mathbf{R})$, *a fortiori* Zariski-denses dans le groupe algébrique $\mathrm{PSL}_d(\mathbf{C})$, et sont donc indécomposables en vertu de la proposition 3 ci-dessous.

(x) Soit Γ un groupe tel que, pour toute paire N_1, N_2 de sous-groupes normaux non réduits à un élément, l'intersection $N_1 \cap N_2$ ne l'est pas non plus; alors Γ est évidemment indécomposable. Dans ce cas, les sous-groupes normaux non réduits à un élément forment une base de voisinage de 1 pour une topologie sur Γ qu'on appelle la *topologie pro-normale* et qui est étudiée dans [GeGl].

Un sous-groupe Zariski-dense Γ d'un groupe de Lie G connexe simple à centre trivial possède cette propriété. En effet, soient N_1 et N_2 deux sous-groupes non réduits à un élément de Γ. Désignons par $\overline{N}_1$ et $\overline{N}_2$ leurs adhérences de Zariski; ce sont des sous-groupes normaux du groupe simple G, de sorte qu'ils coïncident

---

[4]Notons la conséquence suivante pour le problème de décision relatif à la décomposition en produit direct: le problème de savoir si un groupe donné par une présentation finie est décomposable ou non n'est pas algorithmiquement résoluble. En effet, soient A et B deux groupes de présentation finie non réduits à un élément (par exemple $A = B = \mathbf{Z}$) et Δ un groupe de présentation finie. Le groupe $\Gamma = (A \times B) * \Delta$ est décomposable si et seulement si Δ est réduit à un élément, et il est bien connu qu'il n'existe pas d'algorithme permettant de savoir si un groupe donné par une présentation finie est ou n'est pas réduit à un élément (voir par exemple le corollaire 12.33 de [Rotm]).

avec G. Si nous avions $N_1 \cap N_2 = \{1\}$, nous aurions aussi $[N_1, N_2] = \{1\}$ et $[\overline{N}_1, \overline{N}_2] = [G, G] = \{1\}$, ce qui est absurde.

Le «groupe de Grigorchuk» possède aussi cette propriété. C'est le 2-groupe infini de type fini qui apparaît dans [Grig]; voir aussi, par exemple, le théorème VIII.42 de [Harp]. Ceci s'étend à tout groupe *juste infini*, c'est-à-dire à tout groupe infini dont tous les quotients propres sont finis; voir [SaSS], en particulier le chapitre rédigé par J. Wilson.

De même pour le «groupe F de Thompson», dont on sait que le groupe dérivé est d'une part simple et d'autre part contenu dans tout sous-groupe normal non réduit à un élément (théorème 4.5 et preuve du théorème 4.3 dans [CaFP]).

(xi) Nous démontrons ci-dessous l'indécomposabilité des groupes de Coxeter de type fini qui sont irréductibles et infinis (corollaire de la proposition 1 et proposition 8).

(xii) Soit Γ un groupe non réduit à un élément qui est de présentation finie et de dimension cohomologique au plus 2. Alors Γ est ou bien indécomposable, ou bien un produit direct de deux groupes libres. En effet, s'il existe deux sous-groupes Γ_1, Γ_2 de Γ tels que $\Gamma = \Gamma_1 \times \Gamma_2$, un résultat de Bieri implique que Γ_1 et Γ_2 sont de dimension cohomologique au plus un [Bier, corollaire 8.6], de sorte que Γ_1 et Γ_2 sont libres par un théorème de Stallings (voir par exemple [Bier, théorème 7.6]).

Voici deux familles d'exemples de groupes de dimension cohomologique au plus 2: les sous-groupes sans torsion des groupes à un relateur ([Bier, théorème 7.7], résultat dû a Lyndon pour un groupe à un relateur sans torsion), et les groupes fondamentaux des variétés non compactes de dimension 3, en particulier les groupes de noeuds (voir par exemple [Serr], no 1.5 et lemme 5 du no 2.1).

Bagherzadeh a montré un résultat plus précis: les seuls sous-groupes décomposables des groupes à un relateur sans torsion sont des produits directs d'un groupe cyclique infini et d'un groupe libre (voir le corollaire 4.9 de [Bagh]).

(xiii) Pour toute paire k, l d'entiers, $k, l \geq 2$, le produit amalgamé

$$A_{k,l} = \langle a_1, a_2 \mid a_1^k = a_2^l \rangle$$

est indécomposable; répétons l'argument simple du livre de Kurosh.

Le groupe $A_{k,l}$ est sans torsion, et son centre est cyclique infini engendré par $a_1^k = a_2^l$ (ce sont des conséquences immédiates de résultats concernant les formes normales dans les produits amalgamés, voir par exemple le §4.2 de [MaKS]). Si $A_{k,l} = X \times Y$ est une décomposition en produit direct et si $(x, y) \in X \times Y$ est l'écriture du générateur a_1^k du centre de $A_{k,l}$, il en résulte que l'un de x, y est 1; sans restreindre la généralité de ce qui suit, nous pouvons supposer que $y = 1$. Notons $a_1 = (x_1, y_1)$, $a_2 = (x_2, y_2)$ les écritures dans le produit $X \times Y$ des générateurs a_1, a_2 de $A_{k,l}$; remarquons que le groupe X [respectivement le groupe Y] est engendré par x_1 et x_2 [resp. par y_1 et y_2]. Comme Y est sans torsion et comme $y_1^k = y_2^l = y = 1$, nous avons $y_1 = y_2 = 1$. Il en résulte que Y est réduit à un élément.

(xiv) Avec les notations de (xiii), considérons un entier $k \geq 2$ et le produit direct

$$\Gamma = A_{k,k} \times A_{k+1,k+1} = \langle a_1, a_2 \mid a_1^k = a_2^k \rangle \times \langle b_1, b_2 \mid b_1^{k+1} = b_2^{k+1} \rangle.$$

Posons $a = a_1^k = a_2^k$, qui est un générateur du centre de $A_{k,k}$, et $b = b_1^{k+1} = b_2^{k+1}$, qui est un générateur du centre de $A_{k+1,k+1}$. Dans Γ, posons

$$c_1 = ab^{-1}a_1, \quad c_2 = ab^{-1}a_2, \quad c_3 = ab^{-1}b_1, \quad c_4 = ab^{-1}b_2$$
$$c = c_1^k = c_2^k = c_3^{k+1} = c_4^{k+1}, \qquad d = ab^{-1}.$$

On vérifie que le sous-groupe C de Γ engendré par c_1, c_2, c_3, c_4 a un centre cyclique infini engendré par c, et on montre comme en (xiii) que C est indécomposable.

L'intérêt de cet exemple est le suivant: le groupe $A_{k,k} \times A_{k+1,k+1}$ est isomorphe au produit direct du groupe C et du groupe cyclique infini D engendré par d; et les groupes $A_{k,k}, A_{k+1,k+1}, C, D$, tous indécomposables, sont non isomorphes deux à deux. En particulier, le groupe de présentation finie $A_{k,k} \times A_{k+1,k+1}$ n'est pas uniquement directement décomposable. Tout ceci apparaît dans [Kuro, § 42], pour $k = 2$.

Cet exemple montre aussi qu'un réseau dans un groupe de Lie réel semi-simple peut ne pas être uniquement directement décomposable. En effet, pour $k \geq 3$, le quotient $\langle a_1, a_2 \mid a_1^k = a_2^k = 1 \rangle$ de $A_{k,k}$ par son centre est un réseau dans le groupe de Lie $\mathrm{PSL}_2(\mathbf{R})$; c'est un groupe fuchsien qui possède un quadrilatère fondamental dans le demi-plan de Poincaré ayant deux sommets opposés d'angle π/k et les deux autres sommets à l'infini. Le groupe $A_{k,k}$ lui-même est un réseau dans le revêtement universel $\tilde{G}$ de $\mathrm{PSL}_2(\mathbf{R})$. Par suite, $A_{k,k} \times A_{k+1,k+1}$ est un réseau réductible dans le groupe de Lie semi-simple qui est produit direct de deux copies de $\tilde{G}$.

Notons enfin que les groupes $A_{k,k} \times A_{k+1,k+1}$ sont linéaires. En effet, bien que le groupe $\tilde{G}$ ne soit pas linéaire, ses réseaux $A_{k,k}$ le sont (comparer avec la proposition de Toledo, Millson et Gersten qui apparaît au no IV.48 de [Harp]).

(xv) L'exemple suivant, dû à W.R. Scott [Walk], montre *qu'il n'y a pas simplification* pour les décompositions par produit direct. Considérons le produit semi-direct défini par la présentation

$$\Gamma = \langle x, y, z \mid x^{11} = 1, \quad y^{-1}xy = x^2, \quad z^{-1}xz = x^8, \quad yz = zy \rangle \simeq (\mathbf{Z}/11\mathbf{Z}) \rtimes \mathbf{Z}^2,$$

ainsi que les sous-groupes

$$A = \langle x, y \rangle \simeq (\mathbf{Z}/11\mathbf{Z}) \rtimes_2 \mathbf{Z}$$
$$B = \langle x, z \rangle \simeq (\mathbf{Z}/11\mathbf{Z}) \rtimes_8 \mathbf{Z}$$
$$C = \langle y^7 z \rangle \simeq \mathbf{Z}$$
$$C' = \langle yz^3 \rangle \simeq \mathbf{Z}$$

de Γ. D'une part, les restrictions au sous-groupe $\langle x \rangle \simeq \mathbf{Z}/11\mathbf{Z}$ des conjugaisons par $y^7 z$ et yz^3 coïncident avec l'identité, car $2^7 \times 8 = 2 \times 8^3 \equiv 1 \pmod{11}$; d'autre part, nous avons des décompositions en produit direct

$$\mathbf{Z}^2 \;=\; \langle y \rangle \times \langle y^7 z \rangle \;=\; \langle yz^3 \rangle \times \langle z \rangle$$

car les matrices $\begin{pmatrix} 1 & 0 \\ 7 & 1 \end{pmatrix}$ et $\begin{pmatrix} 0 & 1 \\ 1 & 3 \end{pmatrix}$ sont dans $GL_2(\mathbf{Z})$; il en résulte que

$$\Gamma \;=\; A \times C \;=\; C' \times B.$$

Notons que les deux générateurs de l'abélianisé $A/[A, A] \simeq \mathbf{Z}$ agissent sur le groupe dérivé $[A, A] = \langle x \mid x^{11} = 1 \rangle \simeq \mathbf{Z}/11\mathbf{Z}$ par $x \longmapsto x^2$ et $x \longmapsto x^6$. Les automorphismes analogues pour B sont $x \longmapsto x^8$ et $x \longmapsto x^7$. Il en résulte que les groupes A et B ne sont pas isomorphes.

Citons plus brièvement quelques autres exemples illustrant la notion d'unique décomposabilité directe.

(xvi) Un groupe indécomposable est évidemment uniquement directement décomposable. Avec nos conventions, le groupe à un élément est uniquement directement décomposable.

(xvii) Tout groupe fini est uniquement directement décomposable. C'est une conséquence immédiate du théorème de Wedderburn–Remak–Krull–Schmidt, mais un résultat non banal! (Voir aussi ci-dessous la proposition 9.)

(xviii) Un groupe abélien de type fini est uniquement directement décomposable (voir par exemple [BouA'], chapitre VII, §4, no 8). Un groupe abélien libre est uniquement directement décomposable; par exemple, le groupe multiplicatif $\mathbf{Q}^*_+$, isomorphe à la somme restreinte de copies de $\mathbf{Z}$ indexées par les nombres premiers, est uniquement directement décomposable. Une somme restreinte de groupes abéliens de rang 1, c'est-à-dire de sous-groupes de $\mathbf{Q}$, est uniquement directement décomposable. (C'est un résultat de Baer; voir par exemple [Fuch, Section 86].)

En revanche, les groupes abéliens de rang fini sans torsion sont loins d'être tous uniquement directement décomposables. Par exemple, pour tout entier $n \geq 2$, il existe des groupes abéliens de rang fini sans torsion $A, A', A_1, \ldots, A_n$, indécomposables et non isomorphes deux à deux, tels que $A \oplus A'$ et $A_1 \oplus \cdots \oplus A_n$ sont isomorphes. Pour ceci et d'autres exemples de décompositions non isomorphes, voir les §90–91 de [Fuch]; voir aussi [Baum], déjà cité dans l'introduction.

(xix) Soit Γ_U l'un des groupes construits par B.H. Neumann pour montrer qu'il existe une infinité non dénombrable de groupes à deux générateurs [Neum]. Ici, U désigne une suite infinie strictement croissante de nombres impairs supérieurs ou égaux à 5; chaque groupe Γ_U est engendré par un élément d'ordre infini et un élément d'ordre 3; le centre de Γ_U est réduit à un élément. Deux groupes $\Gamma_U, \Gamma_{U'}$ sont isomorphes si et seulement si $U = U'$.

Pour toute sous-suite finie $(u_1, \ldots, u_k)$ de U, c'est une conséquence immédiate de l'analyse de Neumann qu'il existe une décomposition en somme directe

$$\Gamma_U \;=\; \mathrm{Alt}(u_1) \times \cdots \times \mathrm{Alt}(u_k) \times \Gamma_{U'}$$

où $\mathrm{Alt}(u)$ désigne le groupe alterné simple d'ordre $\frac{1}{2}u!$ et U' la suite obtenue à partir de U en supprimant $u_1, \ldots, u_k$; en particulier, Γ_U admet des décompositions en somme directe ayant un nombre arbitrairement grand de facteurs. En revanche, le groupe Γ_U étant de type fini, il n'est pas isomorphe à une somme restreinte d'un nombre infini de groupes non réduits à un élément; *a fortiori*, le groupe Γ_U n'est pas somme restreinte de groupes indécomposables.

Nous revenons à des exemples indécomposables, et en particulier à certains groupes de Coxeter.

Proposition 1. *Soit Γ un groupe possédant un sous-groupe normal abélien T avec les propriétés suivantes:*
  (i) *il existe un entier $d \geq 1$ tel que T est isomorphe à $\mathbf{Z}^d$;*
  (ii) *l'action par conjugaison de Γ/T sur T est fidèle;*
  (iii) *la représentation associée de Γ/T sur $T \otimes_{\mathbf{Z}} \mathbf{Q} \simeq \mathbf{Q}^d$ est irréductible.*
Alors Γ est indécomposable.

Démonstration. Montrons d'abord que tout sous-groupe normal abélien N de Γ est contenu dans T.

Le sous-groupe $[N, T]$ de Γ est dans T et Γ/T-invariant. La propriété (iii) implique qu'il est ou bien d'indice fini dans T ou bien réduit à un élément. S'il était d'indice fini, il existerait un entier $k \geq 1$ tel que $[N, T]$ contienne un sous-groupe $kT \simeq k\mathbf{Z}^d$; par la propriété (ii), N agirait non trivialement sur $[N, T]$ (qui est dans N), ce qui est impossible puisque N est abélien. Donc $[N, T] = \{1\}$; il en résulte que l'action de N sur T est triviale, de sorte que $N \subset T$ par la propriété (ii).

Soient Γ_1, Γ_2 des sous-groupes de Γ tels que $\Gamma = \Gamma_1 \times \Gamma_2$. Posons

$$N_1 = \{a \in \Gamma_1 \mid \text{il existe} \quad b \in \Gamma_2 \quad \text{tel que} \quad (a, b) \in T\},$$
$$N_2 = \{b \in \Gamma_2 \mid \text{il existe} \quad a \in \Gamma_1 \quad \text{tel que} \quad (a, b) \in T\},$$
$$N = N_1 \times N_2.$$

Alors N_1 est un sous-groupe normal abélien de Γ_1 et N_2 un sous-groupe normal abélien de Γ_2, donc N est un sous-groupe normal abélien de Γ; de plus, N contient T. Il résulte de la maximalité de T établie plus haut que $N = T$. La propriété (iii) implique que l'un des facteurs N_1, N_2 est réduit à un élément; convenons que $N_2 = \{1\}$. Comme Γ_2 centralise N_1, la propriété (ii) implique que $\Gamma_2 = \{1\}$. $\quad\square$

Corollaire. Un groupe de Coxeter de type fini qui est irréductible et de type affine est indécomposable.

Démonstration. Un groupe de Coxeter de type fini W_a qui est de type affine est un produit semi-direct $Q \rtimes W$, où W est un groupe de Weyl, en particulier un groupe fini, et Q le groupe des poids radiciels correspondant, en particulier un groupe abélien libre de type fini ([BouL], chapitre 6, § 2, no 1, proposition 1). De plus, les conditions suivantes sont équivalentes: (i) W_a est irréductible comme groupe de Coxeter, (ii) W est irréductible comme groupe de Coxeter, (iii) la représentation de W dans $Q \otimes_{\mathbf{Z}} \mathbf{Q}$ est irréductible ([BouL], chapitre 6, § 1, no 1 et chapitre 5, § 3, no 7).

La proposition 1 s'applique donc à la situation du corollaire. $\quad\square$

3. Décompositions de sous-groupes de groupes algébriques

Il est bien connu qu'un groupe dont le centre est réduit à un élément possède au plus une décomposition en produit direct d'un nombre fini de sous-groupes indécomposables. Nous commençons par rappeler ce résultat et quelques-unes de ses conséquences, notamment le fait que la propriété d'indécomposabilité convenablement formulée passe de certains groupes topologiques à leurs sous-groupes denses.

Nous notons $Z(H)$ le centre d'un groupe H. Si H est sous-groupe d'un groupe G, nous écrivons H' son centralisateur dans G. Lorsque G est un produit direct $A \times B$, nous avons $A' = Z(A) \times B$. Si de plus $Z(G) = \{1\}$, alors $A' = B$ et $A = B'$.

Proposition 2. *Soient G un groupe de centre réduit à un élément et $G_1, \ldots, G_n$ des sous-groupes indécomposables de G tels que $G - G_1 \times \cdots \times G_n$.*

Si A, B sont deux sous-groupes de G tels que $G = A \times B$, il existe une renumérotation des G_i et un entier $m \in \{0, \ldots, n\}$ tels que

$$A = G_1 \times \cdots \times G_m \qquad et \qquad B = G_{m+1} \times \cdots \times G_n.$$

Remarque. Nous citons un résultat plus général à la proposition 9.

Démonstration. Notons $g = (g_1, \ldots, g_n)$ l'écriture d'un élément $g \in G$ selon la décomposition $G = G_1 \times \cdots \times G_n$. Pour tout $i \in I \doteq \{1, \ldots, n\}$, notons $\pi_i : G \longrightarrow G_i, g \longmapsto g_i$ la projection canonique. Posons $A_i = A \cap G_i$ et $B_i = B \cap G_i$.

Nous affirmons que $A = A_1 \times \cdots \times A_n$, de sorte que $A_i = \pi_i(A)$ pour tout $i \in I$. En effet, soit $a = (a_1, \ldots, a_n) \in A$. Comme $a \in B'$, nous avons

$$[(a_1, \ldots, a_n), (b_1, \ldots, b_n)] = ([a_1, b_1], \ldots, [a_n, b_n]) = (1, \ldots, 1)$$

pour tout $(b_1, \ldots, b_n) \in B$. Il en résulte que, pour $i \in I$, nous avons aussi $[a_i, b_j] = 1$ pour tous $j \in J$ et $b_j \in B_j$, et par suite $a_i \in B' = A$. L'affirmation en résulte. De même $B = B_1 \times \cdots \times B_n$ et $B_i = \pi_i(B)$ pour tout $i \in I$.

Soit $i \in I$; vu que $G = A \times B$ et $[A, B] = 1$, nous avons $G_i = A_i \times B_i$. De plus, l'un des groupes A_i, B_i est réduit à un élément parce que G_i est indécomposable. La proposition en résulte. $\qquad\square$

Conséquence. Un groupe G qui satisfait aux hypothèses de la proposition 2 est bien sûr uniquement directement décomposable. De plus, les sous-groupes indécomposables dont G est le produit direct sont uniquement déterminés comme sous-groupes (et non pas seulement à isomorphisme près).

L'argument de la preuve précédente est suffisamment robuste pour s'adapter à d'autres cas. Nous considérons ci-dessous un corps algébriquement clos K et des groupes algébriques définis sur K. Un tel groupe G est *indécomposable* s'il n'est pas produit direct *de sous-groupes algébriques* de manière non banale. Pour des raisons de dimension, tout G est produit direct d'une famille finie de sous-groupes algébriques indécomposables; lorsque de plus $Z(G) = \{1\}$, ces sous-groupes sont uniquement déterminés à l'ordre près.

Proposition 3. *Soient G un groupe algébrique de centre réduit à un élément et $G_1, \ldots, G_n$ des sous-groupes algébriques indécomposables de G tels que $G = G_1 \times \cdots \times G_n$. Soit Γ un sous-groupe Zariski-dense de G.*

Si A, B sont deux sous-groupes de Γ tels que $\Gamma = A \times B$, il existe une renumérotation des G_i et un entier $m \in \{0, \ldots, n\}$ tels que

$$A \subset G_1 \times \cdots \times G_m \qquad et \qquad B \subset G_{m+1} \times \cdots \times G_n.$$

En particulier, si G est de plus indécomposable comme groupe algébrique, c'est-à-dire si $n = 1$, tout sous-groupe Zariski-dense de G est indécomposable.

Démonstration (voir aussi l'exemple (x) *du chapitre 2).* Notons $\overline{A}$ et $\overline{B}$ les adhérences de Zariski de A et B. Alors $[\overline{A}, \overline{B}] = \{1\}$ ([Bore], no 2.4); de plus, $\overline{A}\,\overline{B}$ est fermé dans G ([Bore], no 1.4), de sorte que $G = \overline{A}\,\overline{B}$. Comme $\overline{A} \cap \overline{B}$ est central dans G, cette intersection est réduite à un élément et G est produit direct de $\overline{A}$ et $\overline{B}$.

L'argument de la démonstration précédente montre que, après renumérotation éventuelle des G_j, il existe un entier $m \in \{0, \ldots, n\}$ tel que $A \subset \overline{A} = G_1 \times \cdots \times G_m$ et $B \subset \overline{B} = G_{m+1} \times \cdots \times G_n$. $\qquad\square$

Exemple. Considérons le groupe $\Gamma = \mathbf{Z} \ltimes \mathbf{Z}[1/pq]$, où $p \neq 0$, $q \geq 2$ sont deux entiers premiers entre eux, et où le générateur 1 de $\mathbf{Z}$ agit sur $\mathbf{Z}[1/pq]$ par multiplication par p/q. C'est un groupe métabélien de type fini; dans le cas où $|p| = 1$, c'est aussi un groupe de Baumslag–Solitar. Notons G le groupe des matrices triangulaires de la forme $\begin{pmatrix} a & b \\ 0 & 1 \end{pmatrix}$, avec $a \in \mathbf{C}^*$ et $b \in \mathbf{C}$; c'est un groupe algébrique connexe indécomposable (parce que son algèbre de Lie l'est, puisque c'est l'algèbre de Lie résoluble non abélienne de dimension 2, voir ci-dessous avant la proposition 6). Comme Γ est Zariski-dense dans G, il résulte de la proposition 3 que Γ est indécomposable. En particulier, tout groupe de Baumslag–Solitar résoluble qui n'est pas abélien libre de rang deux est indécomposable. (Les autres groupes de Baumslag–Solitar sont également indécomposables; voir l'exemple (xii) du chapitre 2.)

Avant de généraliser ces propositions à des cas avec centres, nous rappelons les points suivants concernant la notion d'hypercentre.

Considérons à nouveau un groupe G, sans autre structure. La *suite centrale ascendante* de G est la suite $(\zeta^\alpha(G))_\alpha$, indexée par les ordinaux α, définie par récurrence transfinie comme suit (où π_β désigne la projection canonique de G sur $G/\zeta^\beta(G)$):

si $\alpha = 0$, alors $\zeta^0(G) = \{1\}$;

si $\alpha = \beta + 1$, alors $\zeta^\alpha(G) = \pi_\beta^{-1}\big(Z(G/\zeta^\beta(G))\big)$;

si α est un ordinal limite, alors $\zeta^\alpha(G) = \bigcup_{\beta < \alpha} \zeta^\beta(G)$.

Chaque $\zeta^\alpha(G)$ est un sous-groupe caractéristique de G. *L'hypercentre* de G est la réunion $\zeta^\uparrow(G)$ des $\zeta^\alpha(G)$; il suffit de prendre la réunion sur l'ensemble des ordinaux dont le cardinal ne dépasse pas celui de G. L'hypercentre de G est un sous-groupe normal tel que $Z(G/\zeta^\uparrow(G)) = \{1\}$, et qui est minimal pour cette propriété. En particulier, $\zeta^\uparrow(G) = \{1\}$ si et seulement si $Z(G) = \{1\}$.

Soit maintenant G un groupe algébrique *connexe*. Son centre $Z(G) = \zeta^1(G)$ est ou bien de dimension strictement positive ou bien fini. Dans le second cas, $\zeta^2(G)$ est ou bien de dimension strictement positive ou bien fini, et alors égal à $Z(G)$ puisque tout sous-groupe normal fini d'un groupe connexe est central. Plus généralement, pour tout entier $k \geq 1$, l'une au moins des deux relations

$$\dim\left(\zeta^k(G)\right) > \dim\left(\zeta^{k-1}(G)\right), \qquad \zeta^{k+1}(G) = \zeta^k(G)$$

est vraie. Il en résulte qu'il existe un entier h tel que $\zeta^\uparrow(G) = \zeta^h(G)$; en particulier l'hypercentre de G est un sous-groupe algébrique de G. De même, dans un groupe de Lie réel ou complexe connexe, l'hypercentre est un sous-groupe fermé.

Proposition 4. *Soient G un groupe, $H = G/\zeta^\uparrow(G)$ le quotient de G par son hypercentre et $\pi : G \longrightarrow H$ la projection canonique. Soient $H_1, \ldots, H_n$ des sous-groupes indécomposables de H tels que $H = H_1 \times \cdots \times H_n$; posons $G_i = \pi^{-1}(H_i)$, de sorte que $G = G_1 \cdots G_n$ et $G_i \cap \prod_{j \neq i} G_j = \zeta^\uparrow(G)$ pour tout $i \in \{1, \ldots, n\}$.*

Si A, B sont deux sous-groupes de G tels que $G = A \times B$, il existe une renumérotation des G_i et un entier $m \in \{0, \ldots, n\}$ tels que

$$A \subset G_1 \cdots G_m \qquad et \qquad B \subset G_{m+1} \cdots G_n.$$

Démonstration. Il est évident que $H = \pi(A)\pi(B)$. Par ailleurs, tout élément de $\pi(A)$ commute à tout élément de $\pi(B)$; par suite, tout élément de $\pi(A) \cap \pi(B)$ commute à tout élément de $\pi(A)\pi(B)$. Comme $Z(H) = \{1\}$, il en résulte que H est produit direct de ses sous-groupes $\pi(A)$ et $\pi(B)$.

La proposition 4 est donc une conséquence immédiate de la proposition 2. $\square$

Proposition 5. *Soient G un groupe algébrique, $H = G/\zeta^\uparrow(G)$ le quotient de G par son hypercentre et $\pi : G \longrightarrow H$ la projection canonique. Soit $H = H_1 \times \cdots \times H_n$ la décomposition canonique de H en produit de sous-groupes algébriques indécomposables; posons $G_i = \pi^{-1}(H_i)$, de sorte que $G = G_1 \cdots G_n$ et $G_i \cap \prod_{j \neq i} G_j = \zeta^\uparrow(G)$ pour tout $i \in \{1, \ldots, n\}$. Soit Γ un sous-groupe Zariski-dense de G.*

Si A, B sont deux sous-groupes de Γ tels que $\Gamma = A \times B$, il existe une renumérotation des G_i et un entier $m \in \{0, \ldots, n\}$ tels que

$$A \subset G_1 \cdots G_m \qquad et \qquad B \subset G_{m+1} \cdots G_n.$$

En particulier, si de plus H est indécomposable comme groupe algébrique, l'un des facteurs A, B est contenu dans l'hypercentre $\zeta^\uparrow(G)$.

Démonstration. Nous avons $\pi(\overline{A}) = \overline{\pi(A)}$, $\pi(\overline{B}) = \overline{\pi(B)}$ et $[\overline{\pi(A)}, \overline{\pi(B)}] = \{1\}$ ([Bore], numéros 1.4 et 2.4). L'argument de la démonstration de la proposition 3 montre que $H = \pi(A) \times \pi(B)$.

La proposition 5 est alors une conséquence immédiate de la proposition 3. $\square$

Exemple. La proposition 5 s'applique à un groupe réductif G tel que $G/Z(G)$ soit simple, donc en particulier au groupe $\mathrm{GL}_n(K)$; c'est alors la proposition 10 de [Mari].

Exemple. Un groupe algébrique connexe indécomposable dont le centre n'est pas réduit à un élément peut contenir un sous-groupe Zariski-dense décomposable. Considérons en effet le groupe $G = \mathrm{SL}_4(\mathbf{C})$. Notons A le noyau de la réduction $\mathrm{SL}_4(\mathbf{Z}) \longrightarrow \mathrm{SL}_4(\mathbf{Z}/3\mathbf{Z})$ modulo 3 et B le centre $\{\pm 1\}$ de $\mathrm{SL}_4(\mathbf{Z})$. Alors le produit direct $\Gamma = A \times B$ est un sous-groupe Zariski-dense de G. (Voir néanmoins le chapitre 5 sur la c-indécomposabilité.)

Pour appliquer les propositions 3 et 5, il convient de disposer du critère d'indécomposabilité suivant pour les groupes algébriques. Rappelons qu'une algèbre de Lie $\mathfrak{g}$ est *indécomposable* si $\mathfrak{g} \neq \{0\}$ et si, pour tout isomorphisme de $\mathfrak{g}$ avec un produit d'algèbres de Lie $\mathfrak{a} \times \mathfrak{b}$, l'une de $\mathfrak{a}, \mathfrak{b}$ est réduite à zéro.

Critère. Pour qu'un groupe algébrique connexe G soit indécomposable, il suffit que son algèbre de Lie le soit.

Stratégie. Soient K un corps algébriquement clos, G un groupe algébrique connexe à centre fini dont l'algèbre de Lie $\mathfrak{g}$ est indécomposable, Γ un sous-groupe de G et $\Gamma = A \times B$ une décomposition en produit direct.

Il résulte du critère ci-dessus et des propositions qui précèdent que, si Γ est Zariski-dense dans G, l'un des groupes A, B est central d'ordre majoré par l'ordre du centre de G.

Supposons de plus que G est un L-groupe, où L est un sous-corps de K, et que Γ est un sous-groupe du groupe $G(L)$ des points rationnels. Supposons aussi que L est un corps parfait, de sorte que $G(L)$ est Zariski-dense dans G (corollaire 18.3 de [Bore]). Si Γ est Zariski-dense dans $G(L)$, alors de même l'un des groupes A, B est central d'ordre majoré par l'ordre du centre de $G(L)$.

Il y a des exemples immédiats d'algèbres de Lie indécomposables: une algèbre de Lie simple, une algèbre de Lie non abélienne de dimension 2, une algèbre de Lie nilpotente à centre de dimension 1. Voici deux autres familles: pour tout entier $n \geq 1$, le produit semi-direct standard $\mathbf{C}^n \rtimes \mathfrak{gl}_n(\mathbf{C})$ est indécomposable; une algèbre de Lie isomorphe à une sous-algèbre de Lie parabolique d'une algèbre de Lie complexe simple est indécomposable; voir les théorèmes 4.2 et 4.7 de [Meng].

Nous verrons au numéro suivant une famille d'exemples d'algèbres de Lie réelles indécomposables permettant d'appliquer aux groupes de Coxeter la stratégie ci-dessus.

Digression. Une algèbre de Lie $\mathfrak{g}$ de dimension finie (sur un corps arbitraire) s'écrit comme produit direct d'algèbres indécomposables (la vérification par récurrence sur la dimension est de pure routine). De plus, il y a unicité au sens de la proposition suivante, du type Wedderburn–Remak–Krull–Schmidt–Azumaya. Bien que ce soit un résultat classique, nous n'avons pas su en trouver dans la littérature la formulation qui nous convient.

Proposition 6. *Soient $\mathfrak{g}$ une algèbre de Lie de dimension finie sur un corps K et*

$$\mathfrak{g} = \mathfrak{a}_1 \oplus \cdots \oplus \mathfrak{a}_m = \mathfrak{b}_1 \oplus \cdots \oplus \mathfrak{b}_n \qquad (*)$$

deux décompositions de $\mathfrak{g}$ en produits directs d'idéaux indécomposables.

Alors $m = n$ et il existe une permutation σ de $\{1, \ldots, m\}$ telle que $\mathfrak{a}_j$ et $\mathfrak{b}_{\sigma(j)}$ sont isomorphes pour tout $j \in \{1, \ldots, m\}$.

Démonstration. C'est un résultat tout à fait standard qu'il y a unicité de la décomposition de $\mathfrak{g}$ en *sous-$\mathfrak{g}$-modules* indécomposables (appliquer le «théorème de Krull–Schmidt–Azumaya», no 19.21 dans [Lam], à l'algèbre enveloppante de $\mathfrak{g}$).

Pour une application linéaire entre deux idéaux de $\mathfrak{g}$, les conditions d'être un morphisme de $\mathfrak{g}$-modules et d'être un morphisme d'algèbres de Lie sont *différentes*. Toutefois, les isomorphismes fournis par *la preuve* du théorème invoqué sont des compositions d'injections et de projections canoniques associées aux décompositions de $(*)$; ce sont donc *à la fois* des morphismes de $\mathfrak{g}$-modules et des morphismes d'algèbres de Lie. Par suite, le théorème standard fournit bien des isomorphismes d'algèbres de Lie $\mathfrak{a}_j \longrightarrow \mathfrak{b}_{\sigma(j)}$. $\qquad\qquad\square$

4. Groupes de Coxeter et groupes d'Artin

Soient E un espace vectoriel réel de dimension finie et B une forme bilinéaire symétrique sur E. Notons r_B la dimension du noyau

$$\mathrm{Ker}(B) = \{v \in E \mid B(v, w) = 0 \text{ pour tout } w \in E\}$$

de B; soit p_B [respectivement q_B] la dimension maximale d'un sous-espace U de E tel que $B(u, u) > 0$ [resp. $B(u, u) < 0$] pour tout $u \in U$, $u \neq 0$. On sait que $p_B + q_B + r_B$ est la dimension de E, désormais notée n_B, et nous appelons *signature* de B le triplet (p_B, q_B, r_B). Considérons le groupe algébrique

$$\mathrm{Of}(B) = \left\{ g \in \mathrm{GL}(E) \,\middle|\, \begin{array}{l} B(gv, gw) = B(v, w) \text{ pour tout } v, w \in E \text{ et} \\ gv = v \text{ pour tout } v \in \mathrm{Ker}(B) \end{array} \right\},$$

qui est un produit semi-direct de la forme $(\mathbf{R}^{p_B + q_B})^{r_B} \rtimes O(p_B, q_B)$.

Son algèbre de Lie $\mathfrak{of}(B)$ est isomorphe à l'algèbre de Lie des $(n_B \times n_B)$-matrices d'écriture par blocs

$$\begin{pmatrix} a & b & 0 \\ c & d & 0 \\ x & y & 0 \end{pmatrix} \qquad\qquad (*)$$

relativement à la décomposition $n_B = p_B + q_B + r_B$, avec

$$\begin{pmatrix} {}^t a & {}^t c \\ {}^t b & {}^t d \end{pmatrix} \begin{pmatrix} I_p & 0 \\ 0 & -I_q \end{pmatrix} + \begin{pmatrix} I_p & 0 \\ 0 & -I_q \end{pmatrix} \begin{pmatrix} a & b \\ c & d \end{pmatrix} = 0$$

(les préfixes ^{t} indiquent des transpositions, et I_p la $(p \times p)$-matrice unité).

Lorsque $p_B + q_B \geq 3$, on vérifie que le centre de $\mathrm{Of}(B)$ est le groupe $\pm \mathrm{Id}_E$ d'ordre 2, et qu'il coïncide avec son hypercentre.

Proposition 7. *Conservons les notations ci-dessus; supposons de plus*

$$p_B + q_B \geq 2 \quad et \quad (p_B, q_B, r_B) \neq (4,0,0),\ (2,2,0),\ (0,4,0).$$

Alors l'algèbre de Lie $\mathfrak{of}(B)$ *est indécomposable.*

Remarque. Si $p_B + q_B \leq 1$, l'algèbre de Lie $\mathfrak{of}(B)$ est abélienne. Par ailleurs, les algèbres de Lie $\mathfrak{so}(4) = \mathfrak{so}(3) \times \mathfrak{so}(3)$ et $\mathfrak{so}(2,2) = \mathfrak{so}(2,1) \times \mathfrak{so}(2,1) \simeq \mathfrak{sl}_2(\mathbf{R}) \times \mathfrak{sl}_2(\mathbf{R})$ sont décomposables. Notons que $\mathfrak{so}(3,1) = \mathfrak{so}(1,3) \simeq \mathfrak{sl}_2(\mathbf{C})$ est simple (il faut voir ici $\mathfrak{sl}_2(\mathbf{C})$ comme une algèbre de Lie *réelle*), même si l'algèbre de Lie complexifiée ne l'est pas.

Démonstration. Si $r_B = 0$ et $(p_B, q_B) \neq (4,0),\ (2,2),\ (0,4)$, l'algèbre de Lie $\mathfrak{of}(B) = \mathfrak{so}(p_B, q_B)$ est ou bien simple (si $p_B + q_B \geq 3$) ou bien de dimension un (si $p_B + q_B = 2$), donc indécomposable dans tous ces cas. Supposons désormais $r_B \geq 1$.

Comme les signatures (q_B, p_B, r_B) et (p_B, q_B, r_B) donnent lieu à des algèbres de Lie isomorphes, nous pouvons aussi supposer $p_B \geq q_B$ sans restreindre la généralité de ce qui suit. Soient $\mathfrak{a}, \mathfrak{b}$ deux idéaux de $\mathfrak{of}(B)$ tels que $\mathfrak{of}(B) = \mathfrak{a} \times \mathfrak{b}$; il s'agit de montrer que $\mathfrak{a} = 0$ ou $\mathfrak{b} = 0$. Nous séparons la discussion en trois cas.

(i) *Cas où* $p_B + q_B \geq 3$ *et* $(p_B, q_B) \neq (4,0), (2,2)$. Soit $\mathfrak{s}$ une sous-algèbre de Levi de $\mathfrak{of}(B)$; c'est une algèbre de Lie simple. Sa projection sur $\mathfrak{a}$ est ou bien isomorphe à $\mathfrak{s}$ ou bien nulle, et de même pour sa projection sur $\mathfrak{b}$. Supposons les notations telles que sa projection sur $\mathfrak{b}$ soit nulle, c'est-à-dire telles que l'algèbre de Lie $\mathfrak{b}$ soit résoluble. Une vérification de routine établit que l'algèbre de Lie $\mathfrak{of}(B)$ est parfaite (rappelons qu'une algèbre de Lie est dite *parfaite* si elle coïncide avec son idéal dérivé). L'algèbre de Lie $\mathfrak{b}$ est donc résoluble et parfaite, c'est-à-dire réduite à zéro.

(ii) *Cas où* $p_B + q_B = 2$. L'algèbre $\mathfrak{of}(B)$, résoluble, correspond aux matrices de la forme $\begin{pmatrix} 0 & c & 0 \\ c & 0 & 0 \\ x & y & 0 \end{pmatrix}$ avec $c \in \mathbf{R}$ et $x, y \in \mathbf{R}^{r_B}$ (voir $(*)$ ci-dessus). Dans $\mathfrak{of}(B)$, le radical nilpotent $\mathfrak{n}$ coïncide avec l'idéal dérivé $[\mathfrak{of}(B), \mathfrak{of}(B)]$; il est constitué des matrices pour lesquelles $c = 0$, de sorte qu'il est de codimension 1. La projection de $\mathfrak{n} = [\mathfrak{of}(B), \mathfrak{of}(B)]$ sur au moins l'un des facteurs $\mathfrak{a}, \mathfrak{b}$ est donc surjective; ce facteur étant à la fois nilpotent et parfait, il est réduit à zéro.

(iii) *Cas où* $(p_B, q_B) = (4,0)$ *ou* $(p_B, q_B) = (2,2)$. Notons à nouveau $\mathfrak{s}$ une sous-algèbre de Levi de $\mathfrak{of}(B)$; elle possède deux idéaux simples isomorphes $\mathfrak{u}$ et $\mathfrak{v}$ tels que $\mathfrak{s} = \mathfrak{u} \times \mathfrak{v}$. Vu l'argument du cas (i), il suffit de considérer ici le cas où la projection de $\mathfrak{s}$ sur $\mathfrak{a}$ serait isomorphe à $\mathfrak{u}$ et sa projection sur $\mathfrak{b}$ isomorphe à $\mathfrak{v}$; nous allons montrer que ce cas ne se produit pas.

En effet, le facteur $\mathfrak{a}$ posséderait une sous-algèbre simple de dimension trois centralisant le radical résoluble du facteur $\mathfrak{b}$, et de même pour le facteur $\mathfrak{b}$ et sa sous-algèbre simple $\mathfrak{v}$ centralisant le radical résoluble de $\mathfrak{a}$. La représentation naturelle (par restriction de la représentation adjointe) de $\mathfrak{s}$ sur le radical résoluble de $\mathfrak{of}(B)$ contiendrait donc des sous-représentations irréductibles non fidèles. Or ceci est absurde, car les sous-espaces irréductibles de l'action de $\mathfrak{s} = \mathfrak{so}(p_B, q_B)$ sur le radical résoluble $(\mathbf{R}^{p_B+q_B})^{r_B}$ sont tous isomorphes au $\mathfrak{so}(p_B, q_B)$-module fidèle $\mathbf{R}^{p_B+q_B}$. $\qquad\square$

Soit (W, S) un système de Coxeter, avec S fini. Notons E l'espace vectoriel $\mathbf{R}^S$ et B la forme de Tits associée à (W, S). Alors W possède une *représentation géométrique* sur E pour laquelle la forme B est invariante. Cette représentation est fidèle, et fournit donc une injection $W \subset \mathrm{Of}(B)$, où $\mathrm{Of}(B)$ est le groupe algébrique introduit plus haut; de plus W est un sous-groupe discret du groupe des points réels de $\mathrm{Of}(B)$.

Le groupe W est dit *irréductible* si le graphe de Coxeter associé au système (W, S) est connexe. Dans ce cas, la représentation géométrique de W est indécomposable; de plus, les trois conditions suivantes sont équivalentes: cette représentation est irréductible, elle est absolument irréductible, le noyau de B est réduit à zéro.

Pour tout ceci, voir [BouL], chapitre 5, § 4.

Rappelons quelques propriétés de la signature (p_B, q_B, r_B) de B lorsque W est irréductible; $n_B = p_B + q_B + r_B$ désigne comme plus haut la dimension de E, c'est-à-dire le cardinal de S.

- $p_B = n_B$ si et seulement si W est fini;
- $p_B = n_B - 1$ et $r_B = 1$ si et seulement si W est infini et contient un sous-groupe abélien libre d'indice fini (W est alors dit *de type affine*);
- si $q_B = 0$, alors $r_B \leq 1$;
- si $n_B \leq 4$, alors $p_B \geq n_B - 1$; en particulier, $(p_B, q_B, r_B) \neq (2, 2, 0)$;
- si $n_B \geq 4$, alors $p_B \geq 3$;

voir [BouL] pour les trois premières propriétés, et [Par1] pour les deux dernières.

La proposition suivante apparaît dans [Par2].

Proposition 8. *Soit (W, S) un système de Coxeter, avec S fini.*
(i) *Si (W, S) est irréductible infini, W est indécomposable.*
(ii) *Si (W, S) est irréductible infini non affine, tout sous-groupe d'indice fini de W est indécomposable.*
(iii) *Dans tous les cas, W est uniquement directement décomposable.*

Démonstration. Notons que l'assertion (i) pour W de type affine est une répétition du corollaire de la proposition 1.

Soit $\mathcal{G}$ le graphe de Coxeter associé à la paire (W, S). Notons $\mathcal{G}_1, \ldots, \mathcal{G}_m$ les composantes connexes de ce graphe et $W_1, \ldots, W_m$ les groupes de Coxeter correspondants, qui sont les *composantes irréductibles* du groupe W, et dont W est produit direct. Ceux des W_i qui sont infinis ont un centre réduit à un élément ([BouL], chapitre 5, § 4, exercice 3). Si W_i est infini et de plus n'est pas de type affine, alors W_i est Zariski-dense dans un groupe du type $\mathrm{Of}(B_i)$ [BeHa]; il en résulte en particulier que le centre de tout sous-groupe d'indice fini de W_i est encore réduit à un élément.

Il suffit donc d'appliquer la proposition 5 pour démontrer les assertions (i) et (ii). L'assertion (iii) résulte de la proposition 2 lorsque les W_i sont des groupes de Coxeter à centres triviaux (par exemple sont tous des groupes de Coxeter infinis); pour le cas général, nous invoquons la proposition 9. $\square$

La proposition suivante est un cas particulier du «théorème fondamental» du § 47 de [Kuro].

Proposition 9. *Considérons un entier $m \geq 1$, des groupes indécomposables $\Gamma_1, \ldots, \Gamma_m$ et le produit direct $\Gamma = \Gamma_1 \times \cdots \times \Gamma_m$; supposons[5] que, pour tout homomorphisme z de Γ dans le centre de Γ, l'image $z(\Gamma)$ soit finie. Alors Γ est uniquement directement décomposable.*

Soient W un groupe de Coxeter fini irréductible d'un des types A, D, E et Γ le groupe d'Artin correspondant, de quotient W. La proposition 5 permet une autre démonstration du résultat suivant de [Mari].

Proposition 10. *Soit Γ_0 un sous-groupe d'indice fini dans un groupe d'Artin Γ comme ci-dessus. Si Γ_0 n'est pas indécomposable, alors il possède une unique décomposition non triviale comme produit direct, avec l'un des facteurs cyclique infini.*

Démonstration. Le centre $Z(\Gamma)$ de Γ est cyclique infini (théorème 7.2 de [BrSa]). I. Marin a montré qu'il existe un entier d et une représentation irréductible de Γ dans $\mathrm{GL}_d(\mathbf{C})$ d'image Zariski-dense; il en résulte que $Z(\Gamma_0) = Z(\Gamma) \cap \Gamma_0$ est aussi cyclique infini. La proposition 5 permet de conclure. $\square$

En utilisant un autre type d'argument, Paris a montré que tout groupe d'Artin irréductible de type sphérique est indécomposable (voir la proposition 4.2 de [Par3]).

Considérons en particulier le cas du groupe d'Artin B_n des tresses à $n \geq 2$ brins. On sait que le centre de B_n est cyclique infini; plus généralement, tout sous-groupe d'indice fini Γ de B_n possède un centre $Z(\Gamma) = Z(B_n) \cap \Gamma$ qui est d'indice fini dans $Z(B_n) \simeq \mathbf{Z}$. Par suite, si un tel groupe Γ n'est pas indécomposable, alors Γ est produit direct d'un groupe indécomposable Γ_1 et de son centre isomorphe à $\mathbf{Z}$; de plus, Γ_1 est isomorphe au quotient de Γ par son centre, et sa classe d'isomorphisme est donc déterminée par celle de Γ. En particulier, Γ est uniquement directement décomposable.

Ceci s'applique au groupe P_n des tresses pures pour tout $n \geq 3$. En effet, nous avons une suite d'extensions scindées

$$\{1\} \longrightarrow F_{n-1} \longrightarrow P_n \overset{\pi_n}{\longrightarrow} P_{n-1} \longrightarrow \{1\} \tag{$\sharp$}$$

---

[5]Cette hypothèse ne peut en aucun cas être omise, comme le montrent plusieurs des exemples déjà cités (nos (xiv), (xvii) et (xviii) du chapitre 2). Le point important est que tout sous-groupe $z(\Gamma)$ possède une série principale de sous-groupes normaux. Il y a deux types de cas dans lesquels cette hypothèse est évidemment satisfaite: les groupes à abélianisés finis, et les groupes à centres finis.

telles que l'image par π_n du centre de P_n coïncide avec le centre de P_{n-1} (le noyau F_{n-1} de π_n est un groupe libre à $n-1$ générateurs). Définissons par récurrence une suite $(Q_n)_{n\geq 2}$ de sous-groupes des P_n en posant $Q_2 = \{1\}$ et $Q_n = \pi_n^{-1}(Q_{n-1})$ pour $n \geq 3$. (Notons que Q_3 est un groupe non abélien libre à deux générateurs. Notons aussi que, pour $n \geq 3$, Q_n dépend du brin choisi pour définir l'extension ($\sharp$) ci-dessus, et n'est donc pas uniquement défini comme sous-groupe de P_n.) Il est facile de vérifier que P_n est produit direct de son centre, cyclique infini, et du groupe Q_n, indécomposable.

5. Variation sur la notion d'indécomposabilité

Un groupe Γ est dit *c-décomposable* s'il existe deux sous-groupes normaux infinis Γ_1, Γ_2 de Γ tels que $\Gamma_1 \cap \Gamma_2$ est fini et $\Gamma_1\Gamma_2$ d'indice fini dans Γ. Un groupe qui n'est pas c-décomposable est dit *c-indécomposable*. (Voir [Marg], chapitre IX, no 2.2; nous écrivons «c-décomposable» où Margulis écrit «décomposable». La lettre «c» est l'initiale de «commensurable».)

Soient Γ un groupe, Γ_0 un sous-groupe d'indice fini, et F un sous-groupe normal fini de Γ. Il est facile de vérifier que les trois conditions suivantes sont équivalentes: Γ est c-indécomposable, Γ_0 est c-indécomposable, Γ/F est c-indécomposable.

Soit A un ensemble fini non vide. Pour tout $\alpha \in A$, soient k_α un corps local et G_α un k_α-groupe algébrique connexe, presque k_α-simple, et tel que le groupe $G_\alpha(k_\alpha)$ n'est pas compact. Soient G le produit direct $\prod_{\alpha \in A} G_\alpha(k_\alpha)$ et Γ un réseau dans G. Alors Γ est c-indécomposable (comme groupe abstrait) si et seulement si Γ est irréductible (comme réseau dans G); voir [Marg], chapitre IX, no 2.3.

Remarque. Soit Δ un groupe résiduellement fini qui possède un sous-groupe d'indice fini Γ héréditairement indécomposable, ce qui veut dire que tout sous-groupe d'indice fini de Γ est indécomposable. Alors Δ est c-indécomposable.

En effet, soient Δ_1, Δ_2 deux sous-groupes normaux de Δ tels que $\Delta_1 \cap \Delta_2$ est fini et $\Delta_1\Delta_2$ d'indice fini dans Δ. Il s'agit de montrer que l'un des groupes Δ_1, Δ_2 est fini. Pour $j = 1, 2$, posons $\Gamma'_j = \Delta_j \cap \Gamma$. Vu l'hypothèse de finitude résiduelle, nous pouvons choisir un sous-groupe d'indice fini Γ_j de Γ'_j ($j = 1, 2$) de telle sorte que que $\Gamma_1 \cap \Gamma_2 = \{1\}$. Quitte à remplacer à nouveau Γ_j par un sous-groupe d'indice fini, nous pouvons supposer de plus que $\Gamma_1 \times \Gamma_2$ est un sous-groupe de Γ. Les hypothèses impliquent alors que l'un des groupes Γ_1, Γ_2 est fini, de sorte que le groupe Δ_j correspondant est aussi fini.

6. Conditions Max et Min

Un groupe satisfait à la *condition Max-n* si toute chaîne ascendante de sous-groupes normaux de Γ est ultimement stationnaire, et à la condition *Min-n* si toute chaîne descendante de sous-groupes normaux est ultimement stationnaire. On définit de même les *conditions Max-fd et Min-fd*, en termes de facteurs directs. Un groupe satisfaisant à l'une de ces quatre conditions est évidemment produit direct d'un nombre fini de groupes indécomposables.

Un groupe satisfaisant[6] aux deux conditions Max-n et Min-n est uniquement directement décomposable: c'est le théorème de Wedderburn–Remak–Krull–Schmidt; voir par exemple le dernier théorème du §47 de [Kuro], ou le théorème 6.36 de [Rotm], ou le théorème 4.8 de [Suzu, chapitre 2]. Les exemples de Baumslag cités dans l'introduction montrent qu'un groupe satisfaisant la seule condition Max-n n'est pas nécessairement uniquement directement décomposable, puisqu'un groupe polycyclique satisfait cette condition (il satisfait même la condition de chaîne ascendante pour les sous-groupes non nécessairement normaux).

Notons qu'un groupe de Coxeter infini ne possède jamais la propriété Min-n. Plus généralement, un groupe infini résiduellement fini ne possède pas cette propriété. Or les groupes de Coxeter sont résiduellement finis: c'est en effet un fait général, connu sous le nom de «lemme de Mal'cev», que tout groupe linéaire de type fini est résiduellement fini; pour l'esquisse d'un argument valant pour les groupes de Coxeter, voir [BouL], chapitre 5, §4, exercice 9.

Nous allons montrer qu'un groupe de Coxeter de type fini qui n'est pas virtuellement abélien ne possède jamais la propriété Max-n. Il résulte néanmoins de la proposition 8 qu'un groupe de Coxeter de type fini possède les propriétés Min-fd et Max-fd.

Lemme.
  (i) Un quotient d'un groupe qui satisfait la condition Max-n la satisfait aussi.
 (ii) Un sous-groupe d'indice fini d'un groupe qui satisfait la condition Max-n la satisfait aussi.
(iii) Un groupe libre non abélien ne satisfait pas la condition Max-n.

Démonstration. L'assertion (i) est banale et l'assertion (ii) est un résultat de [Wil1].

C'est une conséquence immédiate de la définition qu'un groupe Γ satisfait la condition Max-n si et seulement si tout sous-groupe normal de Γ peut être engendré *comme sous-groupe normal* par un ensemble fini. Il résulte de l'existence de groupes à deux générateurs qui ne sont pas de présentation finie [Neum] que le groupe libre de rang deux ne satisfait pas la condition Max-n, et donc de (i) qu'un groupe libre non abélien de rang quelconque ne la satisfait pas non plus. $\square$

Proposition 11. *Un groupe de Coxeter de type fini qui n'est pas virtuellement abélien ne satisfait pas la condition Max-n.*

Démonstration. Selon un résultat établi indépendamment dans [Gonc] et [MaVi] un groupe de Coxeter de type fini qui n'est pas virtuellement abélien possède un sous-groupe d'indice fini qui se surjecte sur un groupe libre non abélien. La proposition est alors une conséquence immédiate du lemme précédent. $\square$

---

[6]Autrement dit: un groupe possédant une suite de composition distinguée principale, selon la terminologie de Bourbaki ([BouA], chapitre I, §ı,4, exercice 17).

Remarque. Soit $\{1\} \longrightarrow \Gamma' \longrightarrow \Gamma \longrightarrow \Gamma'' \longrightarrow \{1\}$ une extension de groupes. Si Γ' et Γ'' satisfont la condition Max-n, alors Γ la satisfait aussi (si nécessaire, voir par exemple le no 3.1.7 de [Robi]). En particulier, un groupe de Coxeter virtuellement abélien satisfait la condition Max-n.

7. Groupes à quotients majorés

Soit n un entier, $n \geq 2$. Convenons qu'un groupe Γ est *à quotients n-majorés*, ou n-*QM*, si tout sous-groupe de type fini $\Delta \neq \{1\}$ de Γ possède un sous-groupe normal propre d'indice au plus n. Un groupe Γ est dit *à quotients majorés* s'il existe un entier $n \geq 2$ pour lequel il est à quotients n-majorés. Nous collectons quelques exemples et propriétés simples relatifs à cette notion.

(i) Un groupe fini est évidemment un groupe à quotients majorés.

Pour un nombre premier p, un groupe qui est résiduellement un p-groupe fini est p-QM (car tout p-groupe fini non réduit à un élément possède un sous-groupe normal d'indice p). En particulier, les groupes abéliens libres et les groupes non abéliens libres sont 2-QM.

Le groupe $\bigoplus_{p \in \mathbf{P}} \mathbf{Z}/p\mathbf{Z}$, où $\mathbf{P}$ désigne l'ensemble des nombres premiers, n'est pas QM. Plus généralement, un groupe contenant des sous-groupes simples d'ordres arbitrairement grands n'est pas QM.

(ii) Dans la définition de la propriété QM, on ne pourrait pas omettre la condition sur Δ d'être de type fini sans changer la notion. En effet, le groupe additif $\mathbf{Q}$ des rationnels est 2-QM, car tout sous-groupe de type fini non réduit à $\{0\}$ dans $\mathbf{Q}$ est cyclique infini et possède donc un sous-groupe d'indice 2. Mais le groupe $\mathbf{Q}$, qui est divisible, ne possède aucun sous-groupe propre d'indice fini.

(iii) Il est évident que tout sous-groupe d'un groupe n-QM est aussi n-QM. Une somme restreinte (finie ou infinie) de groupes est n-QM si et seulement si chaque facteur est n-QM.

(iv) Soit Γ un groupe qui s'insère dans une extension

$$\{1\} \longrightarrow \Gamma' \longrightarrow \Gamma \xrightarrow{\pi} \Gamma'' \longrightarrow \{1\}.$$

Si Γ' est n'-QM et si Γ'' est n''-QM, alors Γ est $\max(n', n'')$-QM.

En effet, soit Δ un sous-groupe de type fini de Γ. Si $\Delta \subset \Gamma'$, alors Δ possède un sous-groupe normal propre d'indice fini au plus n'. Sinon, soit Δ_0'' un sous-groupe normal propre d'indice $k \leq n''$ de $\pi(\Delta)$; alors $\pi^{-1}(\Delta_0'')$ est un sous-groupe normal propre d'indice k dans Δ.

En particulier, si un groupe Γ possède un sous-groupe normal d'indice fini qui est à quotients majorés, alors Γ est également à quotients majorés.

(v) Pour les groupes de type fini, aucune des propriétés «QM» et «résiduellement fini» n'implique l'autre, comme le montrent les considérations suivantes.

D'une part, soit S un groupe non abélien et T un groupe infini. Le produit en couronne $\Gamma = S \wr T$ n'est pas résiduellement fini (théorème 3.2 de [Grue]). Si S

et T sont deux groupes qui sont de type fini et n-QM pour un entier n, il résulte de (iii) et (iv) que le groupe de type fini Γ est aussi n-QM. [Voir également (vi) ci-dessous.]

D'autre part, tout groupe dénombrable résiduellement fini se plonge dans un groupe de type fini et résiduellement fini [Wil2]. En particulier, le groupe $\bigoplus_{p \in \mathbf{P}} \mathbf{Z}/p\mathbf{Z}$ de (i) se plonge dans un groupe de type fini résiduellement fini qui n'est pas QM, par (iii).

(vi) Rappelons un exemple de groupe résoluble de type fini dû à P. Hall [HalP].

Soient R un anneau commutatif avec unité, dont nous notons R^* le groupe des unités, et R_0 un sous-groupe additif de R. Posons

$$G(R) = \begin{pmatrix} 1 & R & R \\ 0 & R^* & R \\ 0 & 0 & 1 \end{pmatrix} \quad \text{et} \quad Z(R_0) = \begin{pmatrix} 1 & 0 & R_0 \\ 0 & 1 & 0 \\ 0 & 0 & 1 \end{pmatrix}.$$

Le groupe $G(R)$ est résoluble de classe 3, son centre s'identifie à $Z(R)$, et $Z(R_0)$ est un sous-groupe central de $G(R)$.

Soit p un nombre premier; le groupe multiplicatif $p^{\mathbf{Z}}$ est un sous-groupe d'indice deux dans le groupe des unités de $\mathbf{Z}[1/p]$. Le groupe

$$G_+(\mathbf{Z}[1/p]) = \begin{pmatrix} 1 & \mathbf{Z}[1/p] & \mathbf{Z}[1/p] \\ 0 & p^{\mathbf{Z}} & \mathbf{Z}[1/p] \\ 0 & 0 & 1 \end{pmatrix},$$

qui est d'indice deux dans $G(\mathbf{Z}[1/p])$, est engendré par les trois matrices

$$\begin{pmatrix} 1 & 0 & 0 \\ 0 & p & 0 \\ 0 & 0 & 1 \end{pmatrix}, \quad \begin{pmatrix} 1 & 1 & 0 \\ 0 & 1 & 0 \\ 0 & 0 & 1 \end{pmatrix} \quad \text{et} \quad \begin{pmatrix} 1 & 0 & 0 \\ 0 & 1 & 1 \\ 0 & 0 & 1 \end{pmatrix}.$$

Notons $\tilde{\alpha}$ l'automorphisme extérieur de $G_+(\mathbf{Z}[1/p])$ obtenu en conjugant les matrices par la matrice diagonale de coefficients diagonaux $p, 1, 1$. On vérifie que $\tilde{\alpha}(Z(\mathbf{Z})) = Z(p\mathbf{Z})$ est d'indice p dans $Z(\mathbf{Z})$.

L'exemple de Hall est le quotient H de $G_+(\mathbf{Z}[1/p])$ par $Z(\mathbf{Z})$. L'automorphisme $\tilde{\alpha}$ de $G_+(\mathbf{Z}[1/p])$ induit un endomorphisme α de H qui est surjectif de noyau $Z(p^{-1}\mathbf{Z})/Z(\mathbf{Z})$, c'est-à-dire de noyau cyclique d'ordre p. En particulier, le groupe H est de type fini, résoluble (c'est même une extension centrale d'un groupe métabélien) et non Hopfien. Nous avons une suite exacte courte

$$\{1\} \longrightarrow \mathbf{Z}[1/p]/\mathbf{Z} \longrightarrow H \longrightarrow G_+(\mathbf{Z}[1/p])/Z(\mathbf{Z}[1/p]) \longrightarrow \{1\}$$

dont le noyau est le groupe p-QM de l'exemple (ii). Le quotient de la suite exacte, isomorphe à $\mathbf{Z} \ltimes (\mathbf{Z}[1/p])^2$, est un groupe linéaire de type fini, et c'est donc aussi un groupe QM (argument direct, ou proposition 12 ci-dessous). Il résulte de (iv) que le groupe de type fini non Hopfien H est un groupe QM.

(vii) Notre intérêt pour la propriété QM vient du fait qu'elle est partagée par les groupes linéaires de type fini, comme le montre la proposition suivante, bien connue.

Proposition 12. *Soit Γ un groupe de type fini qui est linéaire, c'est-à-dire qui est un sous-groupe de $\mathrm{GL}_d(K)$ pour un entier d et un corps K convenables. Alors il existe un nombre premier p tel que Γ possède un sous-groupe d'indice fini qui est résiduellement un p-groupe fini.*

En particulier, Γ est un groupe à quotients majorés.

Démonstration. Soit A le sous-anneau de K engendré par les coefficients matriciels des éléments d'un système fini de générateurs de Γ; c'est un anneau commutatif intègre de type fini. Soit $\mathfrak{m}$ un idéal maximal de A; le quotient $A/\mathfrak{m}$ est un corps fini (voir par exemple [BoAC'], chapitre 5, §3, no 4, corollaire 1 du théorème 3) dont nous notons p la caractéristique. Pour tout entier $k \geq 0$, notons N_k le noyau de l'application naturelle $\mathrm{GL}_d(A) \longrightarrow \mathrm{GL}_d(A/\mathfrak{m}^k)$.

L'intersection des idéaux $\mathfrak{m}^k$ est réduite à zéro (résultat de Krull, voir par exemple [BoAC], chapitre 3, §3, no 2), et donc l'intersection des sous-groupes N_k est réduite à un élément. Par ailleurs, le quotient N_k/N_{k+1} est fini pour tout $k \geq 0$, et c'est un p-groupe abélien élémentaire pour tout $k \geq 1$. Il en résulte que N_1 est résiduellement un p-groupe fini qui est d'indice fini dans $\mathrm{GL}_d(A)$, et par conséquent que $\Gamma \cap N_1$ est de même résiduellement un p-groupe fini qui est d'indice fini dans Γ.

La seconde assertion de la proposition résulte alors des points (i) et (iv) ci-dessus. $\qquad\qquad\square$

8. Majorations de nombres de facteurs directs

L'objet de ce numéro est d'apporter dans certains cas une précision quantitative aux propriétés Min-fd et Max-fd définies au numéro 6.

Soit Γ un groupe. Pour un entier $n \geq 1$, notons $K_n(\Gamma)$ l'intersection de tous les sous-groupes normaux de Γ d'indice au plus n et $k_n(\Gamma)$ l'indice de $K_n(\Gamma)$ dans Γ.

(i) Pour un produit direct, nous avons

$$K_n\left(\prod_{j=1}^m \Gamma_j\right) = \prod_{j=1}^m K_n(\Gamma_j) \quad \text{et} \quad k_n\left(\prod_{j=1}^m \Gamma_j\right) = \prod_{j=1}^m k_n(\Gamma_j).$$

Par exemple: $k_n(\mathbf{Z}^m) = n^m$ pour tout $m \geq 1$.

Si Γ est simple infini, alors $k_n(\Gamma) = 1$ pour tout $n \geq 1$.

(ii) Soit $\pi : \Gamma \longrightarrow \Delta$ un epimorphisme. Alors $K_n(\Gamma) \subset \pi^{-1}(K_n(\Delta))$ et $k_n(\Gamma) \geq k_n(\Delta)$. En particulier $k_n(\Gamma) \leq k_n(F_g)$ pour un groupe Γ à g générateurs.

Pour une majoration grossière de $k_n(F_g)$, notons que $\log_n(k_n(F_g)) \leq s_n^{\triangleleft}(F_g)$, et signalons que les résultats du chapitre 2 de [LuSe] fournissent une majoration explicite du nombre $s_n^{\triangleleft}(F_g)$ des sous-groupes normaux d'indice au plus n dans F_g.

(iii) Soit $\Gamma = \Gamma_1 \times \cdots \times \Gamma_m$; supposons que Γ possède un système de g générateurs. Notons m_0 le nombre des indices $j \in \{1, \ldots, m\}$ tels que $k_n(\Gamma_j) \geq 2$. Il résulte immédiatement des points (i) et (ii) ci-dessus que $m_0 \leq \log_2(k_n(F_g))$.

Théoème 13. *Soit Γ un groupe qui possède un système de g générateurs et qui est un produit $\Gamma = \Gamma_1 \times \cdots \times \Gamma_m$, de groupes non réduits à $\{1\}$. S'il existe un entier $n \geq 2$ tel que Γ est n-QM, alors $m \leq \log_2(k_n(F_g))$.*

En particulier, si Γ est un groupe linéaire de type fini, Γ peut toujours s'écrire comme produit direct d'un nombre fini de groupes indécomposables et il existe une borne sur le nombre de facteurs des décompositions de Γ en produit direct.

Démonstration. Si le groupe de type fini Γ est n-QM, chaque facteur Γ_j l'est ausssi, et la première assertion de la proposition résulte du point (iii) ci-dessus. La seconde assertion résulte alors de la proposition 12. $\qquad\square$

Remarque. Etant donné deux entiers $d, g \geq 1$ et un anneau de type fini A, il résulte des preuves ci-dessus qu'il existe une constante $M = M(d, g, A)$ telle que tout sous-groupe $\Gamma \subset \mathrm{GL}_d(A)$ à au plus g générateurs possède une décomposition $\Gamma = \Gamma_1 \times \cdots \times \Gamma_m$ en produits de groupes indécomposables, avec $m \leq M$.

M. Abert nous a signalé [Aber] une preuve du fait que, si G est un sous-groupe de $\mathrm{GL}_n(K)$, où K est un corps, alors le nombre de facteurs *non abéliens* dans une décomposition de G en produit direct est borné par une constante ne dépendant que de n, indépendante de K. Ce n'est bien sûr pas le cas des facteurs abéliens, comme en témoignent par exemple les sous-groupes finis cycliques de $\mathrm{GL}_1(\mathbf{C})$. Néanmoins, cela permet de donner une autre preuve de la seconde assertion du théorème 13, par réduction au cas abélien.

Voici pour terminer une conséquence de la proposition 9 et du théorème 13.

Proposition 14. *Soit Γ un groupe linéaire de type fini; supposons[7] que, pour tout homomorphisme z de Γ dans le centre de Γ, l'image $z(\Gamma)$ soit finie. Alors Γ est uniquement directement décomposable.*

Nous remercions Yves Benoist, Martin Bridson, Ken Brown, Luis Paris, Alain Valette et Thierry Vust pour plusieurs observations précieuses concernant notre texte.

Abridged English version

Let Γ be a group. A direct product decomposition $\Gamma \simeq \Gamma_1 \times \cdots \times \Gamma_m$ is *non-trivial* if $m \geq 2$ and $\Gamma_j \not\simeq \{1\}$ for $j = 1, \ldots, m$. The group Γ is *indecomposable* if it does not have any non-trivial decomposition. It is *uniquely directly decomposable* if it has a decomposition $\Gamma \simeq \Gamma_1 \times \cdots \times \Gamma_m$ with indecomposable factors Γ_j, if these factors are uniquely determined by Γ up to isomorphism and order, and moreover if any decomposition of Γ as a direct sum can be refined to a decomposition with indecomposable factors.

Indecomposable groups include simple groups, subgroups of $\mathbf{Q}$, non-trivial free products, torsion-free Gromov-hyperbolic groups, and many other standard

---

[7] Voir la note à laquelle renvoie la proposition 9.

examples reviewed in Chapter 2 of the French version. Finite groups are uniquely directly decomposable (theorem of Wedderburn–Remak–Krull–Schmidt). It is a particular case of a theorem of Kurosh that, if Γ is a centre-free group that is a direct product of a finite number of indecomposable groups, then Γ is uniquely directly decomposable ($\S\,47$ of [Kuro]).

There are abelian groups, indeed subgroups of $\mathbf{Q}^N$, which are not uniquely directly indecomposable ($\S\,90$–91 in [Fuch]). There is a class of spectacular examples due to G. Baumslag [Baum]: given $m, n \geq 2$, there exist indecomposable groups $A_1, \ldots, A_m, B_1, \ldots, B_n$ which are finitely generated, nilpotent, and torsion-free, such that the products $A_1 \times \cdots \times A_m$ and $B_1 \times \cdots \times B_n$ are isomorphic.

Let G be an algebraic group with centre reduced to one element; then G has a decomposition $G = G_1 \times \cdots \times G_n$ with the G_j algebraic subgroups which are indecomposable as algebraic subgroups, and such a decomposition is unique up to isomorphism and order. Let Γ be a Zariski dense subgroup of G; if $\Gamma = A \times B$ is a non-trivial direct product decomposition, then there exists a numeration of the G_j and an integer m with $0 \leq m \leq n$ such that $A \subset G_1 \times \cdots \times G_m$ and $B \subset G_{m+1} \times \cdots \times G_n$. In particular, if G is indecomposable as an algebraic group, then Γ is indecomposable. Chapter 3 of the French version contains appropriately modified statements for the case where G has a non-trivial centre.

There are simple applications of these facts. One is that the soluble Baumslag-Solitar groups (except $\mathbf{Z}^2$) are indecomposable.

Here is another one, which is a new proof of a recent result of Paris [Par2], and which has been motivating for our work. Let (W, S) be a Coxeter system, where S is a finite generating set of reflections. Recall that, to each connected component of the Coxeter graph of (W, S), we can associate a Coxeter subgroup W_j of W ($j = 1, \ldots, m$), which is an *irreducible Coxeter group*, so that $W \simeq W_1 \times \cdots \times W_m$. Irreducible Coxeter groups are of three exclusive kinds: the finite groups (spherical case), the infinite groups which contain free abelian subgroups of finite index (affine case), and the other groups (which are known to have non-abelian free subgroups). Some of the irreducible finite Coxeter groups are decomposable; for example, a dihedral group of order $8k + 4$ is the direct product of its centre of order 2 and of a dihedral group of order $4k + 2$.

Proposition (Paris). Let (W, S) be a Coxeter system, with S finite.

  (i) If W is irreducible and infinite, then W is indecomposable.
 (ii) If W is irreducible, infinite, and not affine, then any subgroup of finite index in W is indecomposable.
(iii) In all cases, W is uniquely directly decomposable.

For the proof, we also use a result of [BeHa] according to which, for W infinite and not affine, the Tits representation provides an embedding of W as a *Zariski-dense* subgroup of an appropriate orthogonal group. Details are given in Chapter 4 of the French version.

In the last chapters, we prove finiteness results on decompositions of linear groups. Recall that a group is *linear* if it is isomorphic to a subgroup of $GL_d(\mathbf{K})$ for some positive integer d and for some field $\mathbf{K}$. The following theorem is part of Theorem 13 and Proposition 14 of the French version.

Theorem. *Let Γ be finitely generated linear group. Then there exists an integer $M \geq 1$ with the following property.*

The group Γ can be written as a direct product of a finite number of indecomposable groups, and any direct decomposition of Γ has at most M factors.

If, moreover, any homomorphism from Γ to the centre of Γ has a finite image, then Γ is uniquely directly decomposable.

There are two particular cases in which the hypothesis on central endomorphisms of Γ is obviously satisfied: groups with finite centre, and groups with finite abelianization.

Acknowledgments

Nous remercions le Fonds National Suisse de la Recherche Scientifique pour son soutien.

References

[Aber] M. Abert, *Representing graphs by the non-commuting relation*, Publ. Math. Debrecen **69/3** (2006), 261–269.

[Bagh] G.H. Bagherzadeh, *Commutativity in one-relator groups*, J. London Math. Soc. **13** (1976), 459–471.

[BaMi] G. Baumslag and C.F. Miller, III, *Some odd finitely presented groups*, Bull. London Math. Soc. **20** (1988), 239–244.

[BaSh] U. Bader et Y. Shalom, *Factor and normal subgroup theorems for lattices in products of groups*, Inventiones math. **163** (2006), 415–354.

[Baum] G. Baumslag, *Direct decompositions of finitely generated torsion-free nilpotent groups*, Math. Z. (1975) **145**, 1–10.

[BeHa] Y. Benoist et P. de la Harpe, *Adhérence de Zariski des groupes de Coxeter*, Compositio Math. **140** (2004), 1357–1366.

[Bier] R. Bieri, *Homological dimension of discrete groups*, Queen Mary College Mathematics Notes, 1976.

[BoAC] N. Bourbaki, *Algèbre commutative, chapitres 1 à 4*, Masson, 1985.

[BoAC'] N. Bourbaki, *Algèbre commutative, chapitres 5 et 6*, Hermann, 1966.

[Bore] A. Borel, *Linear algebraic groups, Second enlarged Edition*, Springer, 1991.

[BouA] N. Bourbaki, *Algèbre, chapitres 1 à 3*, Diffusion C.C.L.S., 1970.

[BouA'] N. Bourbaki, *Algèbre, chapitres 4 à 7*, Masson, 1981.

[BouL] N. Bourbaki, *Groupes et algèbres de Lie, chapitres 4, 5 et 6*, Hermann, 1968.

[BrHa] M.R. Bridson et A. Haefliger, *Metric spaces of non-positive curvature*, Springer, 1999.

[BrSa] E. Brieskorn et K. Saito, *Artin-Gruppen und Coxeter-Gruppen*, Inventiones Math. **17** (1972), 245–271.

[CaFP] J.W. Cannon, W.J. Floyd, et W.R. Parry, *Introductory notes on Richard Thompson's groups*, l'Enseignement math. **42** (1996), 215–256.

[Eber] P.B. Eberlein, *Geometry of nonpositively curved manifolds*, The University of Chicago Press, 1996.

[ErPa] P. Erdös et P. Pálfy, *On the orders of directly indecomposable groups*, Discrete Math. (Paul Erdös memorial collection) **200** (1999), 165–179.

[Fuch] L. Fuchs, *Infinite abelian groups, Vol. I & II*, Academic Press, 1970 & 1973.

[GeGl] T. Gelander et Y. Glasner, *Infinite primitive groups*, arXiv:math. GR/0503001 (Geom. Funct. Anal., to appear).

[Gonc] C. Gonciulea, *Virtual epimorphisms of Coxeter groups onto free groups*, PhD thesis, Ohio State University (2000).

[Grig] R. Grigorchuk, *Burnside's problem on periodic groups*, Functional Anal. Appl. **14** (1980), 41–43.

[Grue] L. Gruenberg, *Residual properties of infinite soluble groups*, Proc. London Math. Soc. **7** (1957), 29–62.

[HalP] P. Hall, *The Frattini subgroups of finitely generated groups*, Proc. London Math. Soc. **11** (1961), 327–352 [= Collected Works, 581–608].

[Harp] P. de la Harpe, *Topics in geometric group theory*, The University of Chicago Press, 2000.

[Hi86] R. Hirshon, *Finitely generated groups L with $L \simeq L \times M$, $M \neq 1$, M finitely presented*, J. of Algebra **99** (1986), 232–238.

[Hi90] R. Hirshon, *On uniqueness of direct decompositions of groups into directly indecomposable factors*, J. Pure Appl. Algebra **63** (1990), 155–160.

[Kapl] I. Kaplansky, *Infinite abelian groups*, Univ. of Michigan Press, 1954.

[Kuro] A.G. Kurosh, *The theory of groups (Volumes I & II)*, Chelsea, 1956.

[Lam] T.Y. Lam, *A first course in noncommutative rings*, Springer, 1991.

[LuSe] A. Lubotzky et D. Segal, *Subgroup growth*, Birkhäuser, 2003.

[MaKS] W. Magnus, A. Karras et D. Solitar, *Combinatorial group theory*, Interscience, 1966.

[Marg] G.A. Margulis, *Discrete subgroups of semisimple Lie groups*, Springer, 1991.

[Mari] I. Marin, *Sur les représentations de Krammer génériques*, arXiv:math.RT/0504143 v2 (Ann. Inst. Fourier, to appear).

[MaVi] G.A. Margulis et E.B. Vinberg, *Some linear groups virtually having a free quotient*, J. Lie Theory **10** (2000), 171–180.

[Meie] D. Meier, *Non-Hopfian groups*, J. London Math. Soc. (2) **26** (1982), 265–270.

[Meng] D.J. Meng, *Some results on complete Lie algebras*, Comm. in Algebra **22(13)** (1994), 5457–5507.

[MoSh] N. Monod et Y. Shalom, *Orbit equivalence rigidity and bounded cohomology*, Annals of Math. **164** (2006), 825–878.

[Neum] B.H. Neumann, *Some remarks on infinite groups*, J. London Math. Soc. **12** (1937), 120–127.

[Par1]   L. Paris, *Signatures des graphes de Coxeter*, Thèse, Université de Genève (1989).

[Par2]   L. Paris, *Irreducible Coxeter groups*, arXiv:math/0412214 v2 (Internat. J. Algebra comput., to appear).

[Par3]   L. Paris, *Artin groups of spherical type up to isomorphism*, J. of Algebra **281** (2004), 666–678.

[Robi]   D.J.S. Robinson, *A course in the theory of groups*, Springer, 1982.

[Rotm]   J.J. Rotman, *An introduction to the theory of groups, Fourth Edition*, Springer, 1995.

[SaSS]   M. du Sautoy, D. Segal, et A. Shalev, *New horizons in pro-p-groups*, Birkhäuser, 2000.

[Schr]   V. Schroeder, *A splitting theorem for spaces of nonpositive curvature*, Inventiones Math. **79** (1985), 323–327.

[ScWa]   P. Scott et T. Wall, *Topological methods in group theory*, in "Homological group theory, Durham 1977", C.T.C. Wall Editor, Cambridge Univ. Press (1979), 137–203.

[Serr]   J.-P. Serre, *Cohomologie des groupes discrets*, in "Prospects in mathematics", Annals of Math. Studies **70**, Princeton Univ. Press (1971), 77–169.

[Stal]   J. Stallings, *A topological proof of Grushko's theorem on free products*, Math. Z. **90** (1965), 1–8.

[Suzu]   M. Suzuki, *Group theory I*, Springer, 1982.

[Ty74]   J.M. Tyrer Jones, *Direct products and the Hopf property* (Collection of articles dedicated to the memory of Hanna Neumann, VI), J. Austral. Math. Soc. **17** (1974), 174–196.

[Ty80]   J.M. Tyrer Jones, *On isomorphisms of direct powers*, in "Word problems, II (Conf. on Decision Problems in Algebra, Oxford, 1976)", North-Holland (1980), 215–245.

[Walk]   E.A. Walker, *Cancellation in direct sums of groups*, Proc. Amer. Math. Soc. **7** (1956), 898–902.

[Wil1]   J.S. Wilson, *Some properties of groups inherited by normal subgroups of finite index*, Math. Z. **114** (1970), 19–21.

[Wil2]   J.S. Wilson, *Embedding theorems for residually finite groups*, Math. Z. **174** (1980), 149–157.

[Zimm]   R. Zimmer, *Ergodic theory and semisimple groups*, Birkhäuser, 1984.

Yves de Cornulier
IRMAR
Campus de Beaulieu
F-35042 Rennes Cedex, France
e-mail: `yves.decornulier@univ-rennes1.fr`

Pierre de la Harpe
Section de Mathématiques
Université de Genève, C.P. 64
CH–1211 Genève 4
e-mail: `Pierre.delaHarpe@math.unige.ch`

Geometric Group Theory
Trends in Mathematics, 103–119
© 2007 Birkhäuser Verlag Basel/Switzerland

Limit Groups of Equationally Noetherian Groups

Abderezak Ould Houcine

Abstract. We collect in this paper some remarks and observations about limit groups of equationally noetherian groups. We show in particular, that some known properties of limit groups of a free group or, more generally, of a torsion-free hyperbolic group can be seen as consequences of the fact that such groups are equationally noetherian. Especially, such properties are still true for linear groups and finitely generated abelian-by-nilpotent groups.

1. Introduction

Limit groups of free groups have been introduced by Sela [20] for his study of equations over free groups. They can be seen, geometrically and algebraically, as *limits* of free groups. This class coincides with the class of fully residually-free groups; a class of groups introduced by Baumslag [1] and studied by Kharlampovich and Myasnikov [12, 13, 14], Remeslennikov [14], Sela [20] and by many others [5, 9, 11, 19].

Some properties of limit groups, as the stationarity of any sequence of epimorphisms, can be deduced from the linearity of free groups. In fact they are consequences of the noetherian nature of the Zariski topology of the field of complex numbers; that is any system of equations in a free group is equivalent to a finite subsystem [8]. This last property is very interesting and a group satisfying it is called *equationally noetherian*; a notion introduced by Baumslag, Myasnikov and Remeslennikov [2]. The purpose of this paper is to see that some known properties of limit groups of free groups or of a torsion-free hyperbolic group can be seen as consequences of the fact such groups are equationally noetherian. These properties are still true for limit groups of any equationally noetherian group and in particular of a linear group and a finitely generated abelian-by-nilpotent group.

This article originates from the Geneva conference.

In [2] Baumslag, Myasnikov and Remeslennikov have developed algebraic geometry over groups. We recall here some definitions and results that we require. Let G be a fixed group and $\bar{x} = (x_1, \ldots, x_n)$. We denote by $G[\bar{x}]$ the group $G * F(\bar{x})$ where $F(\bar{x})$ is the free group with basis $\{x_1, \ldots, x_n\}$.

For an element $s(\bar{x}) \in G[\bar{x}]$ and a tuple $\bar{g} = (g_1, \ldots, g_n) \in G^n$ we denote by $s(\bar{g})$ the element of G obtained by replacing each x_i by g_i $(1 \leq i \leq n)$. Let S be a subset of $G[\bar{x}]$. Then the set

$$V(S) = \{\bar{g} \in G^n \mid s(\bar{g}) = 1 \text{ for all } s \in S\}$$

is called the *algebraic set* over G defined by S. For instance the centralizer of any element in G is an algebraic set. Subsets of $G[\bar{x}]$ will be also seen as *system of equations* with parameters from G. A group G is called *equationally noetherian* if for every $n \geq 1$ and every subset S of $G[\bar{x}]$ there exists a finite subset $S_0 \subseteq S$ such that $V(S) = V(S_0)$. A subset of G^n is *closed* if it is the intersection of an arbitrary number of finite unions of algebraic sets. This defines a topology on G^n, called the *Zariski topology*. Then a group G is equationally noetherian if and only if for each $n \geq 1$ the Zariski topology on G^n is noetherian [2, Theorem D1].

Guba has shown that free groups are equationally noetherian using the fact that a free group is linear [8]. Using the same argument Baumslag, Myasnikov and Remeslennikov have proved that linear groups over a commutative, noetherian, unitary ring, e.g., a field are equationally noetherian [2, Theorem B1]. They have also shown that abelian groups are equationally noetherian. Bryant proved that finitely generated abelian-by-nilpotent groups are equationally noetherian [4]. Since the wreath product of a nontrivial finitely generated abelian group G by a finitely generated nilpotent group H is linear if and only if H is virtually abelian [22], this provides examples of equationally noetherian groups that are not linear [2, p. 38]. Sela has shown that torsion-free hyperbolic groups are equationally noetherian [21, Theorem 1.22]. This answers a question in [2] and gives another proof of the fact that a free group is equationally noetherian. All of this shows that the class of equationally noetherian groups is very large.

We recall the definition of an H-*limit* group. Let H and G be groups. Let $(f_{i \in \mathbb{N}} | f_i : G \to H)$ be a sequence of morphisms from G to H. We say that $f_{i \in \mathbb{N}}$ is *convergent* or *stable* if for every $g \in G$ one of the following sets is finite

$$\{i \in \mathbb{N} | f_i(g) = 1\}, \{i \in \mathbb{N} | f_i(g) \neq 1\}.$$

We let $\underrightarrow{\ker}(f_i) = \{g \in G | \text{ the set } \{i \in \mathbb{N} | f_i(g) \neq 1\} \text{ is finite }\}$. A group K is said to be H-*limit* if there exist a group G and a converging sequence $(f_{i \in \mathbb{N}} | f_i : G \to H)$ such that $K = G/\underrightarrow{\ker}(f_i)$. In general limit groups are supposed finitely generated, but since some results do not depend on the finite generation property, we do not suppose that and we work in the general case. Thus, in our definition, the free group $F_{\aleph_0}$ of rank $\aleph_0$ is a limit group of free groups, though it is not finitely generated.

Limit groups are heavily connected to *residual* properties. Let G be a group and $\mathcal{K}$ a class of groups. We say that G is *residually-$\mathcal{K}$* (or that $\mathcal{K}$ *separates* G)

if for every $g \in G \setminus \{1\}$ there exist $K \in \mathcal{K}$ and a morphism $f : G \to K$ such that $f(g) \neq 1$. If $\mathcal{K}$ consists of the singleton K, then we say simply that G is *residually-K*. We say that G is *fully residually-K* (or that $\mathcal{K}$ *discriminates* G) if for every finite subset $X \subseteq G \setminus \{1\}$ there exist $K \in \mathcal{K}$ and a morphism $f : G \to K$ such that $1 \notin f(X)$. As before, if $\mathcal{K}$ consists of the singleton K, then we say simply that G is *fully residually-K*.

We finish with some notions from Model Theory needed in the sequel. For more details the reader is referred to [6, 10, 15].

The *universal theory* of a group H, denoted $\mathrm{Th}_\forall(H)$, is the set of all *universal* sentences true in H; i.e., sentences of the form

$$\forall \bar{x} (\bigvee_{1 \leq i \leq m} (\bigwedge_{w \in P_i} w(\bar{x}) = 1 \wedge \bigwedge_{v \in N_i} v(\bar{x}) \neq 1)),$$

(true in H), where P_i, N_i are finite sets of words on the variables $\{x_1, \ldots, x_n\}$ and their inverses. An universal sentence *with parameters from H* is defined analogously by allowing in the above words parameters from H.

The *universal Horn theory* of a group H, is the set of all sentences of the form

$$\forall \bar{x} ((\bigwedge_{w \in P} w(\bar{x}) = 1) \Rightarrow v(\bar{x}) = 1),$$

true in H, where P is a finite set of words on the variables $\{x_1, \ldots, x_n\}$ and their inverses.

When ϕ is a sentence, we write $G \models \phi$ to mean that ϕ is true in G and we say that G is a *model* of ϕ. $G \models \mathrm{Th}_\forall(H)$ means that G satisfies every sentence $\phi \in \mathrm{Th}_\forall(H)$ and we say that G is a *model* of $\mathrm{Th}_\forall(H)$ or that G is a model of the universal theory of H.

An *ultrafilter* on a set I is a finitely additive probability measure $\mu : \mathcal{P}(I) \to \{0, 1\}$. An ultrafilter μ is called *nonprincipal* if $\mu(X) = 0$ for every finite subset $X \subseteq I$.

Given an ultrafilter μ on I and a sequence of groups $(G_i)_{i \in I}$ we define an equivalence relation $\sim_\mu$ on $\prod_{i \in I} G_i$ by

$$\hat{a} = (a_i \in G_i)_{i \in I} \sim_\mu \hat{b} = (b_i \in G_i)_{i \in I} \text{ if and only if } \mu(\{i \in I | a_i = b_i\}) = 1.$$

The set of equivalence classes $(\prod_{i \in I} G_i) / \sim_\mu$ is endowed with a structure of group by defining

$$\hat{a}.\hat{b} = \hat{c} \text{ if and only if } \mu(\{i \in I | a_i.b_i = c_i\}) = 1.$$

The group $(\prod_{i \in I} G_i) / \sim_\mu$ is called the *ultraproduct* of the family $(G_i)_{i \in I}$. When $G_i = G$ for all $i \in I$, $(\prod_{i \in I} G_i) / \sim_\mu$ is called an *ultrapower* and it is denoted by $^*G$. If μ is nonprincipal, then $^*G$ is called a *nonprincipal ultrapower*.

Define $\pi : G \to {}^*G$ by $\pi(g) = (g_i = g | i \in I)$. Then π is an embedding. Moreover, a theorem of Łos [6, Theorem 4.1.9] claims that G is an *elementary subgroup* of $^*G$; that is, any sentence with parameters from G which is true in G is also true in $^*G$. As consequence, we get the property, needed in the sequel, that if $G \models \phi$, where ϕ is an universal sentence with parameters from G, then $^*G \models \phi$. In particular $^*G$ and G have the same universal theory.

Another theorem in Model theory, due to Keisler ([6, Theorem 6.1.1] or [15, Exercice 4.5.37]), claims that when I is countable, a *nonprincipal* ultrapower $^*G$ of G is $\aleph_1$-*saturated*. In what follows, we need only a special case of that saturation property (compactness): if

$$^*G \models \forall \bar{x}\left(\bigvee_{i \in \mathbb{N}} \phi_i(\bar{x})\right),$$

where each ϕ_i is of the form $\left(\bigwedge_{1 \leq j \leq p} w_j(\bar{x}) = 1 \wedge \bigwedge_{1 \leq j \leq q} v_j(\bar{x}) \neq 1\right)$, then there exists $n \in \mathbb{N}$ such that

$$^*G \models \forall \bar{x}\left(\bigvee_{0 \leq i \leq n} \phi_i(\bar{x})\right).$$

(The words $w_j(\bar{x}), v_j(\bar{x})$ are allowed to have parameters from $^*G$.)

Conventions

- In what follows, for a finitely generated group G we write $G = \langle \bar{x} | P(\bar{x}) \rangle$ to mean that G is generated by $\bar{x} = (x_1, \ldots, x_n)$ and $P(\bar{x})$ is a presentation *closed by deduction*; that is, if $G \models w(\bar{x}) = 1$ then $w \in P$. If $S(\bar{x})$ is a set of words we write $S(\bar{x}) = 1$ as abbreviation of the formula $\bigwedge_{w \in S} w(\bar{x}) = 1$.

- Let G be an equationally noetherian group and let $S \subseteq G[\bar{x}]$. A subset $S_0 \subseteq S$ is called a G-*witness* of S if S_0 is finite and $V(S) = V(S_0)$. Notice that we have

$$G \models \forall \bar{x}(S_0(\bar{x}) = 1 \Leftrightarrow S(\bar{x}) = 1).$$

- In all of the rest of this paper, all the ultrafilter considered will be on $\mathbb{N}$.

2. First properties

We begin this section with the following theorem which regroups several relations between the different notions given in the introduction.

Theorem 2.1. *Let H be a group.*

(1) *A countable fully residually-H group is an H-limit group.*

(2) *An H-limit group is embeddable in all nonprincipal ultrapowers of H; in particular it is a model of $Th_\forall(H)$.*

(3) *Let G be a finitely generated group. The following properties are equivalents:*
    (i) *G is a model of $Th_\forall(H)$,*
    (ii) *G is embeddable in some nonprincipal ultrapower of H,*
    (iii) *G is embeddable in all nonprincipal ultrapowers of H.*

(4) *If H is equationally noetherian then for every finitely generated group G the following properties are equivalent:*

> (i) G is a model of $Th_\forall(H)$,
>
> (ii) G is fully residually-H,
>
> (iii) G is an H-limit group.

(5) *If H is equationally noetherian then for every finitely generated group G the following properties are equivalent:*

> (i) G is a model of the universal Horn theory of H,
>
> (ii) G is residually-H.

The following lemma is classical in Model Theory, but for completeness we provide a proof.

Lemma 2.2. *Let $G = \langle \bar{x} | P(\bar{x}) \rangle$ be a finitely generated group and H a group. Let $N(\bar{x})$ be the set of all words $w(\bar{x})$ such that $G \models w(\bar{x}) \neq 1$. Suppose that for every finite subset $X \subseteq P(\bar{x}) \cup N(\bar{x})$, there exists a tuple $\bar{a}$ in H such that $H \models \bigwedge_{w \in X \cap P} w(\bar{a}) = 1 \wedge \bigwedge_{v \in X \cap N} v(\bar{a}) \neq 1$. Then G is embeddable in all nonprincipal ultrapowers of H. In particular G is a model of the universal theory of H.*

Proof. Let μ be a nonprincipal ultrafilter on $\mathbb{N}$ and let $^*H$ be the ultrapower of H relatively to μ.

Write $P(\bar{x}) \cup N(\bar{x}) = \bigcup_{i \in \mathbb{N}} S_i$, where each S_i is finite and $S_i \subseteq S_{i+1}$. For every $i \in \mathbb{N}$, pick $\bar{a}_i \in H$ such that $H \models \bigwedge_{w \in S_i \cap P} w(\bar{a}_i) = 1 \wedge \bigwedge_{v \in S_i \cap N} v(\bar{a}_i) \neq 1$.

We claim that the map $f : G \to {}^*H$ defined by $f(\bar{x}) = (\bar{a}_i | i \in \mathbb{N})$ extends to an embedding.

If $w(\bar{x}) \in P$, then there exists $n \in \mathbb{N}$ such that $w \in S_n$ and thus $H \models w(\bar{a}_i) = 1$ for every $i \geq n$. Therefore $\mu(\{i \in \mathbb{N} | H \models w(\bar{a}_i) = 1\}) = 1$ and thus $^*H \models w((\bar{a}_i | i \in \mathbb{N})) = 1$.

Similarly, if $v(\bar{x}) \in N$, then there exists $n \in \mathbb{N}$ such that $v \in S_n$ and thus $H \models v(\bar{a}_i) \neq 1$ for every $i \geq n$. As before, $\mu(\{i \in \mathbb{N} | H \models v(\bar{a}_i) \neq 1\}) = 1$ and thus $^*H \models v((a_i | i \in \mathbb{N})) \neq 1$. This ends the proof of our claim.

Now, since universal sentences are conserved by taking subgroups, G is a model of the universal theory of $^*H$. As H and $^*H$ have the same universal theory, G is a model of the universal theory of H. $\qquad\square$

Proof of Theorem 2.1. (1). Let G be a countable fully residually-H group and write $G \setminus \{1\} = \bigcup_{i \in \mathbb{N}} S_i$ where each S_i is finite and $S_i \subseteq S_{i+1}$. Then there exists a convergent sequence $(f_{i \in \mathbb{N}} | f_i : G \to H)$ such that $1 \notin f_i(S_i)$. Clearly $\underrightarrow{\ker}(f_i) = 1$ and thus G is an H-limit group.

(2). Let K be an H-limit group and G be a group for which there exists a convergent sequence $(f_{i \in \mathbb{N}} | f_i : G \to H)$ such that $K = G/\underrightarrow{\ker}(f_i)$. Let μ be a nonprincipal ultrafilter on $\mathbb{N}$, and let $^*G$ (resp. $^*H$) be the ultrapower of G (resp. H) relatively to μ. Let $\pi : G \to {}^*G$ be the natural embedding defined by $g \to \hat{g} = (g_i = g | i \in \mathbb{N})$. The map $f : {}^*G \to {}^*H$ defined by $f((g_i | i \in \mathbb{N})) = (f_i(g_i) | i \in \mathbb{N})$ is a morphism and satisfies

$$f((g_i | i \in \mathbb{N})) = 1 \Leftrightarrow \mu(\{i \in \mathbb{N} | f_i(g_i) = 1\}) = 1.$$

Therefore, the morphism $\varphi = f \circ \pi : G \to {}^{*}H$ satisfies

$$\varphi(g) = 1 \Leftrightarrow \mu(\{i \in \mathbb{N} | f_i(g) = 1\}) = 1.$$

Since $(f_{i\in\mathbb{N}} | f_i : G \to H)$ is a convergent sequence and μ is nonprincipal we get

$$\varphi(g) = 1 \Leftrightarrow \{i \in \mathbb{N} | f_i(g) \neq 1\} \text{ is finite.}$$

Hence $\ker(\varphi) = \underrightarrow{\ker}(f_i)$. Thus $K = G/\ker(\varphi)$ and K is embeddable in ${}^{*}H$. In particular $K \models \mathrm{Th}_\forall(H)$.

$(3)\,(\mathrm{iii}) \Rightarrow (\mathrm{ii})$ is obvious, $(3)\,(\mathrm{ii}) \Rightarrow (\mathrm{i})$ is also obvious as universal sentences are preserved by subgroups and $\mathrm{Th}_\forall(H){=}\mathrm{Th}_\forall({}^{*}H)$. The implication $(3)\,(\mathrm{i}) \Rightarrow (\mathrm{ii})$ is a consequence of $(3)\,(\mathrm{i}) \Rightarrow (\mathrm{iii})$ and of the existence of nonprincipal ultrafilters on $\mathbb{N}$; so it remains to show $(3)\,(\mathrm{i}) \Rightarrow (\mathrm{iii})$.

Write $G = \langle \bar{x} | P(\bar{x}) \rangle$ and let $N(\bar{x})$ be the set of words $v(\bar{x})$ such that $G \models v(\bar{x}) \neq 1$. Since G is a model of the universal theory of H, for every finite subset $X \subseteq P(\bar{x}) \cup N(\bar{x})$, there exists a tuple $\bar{a}$ in H such that $H \models \bigwedge_{w \in X \cap P} w(\bar{a}) = 1 \wedge \bigwedge_{v \in X \cap N} v(\bar{a}) \neq 1$. Therefore, by Lemma 2.2, G is embeddable in all nonprincipal ultrapowers of H.

$(4)\,(\mathrm{ii}) \Rightarrow (\mathrm{iii})$ is a consequence of (1) and $(4)\,(\mathrm{iii}) \Rightarrow (\mathrm{i})$ is a consequence of (2), it remains to show $(4)\,(\mathrm{i}) \Rightarrow (\mathrm{ii})$. Write $G = \langle \bar{x} | P(\bar{x}) \rangle$ and let $P_0(\bar{x})$ be an H-witness of $P(\bar{x})$. Since $G \models \mathrm{Th}_\forall(H)$, for any words $v_1(\bar{x}), \ldots, v_n(\bar{x})$ such that $G \models \bigwedge_{1 \leq i \leq n} v_i(\bar{x}) \neq 1$, H is a model of

$$\exists \bar{x}(P_0(\bar{x}) = 1 \wedge \bigwedge_{1 \leq i \leq n} v_i(\bar{x}) \neq 1).$$

Therefore there exists $\bar{a} \in H$ satisfying $P(\bar{a}) = 1 \wedge \bigwedge_{1 \leq i \leq n} v_i(\bar{a}) \neq 1$. Since $H \models \forall \bar{x}(P_0(\bar{x}) = 1 \Leftrightarrow P(\bar{x}) = 1)$ there exists a morphism $f : G \to H$ such that $f(v_i(\bar{x})) = v_i(\bar{a}) \neq 1$, $1 \leq i \leq n$. Thus G is fully residually-H.

$(5)\,(\mathrm{i}) \Rightarrow (\mathrm{ii})$. Write $G = \langle \bar{x} | P(\bar{x}) \rangle$ and let $P_0(\bar{x})$ be an H-witness of $P(\bar{x})$. Since G is a model of the universal Horn theory of H, for any word $v(\bar{x})$ such that $G \models v(\bar{x}) \neq 1$, H is a model of $\exists \bar{x}(P_0(\bar{x}) = 1 \wedge v(\bar{x}) \neq 1)$. As before, there exists $\bar{a} \in H$ satisfying $P(\bar{a}) = 1 \wedge v(\bar{a}) \neq 1$. Thus there exists a morphism $f : G \to H$ such that $f(v(\bar{x})) \neq 1$. Thus G is residually-H.

$(5)\,(\mathrm{ii}) \Rightarrow (\mathrm{i})$. Clearly if G is residually-H, then G is embeddable in some product $\prod_{i \in I} H$ and thus G is a model of the universal Horn theory of H. $\qquad \square$

Remarks 2.3. (1). The implications $(4)\,(\mathrm{ii}) \Rightarrow (\mathrm{i})$ and $(5)\,(\mathrm{ii}) \Rightarrow (\mathrm{i})$ do not depend on the equationally noetherian nature of H neither on the finite generation property of G; that is $(4)\,(\mathrm{ii}) \Rightarrow (\mathrm{i})$ and $(5)\,(\mathrm{ii}) \Rightarrow (\mathrm{i})$ are still true for any group H and any group G. The implication $(5)\,(\mathrm{ii}) \Rightarrow (\mathrm{i})$ is very clear. Let us show that $(4)\,(\mathrm{ii}) \Rightarrow (\mathrm{i})$ is true for any group H and any group G. Suppose that G is fully residually-H. Suppose

$$(*) \qquad H \models \forall \bar{x}(\bigvee_{1 \leq i \leq m} (\bigwedge_{w \in P_i} w(\bar{x}) = 1 \wedge \bigwedge_{v \in N_i} v(\bar{x}) \neq 1)),$$

where P_i, N_i are finite sets of words, and suppose towards a contradiction

$$G \models (\bigwedge_{1 \leq i \leq m} (\bigvee_{w \in P_i} w(\bar{a}) \neq 1 \vee \bigvee_{v \in N_i} v(\bar{a}) = 1)), \text{ for some tuple } \bar{a} \text{ in } G.$$

Let

$$X = \{w(\bar{a}) | w(\bar{x}) \in P_1 \cup \cdots \cup P_m, \; G \models w(\bar{a}) \neq 1\}.$$

Since G is fully residually-H, there exists a morphism $f : G \to H$ such that $1 \notin f(X)$. Then

$$H \models (\bigwedge_{1 \leq i \leq m} (\bigvee_{w \in P_i} w(f(\bar{a})) \neq 1 \vee \bigvee_{v \in N_i} v(f(\bar{a})) = 1)),$$

contradicting $(*)$.

(2). Note that the equivalences (4)(i)–(iii) are not true if G is not finitely generated: $\mathbb{Q}$ is a model of the universal theory of $\mathbb{Z}$ but $\mathbb{Q}$ is not residually-$\mathbb{Z}$. Furthermore, $\mathbb{Q}$ is embeddable in all nonprincipal ultrapowers of $\mathbb{Z}$.

Corollary 2.4. *Let H be an equationally noetherian group. Then a group G is a model of $Th_\forall(H)$ if and only if G is locally fully residually-H.* $\qquad\square$

Definition 2.5. *A group G is said H-pseudo-limit if G is a model of the universal Horn theory of H.*

In order to prove the next theorem we need the following definitions from [2]. Let $Y \subseteq G^n$. We define

$$I(Y) = \{w(\bar{x}) \in G[\bar{x}] | w(\bar{y}) = 1 \text{ for all } \bar{y} \in Y\}.$$

Notice that when $S \subseteq G[\bar{x}]$, then

$$w(\bar{x}) \in I(V(S)) \Leftrightarrow G \models \forall \bar{x}(S(\bar{x}) = 1 \Rightarrow w(\bar{x}) = 1).$$

For $S \subseteq G[\bar{x}]$ the *coordinate group* of S, denoted G_S, is $G[\bar{x}]/I(V(S))$. The proof of the following Lemma can certainly be extracted from [16] but for completeness we provide a proof.

Lemma 2.6. *For any group H and any $S \subseteq H[\bar{x}]$, H_S is residually-H. In particular H_S is an H-pseudo-limit group.*

Proof. Let $v(\bar{x})$ be a word such that $H_S \models v(\bar{x}) \neq 1$. Clearly we have $v(\bar{x}) \notin I(V(S))$ and thus

$$H \models \exists \bar{x}(S(\bar{x}) = 1 \wedge v(\bar{x}) \neq 1).$$

Thus there exists a morphism $f : H_S \to H$ which fixes every element of H and $f(v(\bar{x})) \neq 1$. Therefore H_S is residually-H as desired.

By Remarks 2.3 (1), H_S is an H-pseudo-limit group. $\qquad\square$

Theorem 2.7. *If H is equationally noetherian then any sequence of epimorphisms of finitely generated H-pseudo-limit groups*

$$G_1 \to_{\phi_{1,2}} G_2 \to_{\phi_{2,3}} \cdots$$

terminates after finitely many steps.

Conversely, if H is finitely generated and any sequence of epimorphisms of finitely generated residually-H groups terminates after finitely many steps, then H is equationally noetherian.

Proof. Suppose that H is equationally noetherian and write $\bar{x} = (x_1, \ldots, x_n)$, $G_i = \langle \bar{x} | P_i(\bar{x}) \rangle$, $\phi_{i,i+1}(x_k) = x_k$. Since H is equationally noetherian there exists a finite subset $S(\bar{x}) \subseteq P(\bar{x}) = \bigcup_i P_i(\bar{x})$ such that

$$H \models \forall \bar{x}(S(\bar{x}) = 1 \Rightarrow w(\bar{x}) = 1)) \text{ for every } w \in P.$$

Let $m \in \mathbb{N}$ such that $S(\bar{x}) \subseteq P_m(\bar{x})$. We claim that for every $k \geq m$, the epimorphism $\phi_{k,k+1} : G_k \to G_{k+1}$ is an isomorphism. Let $v(\bar{x})$ be a word on $\bar{x}$ such that $G_{k+1} \models v(\bar{x}) = 1$. Then there exists a finite subset $A \subseteq P_{k+1}$ such that $\forall \bar{x}(A(\bar{x}) = 1 \Rightarrow v(\bar{x}) = 1)$ is true in any group. Since $k + 1 \geq m$ we have

$$H \models \forall \bar{x}(S(\bar{x}) = 1 \Rightarrow A(\bar{x}) = 1)),$$

thus we get

$$H \models \forall \bar{x}(S(\bar{x}) = 1 \Rightarrow v(\bar{x}) = 1)).$$

Since G_k is a model of the universal Horn theory of H, we get $G_k \models v(\bar{x}) = 1$.

Now suppose that H is finitely generated and that any sequence of epimorphisms of finitely generated residually-H groups terminates after finitely many steps. Let $S(\bar{x})$ be a system of equations with parameters from H. Since H is finitely generated $S(\bar{x})$ is countable. Write $S(\bar{x}) = \bigcup_{i \in \mathbb{N}} S_i(\bar{x})$ where S_i is finite and $S_i \subseteq S_{i+1}$. Let $P_i(\bar{x})$ be the set of words $w(\bar{x})$ with parameters from H such that

$$H \models \forall \bar{x}(S_i(\bar{x}) = 1 \Rightarrow w(\bar{x}) = 1),$$

i.e., $P_i(\bar{x}) = I(V(S_i))$. Let $H_i = \langle H * F(\bar{x}) | P_i(\bar{x}) \rangle = H_{S_i}$. By Lemma 2.6, H_i is residually-H. Now the sequence

$$H_1 \to H_2 \to H_3 \cdots$$

is a sequence of epimorphisms of finitely generated residually-H groups, and thus stabilizes. Therefore there exists n such that for every $i, j \geq n$, $P_i(\bar{x}) = P_j(\bar{x})$. Hence for every $i \geq n$ we have

$$H \models \forall \bar{x}(S_n(\bar{x}) = 1 \Rightarrow S_i(\bar{x}) = 1).$$

Thus $S(\bar{x})$ is equivalent to a finite subsystem as desired. $\qquad\square$

By Theorem 2.1 (2) an H-limit group is a model of the universal theory of H and thus it is H-pseudo-limit. Thus we get:

Corollary 2.8. *Let H be an equationally noetherian group. Then any sequence of epimorphisms of finitely generated H-limit groups*

$$G_1 \to G_2 \to \cdots$$

terminates after finitely many steps. $\qquad\square$

Corollary 2.9. *Let H be an equationally noetherian group. Then every finitely generated H-pseudo-limit group is Hopfian. In particular every H-limit group is Hopfian.*

Proof. Let G be a finitely generated H-pseudo-limit group and $\phi : G \to G$ be a surjective morphism. Then the sequence $G \to_\phi G \to_\phi G \to_\phi G \cdots$ terminates after finitely many steps. Thus ϕ is an isomorphism. $\square$

Corollary 2.10. *Let H be an equationally noetherian group. Then any finitely generated H-pseudo-limit group is equationally noetherian. In particular any finitely generated H-limit group is equationally noetherian.*

Proof. Let G be finitely generated H-pseudo-limit group. Then any sequence of epimorphisms of finitely generated residually-G groups is also a sequence of epimorphisms of finitely generated H-pseudo-limit groups; thus terminates after finitely many steps, by Theorem 2.7. Again by Theorem 2.7, G is equationally noetherian. $\square$

Theorem 2.11. *Let H be an equationally noetherian group. Then there exists at most countably many nonisomorphic finitely generated H-pseudo-limit groups. In particular there exists at most countably many nonisomorphic finitely generated H-limit groups.*

Proof. Suppose towards a contradiction that the opposite is true. Then there exists $n \in \mathbb{N}$ such that there exists at least λ nonisomorphic n-generated H-pseudo-limit groups for some $\lambda > \aleph_0$, (n-generated means generated by n elements). (In fact we can assume that $\lambda = 2^{\aleph_0}$, see [18]). Let $(G_i = \langle \bar{x} | P_i(\bar{x}) \rangle | i \in \lambda > \aleph_0)$ be the list of nonisomorphic n-generated H-pseudo-limit groups. For every $i \in \lambda$ there exists a finite subset $S_i \subseteq P_i$ such that $H \models \forall \bar{x}(S_i(\bar{x}) = 1 \Leftrightarrow P_i(\bar{x}) = 1)$.

Since for every $i \in \lambda$ the set S_i is finite, the set $\{S_i | i \in \lambda\}$ is countable. Therefore the map $f : \{P_i | i \in \lambda\} \to \{S_i | i \in \lambda\}$ defined by $P_i \mapsto S_i$ is not injective and thus there exist $i, j \in \lambda, i \neq j$ such that $S_i = S_j$.

Since G_i, G_j are models of the universal Horn theory of H we get $P_i = P_j$, a contradiction with the fact that G_i and G_j are not isomorphic. $\square$

3. Factor sets

Let H be a group equipped with the Zariski topology. Recall that a closed set is called *irreducible* if it is not the union of two proper, nonempty, closed subsets. In our context, if $S \subseteq H[\bar{x}]$ is a system of equations such that $V(S)$ is irreducible, then whenever $S_1, \ldots, S_n \subseteq H[\bar{x}]$ are systems of equations and if

$$H \models \forall \bar{x}(S(\bar{x}) = 1 \Rightarrow S_1(\bar{x}) = 1 \vee \cdots \vee S_n(\bar{x}) = 1),$$

then there exists i such that $H \models \forall \bar{x}(S(\bar{x}) = 1 \Rightarrow S_i(\bar{x}) = 1)$.

Now suppose that H is equationally noetherian. As noticed in the introduction the Zariski topology on H^n is noetherian. This implies that every closed subset of H^n is a finite union of algebraic sets. Recall also that every closed set in a noetherian topological space is a finite union of irreducible closed subsets. Thus, in our context, for any system $S(\bar{x})$ of equations there exist systems of equations $S_1, \ldots, S_n$ such that $(V(S_i)|1 \leq i \leq n)$ are irreducible and

$$H \models \forall \bar{x}(S(\bar{x}) = 1 \Leftrightarrow S_1(\bar{x}) = 1 \vee \cdots \vee S_n(\bar{x}) = 1).$$

Some properties contained in the next lemma, can certainly be extracted from [2].

Lemma 3.1. *Let H be a group and $S \subseteq H[\bar{x}]$ such that $V(S)$ is irreducible. Then H_S is fully residually-H. In particular, H_S is a model of the universal theory of H, and if H is countable then H_S is an H-limit group.*

Proof. Let $v_1(\bar{x}), \ldots, v_p(\bar{x})$ be words such that $H_S \models \bigwedge_{1 \leq j \leq p} v_j(\bar{x}) \neq 1$. Suppose that

$$H \models \forall \bar{x}(S(\bar{x}) = 1 \Rightarrow \bigvee_{1 \leq j \leq p} v_j(\bar{x}) = 1).$$

Then, since $V(S)$ is irreducible, there exists j such that

$$H \models \forall \bar{x}(S(\bar{x}) = 1 \Rightarrow v_j(\bar{x}) = 1),$$

contradicting the fact that $H_S \models v_j(\bar{x}) \neq 1$. Therefore

$$H \models \exists \bar{x}(S(\bar{x}) = 1 \wedge \bigwedge_{1 \leq j \leq p} v_j(\bar{x}) \neq 1).$$

Consequently, there exists a morphism $f : H_S \to H$, which fixes every element of H, such that $f(v_j) \neq 1$ $(1 \leq j \leq p)$. Thus H_S is fully residually-H as desired.

It follows, by Remarks 2.3 (1), that H_S is a model of the universal theory of H. Furthermore, if H is countable then H_S is also countable and by Theorem 2.1 (1), H_S is an H-limit group. $\square$

Factors sets are the first step in the construction of Makanin-Razborov diagrams. Let H be a group, K an H-limit group and G a finitely generated group. A *factor set* of G relatively to K is a finite collection of proper quotients $\{f_i : G \to L_i\}$ of H-limit groups such that any morphism $f : G \to K$ factors through some f_i after precomposition with some automorphism of G.

The following theorem gives a weak version of factor sets.

Theorem 3.2. *Let H be an equationally noetherian group. Then for any finitely generated group G there exists a finite collection of epimorphisms $\{f_i : G \to L_i\}$, where each L_i is an H-limit group, such that for any H-limit group L any morphism $f : G \to L$, factors through some f_i.*

Proof. Put $G = \langle \bar{x}|P(\bar{x}) \rangle$ and let $S(\bar{x})$ be an H-witness of P. Let $S_1, \ldots, S_n$ be systems of equations (maybe with parameters from H) such that each S_i is finite, $(V(S_i)|1 \leq i \leq n)$ are irreducible and

$$H \models \forall \bar{x}(S(\bar{x}) = 1 \Leftrightarrow S_1(\bar{x}) = 1 \vee \cdots \vee S_n(\bar{x}) = 1).$$

Let P_i be the set of words $w(\bar{x})$, *without parameters from H*, such that

$$H \models \forall \bar{x}(S_i(\bar{x}) = 1 \Rightarrow w(\bar{x}) = 1).$$

Let $L_i = \langle \bar{x} | P_i(\bar{x}) \rangle$. We claim that the sequence $L_1, \ldots, L_n$ satisfies the desired properties.

First we claim that L_i is an H-limit group. Let us show that L_i is the subgroup of H_{S_i} generated by $\bar{x}$. Clearly the map $\phi : L_i \to H_{S_i}$ which sends x_i to x_i is a morphism. We claim that it is injective. Indeed if $v(\bar{x})$ is a word, without parameters from H, such that $H_{S_i} \models v(\bar{x}) = 1$ then $v(\bar{x}) \in I(V(S_i))$ and thus $v(\bar{x}) \in P_i$. Therefore $L_i \models v(\bar{x}) = 1$ and thus ϕ is injective.

By Lemma 3.1, H_{S_i} is a model of $\mathrm{Th}_\forall(H)$ and thus L_i is also a model of $\mathrm{Th}_\forall(H)$. As L_i is finitely generated, by Theorem 2.1 $(4)(i) \Leftrightarrow (iii)$, L_i is an H-limit group.

Since L_i is an H-limit group and $L_i \models S(\bar{x}) = 1$, there is an epimorphism $f_i : G \to L_i$ such that $f_i(x_k) = x_k$.

Now let L be an H-limit group and $h : G \to L$ be an epimorphism. Since $L \models \mathrm{Th}_\forall(H)$, by Theorem 2.1 (3) L is embeddable in some nonprincipal ultrapower $^*H$ of H. Without loss of generality we may assume that $L \subseteq {}^*H$. Write $L = \langle \bar{y} | Q(\bar{y}) \rangle$ such that $h(x_i) = y_i$. Then $L \models P(\bar{y}) = 1$, and thus $L \models S(\bar{y}) = 1$. Since every universal sentence with parameters from H is true in $^*H$ (see the introduction) we get

$$^*H \models \forall \bar{x}(S(\bar{x}) = 1 \Leftrightarrow S_1(\bar{x}) = 1 \vee \cdots \vee S_n(\bar{x}) = 1).$$

Therefore there exists j such that $^*H \models S_j(\bar{y}) = 1$. Again, since every universal sentence with parameters from H is true in $^*H$, we get $^*H \models P_j(\bar{y}) = 1$ and thus $L \models P_j(\bar{y}) = 1$. Hence there exists a morphism $\varphi : L_j \to L$ which sends x_i to y_i. Now clearly $h = \varphi \circ f_j$. $\qquad\square$

Note that the proof of the above theorem is slightly different of those presented in [7] for free groups, as here we do not use the fact that some ultrapower of H is equationally noetherian.

Corollary 3.3. *Let H be an equationally noetherian group. Then a finitely generated group G is residually-H if and only if there exist H-limit groups $L_1, \ldots, L_n$ such that G is embeddable in $L_1 \times \cdots \times L_n$.*

Proof. Clearly if G is embeddable in $L_1 \times \cdots \times L_n$ for some H-limit groups $L_1, \ldots, L_n$ then G is a model of the universal Horn theory of H and thus it is residually-H by Theorem 2.1(5).

Suppose that G is residually-H. By Theorem 3.2, there exists a finite collection of epimorphisms $\{f_i : G \to L_i\}$, where each L_i is an H-limit group, such that for any H-limit group L any morphism $h : G \to L$, factors through some f_i. We claim that G is embeddable in $K = L_1 \times \cdots \times L_m$.

Write $G = \langle x_1, \ldots, x_n | P(\bar{x}) \rangle$ and $L_i = \langle y_{i,1}, \ldots, y_{i,n} | P_i(\bar{y}_i) \rangle$ such that $f_i(x_j) = y_{i,j}$. Then the map $f : G \to L_1 \times \cdots \times L_m$ defined by

$$f(x_j) = \hat{y}_j = (y_{1,j}, \ldots, y_{m,j})$$

is a morphism. It remains to show that f is an embedding. Let $w(\hat{y}_1, \ldots, \hat{y}_n)$ be a word on the variables $\{\hat{y}_1, \ldots, \hat{y}_n\}$ such that $w(\hat{y}_1, \ldots, \hat{y}_n) = 1$ in K. Suppose towards a contradiction that $G \models w(x_1, \ldots, x_n) \neq 1$. Since G is residually-H there exists a morphism $\phi : G \to H$ such that $\phi(w(x_1, \ldots, x_n)) \neq 1$. Now there exist f_i and a morphism $h : L_i \to H$ such that $\phi = h \circ f_i$ and thus $h(w(y_{i,1}, \ldots, y_{i,n})) \neq 1$. Therefore $L_i \models w(y_{i,1}, \ldots, y_{i,n}) \neq 1$ and thus $K \models w(\hat{y}_1, \ldots, \hat{y}_n) \neq 1$, contradiction. Hence f is an embedding as claimed. $\qquad\square$

A natural question arises from Theorem 3.2: when can we say that the groups L_i are a proper quotients of G? Clearly if G is not H-limit, then the groups L_i are proper quotients. When G is an H-limit group, to answer the question, we need the following notion.

Definition 3.4. *A finitely generated H-limit group G is said H-determined if there exists a finite subset $X \subseteq G \setminus \{1\}$ such that for any morphism $f : G \to L$, where L is an H-limit group, if $1 \notin f(X)$ then f is an embedding. We denote by $\mathcal{D}_H$ the class of H-determined groups.*

Notice that every H-determined group is a subgroup of H. In particular if H is free then any H-determined group is also free. More precisely, 1, $\mathbb{Z}$, F_2 are the only F_2-determined F_2-limit groups. Thus, in some sense, H-determined groups play the same role in general case, as free groups in the special case of limit groups of free groups. Notions related to determined groups are investigated in [3] and [17].

A *primitive-quantifier-free formula* is a formula $\vartheta(\bar{x})$ of the form

$$\left(\bigwedge_{w \in P} w(\bar{x}) = 1 \wedge \bigwedge_{v \in N} v(\bar{x}) \neq 1 \right),$$

where P, N are finite sets of words on the variables $\bar{x} = \{x_1, \ldots, x_n\}$ and their inverses. We begin with the following proposition.

Proposition 3.5. [17, Proposition 2.3.1] *Let H be a group such that $Th_\forall(H)$ has at most countably many nonisomorphic finitely generated models. Then for any primitive-quantifier-free formula $\vartheta(\bar{x})$ such that $H \models \exists \bar{x}(\vartheta(\bar{x}))$, there exists a primitive-quantifier-free formula $\xi(\bar{x})$ such that $H \models \exists \bar{x}(\vartheta(\bar{x}) \wedge \xi(\bar{x}))$ and for any word $w(\bar{x})$ on the variables $\bar{x} = \{x_1, \ldots, x_n\}$ and their inverses one has*

$$H \models \forall \bar{x}(\vartheta(\bar{x}) \wedge \xi(\bar{x}) \Rightarrow w(\bar{x}) = 1) \ \text{or} \ H \models \forall \bar{x}(\vartheta(\bar{x}) \wedge \xi(\bar{x}) \Rightarrow w(\bar{x}) \neq 1).$$

Proof. Let $\vartheta(\bar{x})$ be a primitive-quantifier-free formula such that $H \models \exists \bar{x}(\vartheta(\bar{x}))$ and suppose towards a contradiction that $\vartheta(\bar{x})$ does not satisfies the conclusions of the proposition. We are going to construct a tree. By hypothesis there exists a word $\alpha_1(\bar{x})$ such that $H \models \exists \bar{x}(\vartheta(\bar{x}) \wedge \alpha_1(\bar{x}) = 1)$ and $H \models \exists \bar{x}(\vartheta(\bar{x}) \wedge \alpha_1(\bar{x}) \neq 1)$ (to simplify notation we omit $\bar{x}$). We can do the same thing with $\vartheta \wedge \alpha_1 = 1$ and $\vartheta \wedge \alpha_1 \neq 1$.

Thus we have:

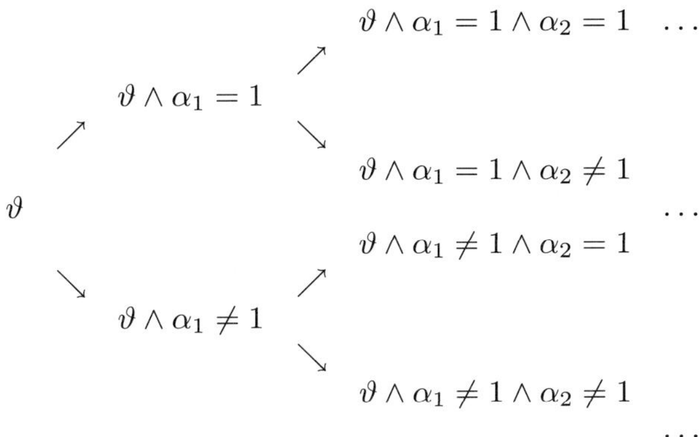

Now every branch B in the above tree defines a finitely generated group G_B, by taking G_B the group generated by $\bar{x}$ and with presentation the set of all words equal to 1 which appear in the branch B. Then G_B satisfies any inequality which occurs in B. Since any finite segment in any branch B is consistent in H, by Lemma 2.2, G_B is a model of the universal theory of H. Since there exists $2^{\aleph_0}$ branch, we get $2^{\aleph_0}$ nonisomorphic finitely generated models of $\mathrm{Th}_\forall(H)$. Contradiction with the property satisfied by H. $\square$

Theorem 3.6. *Let H be an equationally noetherian group and G a nontrivial finitely generated H-limit group. Then for any finite subset $X \subseteq G \setminus \{1\}$ there exists an epimorphism $f : G \to L$ where L is an H-determined group such that $1 \notin f(X)$.*

Proof. Write $G = \langle \bar{x} | P(\bar{x}) \rangle$ and let $S(\bar{x})$ be an H-witness of P. Let $X \subseteq G \setminus \{1\}$ be finite and let $v_1(\bar{x}), \dots, v_n(\bar{x})$ be words representing the elements of X.

Let $\vartheta(\bar{x}) = (S(\bar{x}) = 1 \wedge \bigwedge_{1 \leq i \leq n} v_i(\bar{x}) \neq 1)$. By Theorem 2.11, there exists countably many nonisomorphic H-limit groups and thus by Proposition 3.5 there exists a primitive-quantifier-free formula $\xi(\bar{x})$ such that $H \models \exists \bar{x}(\vartheta(\bar{x}) \wedge \xi(\bar{x}))$ and for any word $w(\bar{x})$ on the variables $\bar{x} = \{x_1, \dots, x_n\}$ and their inverses, one has

$$(1) \qquad H \models \forall \bar{x}(\vartheta(\bar{x}) \wedge \xi(\bar{x}) \Rightarrow w(\bar{x}) = 1) \text{ or } H \models \forall \bar{x}(\vartheta(\bar{x}) \wedge \xi(\bar{x}) \Rightarrow w(\bar{x}) \neq 1).$$

Let $\xi(\bar{x}) = (\bigwedge_{w \in Q} w(\bar{x}) = 1 \wedge \bigwedge_{u \in N} u(\bar{x}) \neq 1))$, where Q, N are finite sets of words. Since $H \models \exists \bar{x}(\vartheta(\bar{x}) \wedge \xi(\bar{x}))$ there exists $\bar{a} \in H^n$ such that

$$H \models (S(\bar{a}) = 1 \wedge \bigwedge_{w \in Q} w(\bar{a}) = 1 \wedge \bigwedge_{1 \leq i \leq n} v_i(\bar{a}) \neq 1 \wedge \bigwedge_{u \in N} u(\bar{a}) \neq 1)).$$

Let $L = \langle \bar{a} \rangle$. Then L is an H-limit group as it is a subgroup of H and since $L \models S(\bar{a}) = 1$ there is an epimorphism $f : G \to L$ which sends x_i to a_i.

We claim that L is an H-determined group. Let C be the set of elements of L which are represented by the words $v_1(\bar{a}), \dots, v_n(\bar{a})$ and $u(\bar{a}), u \in N$. Then for any morphism $h : L \to K$ where K is an H-limit group if $1 \notin h(C)$ then

$$K \models \vartheta(h(\bar{a})) \wedge \xi(h(\bar{a})),$$

and thus, since K is H-limit, by (1) we get

$$L \models w(\bar{a}) = 1 \Leftrightarrow K \models w(h(\bar{a})) = 1.$$

Therefore h is an embedding. Thus L is H-determined. $\qquad\square$

Theorem 3.7. *Let H be an equationally noetherian group. Then a finitely generated group G is H-limit if and only if G is fully residually-$\mathcal{D}_H$. Furthermore, a finitely generated H-limit group G has the same universal theory as H if and only if every H-determined group is embeddable in G.*

Proof. Let G be a finitely generated H-limit group. Then, by Theorem 3.6, G is fully residually-$\mathcal{D}_H$.

Now if G is fully residually-$\mathcal{D}_H$, then clearly it is fully residually-H as any H-determined group is a subgroup of H.

Let G be a finitely generated group having the same universal theory as H. Let $L = \langle \bar{x} | P(\bar{x}) \rangle$ be an H-determined group and let $X \subseteq L \setminus \{1\}$, given by words $v_1(\bar{x}), \ldots, v_n(\bar{x})$, such that for any morphism $f : L \to K$, where K is an H-limit group, if $1 \notin f(X)$ then f is an embedding. Let $S(\bar{x})$ be H-witness of P. Then

$$(1) \qquad H \models \exists \bar{x} (S(\bar{x}) = 1 \wedge \bigwedge_{1 \leq i \leq n} v_i(\bar{x}) \neq 1).$$

Since $\mathrm{Th}_\forall(G) = \mathrm{Th}_\forall(H)$, G satisfies the sentence which appears in (1). Therefore, as L is H-determined and G is H-limit, L is embeddable in G.

Now suppose that every H-determined group is embeddable in G and let us show that H is a model of the universal theory of G. Since G is H-limit, by Corollary 2.10, G is equationally noetherian. Therefore, by Corollary 2.4, it remains to show that every finitely generated subgroup of H is G-limit. By the above result every finitely generated subgroup of H is fully residually-$\mathcal{D}_H$. Therefore, every finitely generated subgroup of H is fully residually-G and thus it is G-limit. $\qquad\square$

Corollary 3.8. *If H is equationally noetherian then $Th_\forall(H) = \bigcap_{K \in \mathcal{D}_H} Th_\forall(K)$.* $\qquad\square$

Corollary 3.9. *If H is equationally noetherian and if every H-determined group is residually-finite then every finitely generated H-limit group is residually-finite.* $\qquad\square$

This justifies the following problem.

Problem. What are the H-determined groups for H a hyperbolic group?

Definition 3.10. *A strong factor set for a group G is a finite collection of proper quotients $\{f_i : G \to G_i\}$ of H-limit groups such that for any H-limit group L and any morphism $f : G \to L$ either f is an embedding or f factors through some f_i.*

Theorem 3.11. *Let H an equationally noetherian group. A nontrivial finitely generated H-limit group G is H-determined if and only if G has a strong factor set.*

Proof. Let G be a nontrivial finitely generated H-limit group which is H-determined and let us prove that G has a strong factor set. If every proper quotient of G which is H-limit is trivial then we have the desired conclusion. So suppose that G has a nontrivial proper quotient which is H-limit. Write $G = \langle \bar{x} | P(\bar{x}) \rangle$. By Theorem 2.11, G has at most countably many nonisomorphic H-limit quotients. Let $(G_i = \langle \bar{x} | P_i(\bar{x}) \rangle \mid i \in \mathbb{N})$ be the list of all nontrivial proper quotients of G which are H-limit. Since H is equationally noetherian there exist finite subsets $S(\bar{x}) \subseteq P(\bar{x})$, $S_i(\bar{x}) \subseteq P_i(\bar{x})$ such that

$$H \models \forall \bar{x}(S(\bar{x}) = 1 \Leftrightarrow P(\bar{x}) = 1).$$

$$H \models \forall \bar{x}(S_i(\bar{x}) = 1 \Leftrightarrow P_i(\bar{x}) = 1).$$

Since G is H-determined there exists a finite number $v_1(\bar{x}), \ldots, v_n(\bar{x})$ of words such that if K is an H-limit group containing $\bar{a}$ which satisfies $S(\bar{a}) = 1 \wedge \bigwedge_{1 \leq i \leq n} v_i(\bar{a}) \neq 1$, then the morphism $f : G \to K$ defined by $f(x_i) = a_i$ is an embedding.

Let $^*H$ be a nonprincipal ultrapower of H. Then every H-limit group is embeddable in $^*H$. Now if $\bar{b} \in {}^*H$ is a tuple such that $^*H \models S(\bar{b}) = 1$, then

$$^*H \models \bigwedge_{1 \leq i \leq n} v_i(\bar{b}) \neq 1 \vee \bigvee_{j \in \mathbb{N}} S_j(\bar{b}) = 1.$$

Since $^*H$ is $\aleph_1$-saturated, there is $m \in \mathbb{N}$ such that for every $\bar{b} \in {}^*H$ satisfying $^*H \models S(\bar{b}) = 1$ we get

$$^*H \models \bigwedge_{1 \leq i \leq n} v_i(\bar{b}) \neq 1 \vee \bigvee_{1 \leq j \leq m} S_j(\bar{b}) = 1.$$

Therefore

$$(1) \qquad H \models \forall \bar{x}(S(\bar{x}) = 1 \Rightarrow \bigwedge_{1 \leq i \leq n} v_i(\bar{x}) \neq 1 \vee \bigvee_{1 \leq j \leq m} S_j(\bar{x}) = 1).$$

We claim that $\{G_1, \ldots, G_m\}$ is a strong factor set where f_i is defined by the obvious manner. Let $f : G \to L$ be a morphism, where L is an H-limit group. Then there exists $\bar{a} \in L$ such that $L \models S(\bar{a}) = 1$ and $f(x_i) = a_i$. Since L is an H-limit group it satisfies the sentence appearing in (1) and thus $\langle \bar{a} \rangle$ is either isomorphic to G or $\langle \bar{a} \rangle \models \bigvee_{1 \leq j \leq m} S_j(\bar{a}) = 1$. Thus if f is not an embedding, there exists j and a morphism $h : G_j \to \langle \bar{a} \rangle$ and we see that $f = f_i \circ h$. Thus $\{G_1, \ldots, G_m\}$ is a strong factor set as claimed.

Suppose now that G has a strong factor set $\{f_i : G \to G_i \mid 1 \leq i \leq m\}$ and let us show that G is H-determined.

Write $(G_i = \langle \bar{x} | P_i(\bar{x}) \rangle \mid 1 \leq i \leq m)$ with $f(x_i) = x_i$. Since every G_i is a proper quotient, there exists words $v_1(\bar{x}), \ldots, v_m(\bar{x})$ such that

$$G_i \models v_i(\bar{x}) = 1, G \models v_i(\bar{x}) \neq 1.$$

Therefore if $f : G \to L$ where L is an H-limit group such that $\bigwedge_{1 \leq i \leq m} f(v_i) \neq 1$ then clearly f is an embedding. $\qquad \square$

References

[1] B. Baumslag. *Residually free groups*. Proc. London Math. Soc. (3) 17 (1967), 402–418.

[2] G. Baumslag, A. Myasnikov, V.N. Remeslennikov. *Algebraic geometry over groups I: Algebraic sets and ideal theory*. J. Algebra, 219 (1999), 16–79.

[3] O. Belegradek. *Finitely determined members of varieties of groups and rings*. J. Algebra, 228 (2000), 586–602.

[4] R. Bryant. *The verbal topology of a group*. J. Algebra 48 (1977), 340–346.

[5] C. Champetier and V. Guirardel. *Limit groups as limits of free groups: compactifying the set of free groups*. Israel Journal of Mathematics, 146 (2005).

[6] C.C. Chang and H.J. Keissler. *Model Theory*. Studies in Logic and the foundations of mathematics, Volume 73, Elsevier Science, 1990.

[7] Z. Chatzidakis. *Limit groups, viewed by a logician*.
Preprint available at http://www.logique.jussieu.fr/www.zoe

[8] V. Guba. *Equivalence of infinite systems of equations in free groups and semigroups to finite subsystems*. Mat. Zametki 40, NO. 3 (1986), 321–324.

[9] V. Guirardel. *Limit groups and groups acting freely on R^n-trees*. Geom. Topol. 8 (2004), 1427–1470.

[10] W. Hodges. *Model theory*, volume 42 of *Encyclopedia of Mathematics and its Applications*. Cambridge University Press, Cambridge, 1993.

[11] I. Kapovich. *Subgroup properties of fully residually free groups*. Trans. Amer. Math. Soc. 354 (2002), 335–362.

[12] O. Kharlampovich and A. Myasnikov. *Irreducible affine varieties over a free group. I. Irreducibility of quadratic equations and Nullstellensatz*. J. Algebra 200 (1998), 472–516.

[13] O. Kharlampovich and A. Myasnikov. *Irreducible affine varieties over a free group. II. Systems in triangular quasi-quadratic form and description of residually free groups*. J. Algebra 200 (1998), 517–570.

[14] O. Kharlampovich, A.G. Myasnikov, V.N. Remeslennikov, D.E. Serbin. *Subgroups of fully residually free groups: algorithmic problems*. In: Group theory, statistics, and cryptography, Contemp. Math., 360 (2004), Amer. Math. Soc., Providence, RI, pp. 63–101.

[15] D. Marker. *Model Theory: An Introduction*. Springer (2000).

[16] A.G. Myasnikov, V.N. Remeslennikov. *Algebraic Geometry over Groups II. Logical Foundations*. J. Algebra 234 (2000), 225–276.

[17] A. Ould Houcine. *Sur quelques problèmes de plongement dans les groupes*. Ph.D. thesis, Université Paris 7, (2003).

[18] A. Ould Houcine. *Some properties of finitely generated models*. Preprint, (2004).

[19] F. Paulin. *Sur la théorie élémentaire des groupes libres (d'après Sela)*. Astérisque 294 (2004), 363–402.

[20] Z. Sela. *Diophantine geometry over groups. I. Makanin-Razborov diagrams*. Publ. Math. Inst. Hautes Études Sci, 93 (2001) 31–105.

[21] Z. Sela. *Diophantine Geometry over Groups VIII: The elementary theory of a hyperbolic groups.* Preprint.

[22] B.A.F. Wehrfrits. *Infinite Linear Groups.* Springer-Verlag, New York – Heidelberg – Berlin, 1973.

Abderezak Ould Houcine
Université de Lyon
Université de Lyon1
CNRS, UMR 5208, Institut Camille Jordan
Bâtiment du Doyen Jean Braconnier
43, blvd du 11 novembre 1918
F-69200 Villeurbanne Cedex, France
e-mail: `ould@math.univ-lyon1.fr`

Geometric Group Theory

Trends in Mathematics, 121–168

© 2007 Birkhäuser Verlag Basel/Switzerland

Solution of the Conjugacy Problem and Malnormality of Subgroups in Certain Relative Small Cancellation Group Presentations

Arye Juhász

Introduction

The three fundamental decision problems posed by Max Dehn in 1912 are the word problem, the conjugacy problem and the isomorphism problem. Let G be a group and let u and v be elements of G. The word problem asks for an algorithm for deciding whether $u = v$. The conjugacy problem asks about the existence of an element $g \in G$ which conjugates u to v, i.e., $v = g^{-1}ug$. A solution of the conjugacy problem clearly contains a solution of the word problem. The isomorphism problem asks whether two group presentations define isomorphic groups. The word and conjugacy problems received much attention in the literature. For a summary of results concerning the conjugacy problem until 1987 see [13] and references therein.

There are several methods to prove the solvability of the word problem and the conjugacy problem in groups. In [11] algebraic methods are used while in [10] and [20] geometric methods are applied. In [19] Lyndon proved that the word problem is solvable for groups with a presentation which satisfies some combinatorial conditions. In [22] Schupp solved the conjugacy problem for groups with the same combinatorial conditions on the presentation.

In this work we consider the word and conjugacy problems in quotients F/N of groups F which are built up from some of their subgroups with solvable word and conjugacy problems. We show that if N is generated as a normal subgroup by elements which satisfy a certain combinatorial condition, similar to those dealt with in [19] and [22], then no matter which methods were used to prove the solvability of the word and conjugacy problems in the subgroups which build the group F, the quotients F/N will have solvable word and conjugacy problems. More precisely, we consider the solvability of the word and conjugacy problems in quotients of free products, amalgamated free products and HNN-extensions with solvable word

---

This article originates from the Geneva conference.

and conjugacy problems. (See [20, Ch. IV].) We assume that the quotients are taken over sets of relations which satisfy the small cancellation condition $V(6)$, which we define now, first via diagrams.

Let M be a map (see [20, Ch. V]). M satisfies the *small cancellation condition* $V(6)$ if for every region D of M one of the following holds:

(i) D has at least four edges and every inner vertex of D
   has valency at least four;

(ii) D has at least six edges.

This condition is a special case of the small cancellation condition $W(6)$ introduced in [14], and it is a common generalisation of the well-known $C(6)$ and $C(4)\&T(4)$ small cancellation conditions (See [20, Ch. V] .)

Let $\mathcal{P} = \langle X|\mathcal{R}\rangle$ be a free group-presentation, where $\mathcal{R}$ is a symmetrized set of words in the free group $F(X)$ freely generated by X, and let N be the normal closure of $\mathcal{R}$ in $F(X)$. We say that $\mathcal{P}$ *satisfies the small cancellation condition* $V(6)$ if the underlying map of every reduced $\mathcal{R}$-diagram over $F(X)$ satisfies the small cancellation condition $V(6)$. In [19] Lyndon solved the word problem and in [22] P.E. Schupp solved the conjugacy problem for free presentations which satisfy the small cancellation condition $C(4)\&T(4)$ and also for free presentations which satisfy the small cancellation condition $C(6)$. G. Huck and S. Rosebrock extended these results considerably in [12] to free presentations which satisfy the most general conditions on the valency of vertices and number of edges of an inner region of M, which guarantee combinatorial non-positive curvature. They use the theory of non-positively curved PE 2-complexes, as developed by M. Bridson in [3] in order to estimate the area of the corresponding diagrams in terms of their boundaries. This approach was taken earlier by S. Gersten in [8]. Their result contains the solution of the word and conjugacy problems for groups with free $V(6)$ presentations.

The situation concerning small cancellation quotients of free products, amalgamated free products and HNN-extensions is not as satisfactory as with small cancellation quotients of free groups. The only work in the literature concerning the conjugacy problem for small cancellation quotients of one of these relative presentations seems to be the result of P.E. Schupp [23] which solves it for free product presentations with the metric small cancellation condition $C'(^1/_6)$ and probably can be deduced from [21] for quotients of free products by the normal closure of relations which make the relevant complex hyperbolic, like the condition $C(7)$. (Recall, that for a real number λ, $0 < \lambda < 1$ a presentation satisfies the small cancellation condition $C'(\lambda)$ if for every piece P of any relation R the metric condition $|P| < \lambda|R|$ is satisfied. (Here $|P|$ is the length of P.) The conjugacy problem was also solved for certain special cases of small cancellation quotients of free products, where the geometrical small cancellation condition $C(6)$ is satisfied. See [4] and references therein. Also, the works [21] and [2] are relevant here. No results seem to appear in the literature on the conjugacy problem for small cancellation quotients of amalgamated free products and HNN-extensions with the geometric small cancellation conditions. The reason for this is expressed perhaps in the remark in [20, p. 288–289]: "One cannot as in the free group or free product case, say much about decision problems for F/N. There are two points of difficulty. First if $r = z_1 \ldots z_n$ is a cyclically reduced element of $\mathcal{R}$ then $ara^{-1} = (az_1) \ldots (z_n a^{-1})$ is also a cyclically reduced element of $\mathcal{R}$ for any $a \in A$. Also, we have no control over which particular normal forms of elements are used". (Here A is the amalgamated subgroup.) We resolved these problems under certain conditions in [15] for amalgamated free products and in [16] for HNN-extensions in the special case when $\mathcal{R}$ consists of a single relator S^n, $n \geq 4$. Basically, we traced back the relative presentations to the corresponding free presentations. In the present paper we show that this idea extends to the general case under the small cancellation condition $V(6)$ (Theorems 0.2 and 0.3). In each of these cases diagrams are defined (see [20, Ch. V] and see also Section 3.1 for more background), hence we may extend the definition of condition $V(6)$ to these groups. Thus if F is a free product, an amalgamated free product or an HNN-extension, we say that the relative presentation $\langle F|\mathcal{R} \rangle$ *satisfies the small cancellation condition* $V(6)$ if the underlying map of every reduced $\mathcal{R}$-diagram over F satisfies the condition $V(6)$. In this case we say that the group F/N satisfies the condition $V(6)$.

Our main results are the following.

Theorem 0.1. *Let A and B be groups with solvable word and conjugacy problems and let $F = A * B$. Let $\mathcal{R}$ be a finite symmetrized set of words in F (see [20, p. 286]) and let $\mathcal{P} = \langle F|\mathcal{R} \rangle$. If $\mathcal{P}$ satisfies the condition $V(6)$, then $\mathcal{P}$ has solvable word and conjugacy problems.*

Although this theorem is formulated for two factors, it easily extends to several factors.

For an amalgamated free product we have the following result. Recall that a subgroup K of G is *malnormal* in G if $g^{-1}Kg \cap K = \{1\}$, for every $g \in G\backslash K$.

Malnormality of subgroups is a well-known and important condition which, for example, makes the various versions of the combination theorem work. See [1], [9], [17], [18] and references therein. It turns out that this is also a sufficient condition for the conjugacy problem to be solvable in $V(6)$-quotients of amalgamated free products and HNN-extensions.

Recall that for a subgroup H of a group G the *double coset problem* asks for an algorithm to decide for given elements x and y in G whether $x \in HyH$. For the special case when $y \in H$ this problem reduces to the membership problem for the subgroup H in G.

Theorem 0.2. *Let A, B and C be groups; let $\alpha : C \to A$ and $\beta : C \to B$ be injections. Let F be the amalgamated free product of the groups A and B amalgamated along C, $F = A *_C B$, and assume that each of the following holds:*

(i) *A and B have solvable word and conjugacy problems;*

(ii) *$\beta \circ \alpha^{-1}\big|_{\alpha(C)}$ and $\alpha \circ \beta^{-1}\big|_{\beta(C)}$ are effectively calculable;*

124 A. Juhász

(iii) *the double coset problems for $\alpha(C)$ and $\beta(C)$ in A and B respectively are solvable;*

(iv) *C is malnormal in both A and in B.*

Let $\mathcal{R}$ be a finite symmetrized set of words in F and let $\mathcal{P} = \langle F|\mathcal{R}\rangle$. If $\mathcal{P}$ satisfies the condition $V(6)$, then $\mathcal{P}$ has solvable word and conjugacy problems.

For HNN-extensions $F = \langle H, t|t^{-1}At = B\rangle$ we have a similar result except that the malnormality condition is more involved. Let $C = A \cup B$. We define below two versions of malnormality. The first version is standard (See [17], [18]), the second seems to be new. Thus, by the first version C is *malnormal* in H if A and B are malnormal in H and $h^{-1}Ah \cap B = 1$, for every $h \in H$. C is *weakly malnormal* in H if A and B are malnormal in H and $h^{-1}(AB)h \cap BA = 1$, for every $h \notin AB$.

Weak malnormality is used in [15] and [16] and is planned to be used in future works to show that relative presentations $\langle F|\mathcal{R}\rangle$ where the elements of $\mathcal{R}$ are of some specific form, satisfy the condition $C(p)$, $p \geq 6$. Finally, we need a definition.

Definition 0.1. Let W be a cyclically reduced cyclic word in F. W is *absolutely reduced* if each of the following holds:

(i) If $t^\varepsilon h t^\varepsilon$ is a subword of W, then $h \notin BA$ for $\varepsilon = 1$ and $h \notin AB$ for $\varepsilon = -1$;

(ii) If $t^\varepsilon h t^{-\varepsilon}$ is a subword of W, then $h \notin B$ for $\varepsilon = 1$ and $h \notin A$ for $\varepsilon = -1$.

We have the following result:

Theorem 0.3. *Let H be a group, A and B isomorphic subgroups, and ψ an isomorphism $\psi : A \to B$. Let $F = \langle H, t|t^{-1}At = B\rangle$, $t^{-1}at = \psi(a)$, $a \in A$. Let $\mathcal{R}$ be the symmetrization of a finite set of words in F such that the presentation $\langle F|\mathcal{R}\rangle$ satisfies the condition $V(6)$. Let G be the group given by this presentation.*

Assume that each of the following holds:

(i) *ψ and ψ^{-1} are effectively computable;*

(ii) *H has solvable word and conjugacy problems;*

(iii) *the double coset problems for A and B (AxA, AxB, BxA and BxB) are solvable in H;*

(iv) *One of the following holds*

    1) *A and B are malnormal in H and $h^{-1}Ah \cap B = \{1\}$, for every h in H.*

    2) *Every element of $\mathcal{R}$ is in absolutely reduced normal form and C is weakly malnormal. $(C = A \cup B)$.*

Then G has solvable word and conjugacy problems.

Notice that while condition (iv)(1) imposes a severe restriction on $A \cap B$ but no condition on the elements of $\mathcal{R}$, condition (iv)(2) imposes no restriction on $A \cap B$ but imposes restriction on the elements of $\mathcal{R}$.

Our results on diagrams with the condition $V(6)$ allow us to show that certain subgroups of G are malnormal. These results are a kind of "conjugacy analogues" of the Freiheitssatz.

Theorem 0.4. *Let $F = A * B$ or $F = A \underset{C}{*} B$ or $F = \langle H, t | t^{-1}At = B \rangle$ and let $\mathcal{R}$ be a symmetrized set of words in F such that the presentation $\mathcal{P} = \langle F | \mathcal{R} \rangle$ satisfies the condition $V(6)$. Let N be the normal closure of $\mathcal{R}$ in F and let $G = F/N$. Let $\nu : F \to G$ and let $\nu_A : A \to G$ and let $\nu_B : B \to G$ be the natural projections. Let $K = A \cup B$ if $F = A * B$ or $F = A \underset{C}{*} B$ and let $K = H$ if $F = \langle H, t | t^{-1}At = B \rangle$. Then*

(a) *ν_K is an injection.*

(b) *Let $c, d \in K$. If $\nu(c)$ is conjugate in G to $\nu(d)$ while c is not conjugate to d in F, then there exists natural numbers n and m and relators $R_{i,j} \in \mathcal{R}$, $1 \leq i \leq n$, $1 \leq j \leq m$ such that $R_{i,j} = w_{i,j} u_{i,j} w_{i+1,j}^{-1} v_{i,j}^{-1}$ $(w_{n+1,j} = w_{1,j})$ with $w_{i,j}$ pieces for $1 \leq i \leq n$, $1 \leq j \leq m$, $v_{i,j}$ pieces for $1 \leq i \leq n$, $1 \leq j \leq m - 1$ and $u_{i,j}$ pieces for $1 \leq i \leq n$, $2 \leq j \leq m$, which satisfy $u_{1,1} u_{2,1} \cdots u_{n,1} = c$, $v_{1,m} v_{2,m} \cdots v_{n,m} = d$, $u_{i,j+1} = v_{i,j}$, $i = 1, \ldots, n$, $j = 1, \ldots, m - 1$ and $\nu(w^{-1}cw) = \nu(d)$, where $w = w_{1,1} \cdots w_{1,m}$.*

(c) *If one of the following holds then $\nu(c)$ is conjugate to $\nu(d)$ in G if and only if c is conjugate to d in F.*

    (1) *There is no sequence of relations $w_1 u_1 w_2^{-1} v_1$, $\ldots$, $w_i u_i w_{i+1}^{-1} v_i$, $\ldots$, $w_n u_n w_1^{-1} v_n$, such that $u_1 \cdots u_n = c$ and $v_1 \cdots v_n = d$, where u_i, v_i, w_i are pieces and $u_i \in A$ for every i or $u_i \in B$ for every i, if F is a free product or an amalgamated free product and $u_i \in H$ if F is an HNN-extension.*

    (2) *$\mathcal{R}$ satisfies the condition $C(5)$ (in addition to the condition $V(6)$).*

(d) *Under the condition (1) or (2) of part (c), if $L \subseteq K$ and L is a malnormal subgroup in A, B or H, as makes sense, then $\nu(L)$ is malnormal in G.*

In what follows we examplify Theorems 0.1–0.4.

Example 1 shows how Theorems 0.1 – 0.4 fit together to prove that groups constructed from smaller groups by free products, amalgamated free products, HNN-extension and taking $V(6)$-quotients have solvable conjugacy problems and certain subgroups are malnormal provided that the same is true for the smaller groups from which the groups are constructed. This example emphasizes the need for malnormality and the interplay between Theorems 0.1–0.3 and Theorem 0.4.

Example 1. First a convention. Let $F = A * B$ or $F = A \underset{C}{*} B$ and let $W = a_1 b_1 \ldots a_n b_n$, $a_i \in A$, $b_i \in B$, $n \geq 2$, be a word in reduced form. We call the a_i and the b_i the *components of* W. They are defined uniquely if $F = A * B$ and up to double cosets of C if $F = A \underset{C}{*} B$. Let $A_0 = \langle x_1, x_2 | x_1 x_2 x_1^{-1} x_2^{-1}, x_1^{13}, x_2^{13} \rangle$ and let $A_1 = \langle a, b | aba^{-1}b^{-1}, a^7, b^7 \rangle$. Let $B_0 = \langle y_1, y_2 | y_1 y_2 y_1^{-1} y_2^{-1}, y_1^{13}, y_2^{13} \rangle$ and let $B_1 = \langle c, d | cdc^{-1}d^{-1}, c^7, d^7 \rangle$. Thus $A_0 \cong B_0 \cong \mathbb{Z}_{13} \oplus \mathbb{Z}_{13}$ and $A_1 \cong B_1 \cong \mathbb{Z}_7 \oplus \mathbb{Z}_7$.

Let $R_1 = x_1^5 a^2 x_2^8 b^{-3} x_1^{-4} x_2^{-8} a \in A_0 * A_1$ and let $R_2 = y_1^7 cd^{-1} c^3 y_2^{-4} cdy_1^9 c^2 d^3 \in B_0 * B_1$. Let $\mathcal{R}_i$, $i = 1, 2$ be the symmetric closure of R_i in $A_0 * A_1$ and $B_0 * B_1$, respectively. Let $A = A_0 * A_1 / \langle\!\langle R_1 \rangle\!\rangle$ and let $B = B_0 * B_1 / \langle\!\langle R_2 \rangle\!\rangle$. Here $\langle\!\langle R \rangle\!\rangle$ denotes the normal closure of R in the relevant group. Then

(1) $\mathcal{R}_1$ and $\mathcal{R}_2$ satisfy the condition $C(6)$.
   (We shall prove (1), (2) and (3) below in Section 3.5.)
   Therefore, by Theorem 0.1.
     (i) A and B have solvable word and conjugacy problems.
   Also, by Theorem 0.4(a)
     (ii) A_0 and A_1 are embedded in A and B_0 and B_1 are embedded in B.
   Denote by $\bar{A}_0$, $\bar{A}_1$, $\bar{B}_0$ and $\bar{B}_1$ the images of A_0, A_1, B_0 and B_1 in A and B,
   respectively. It follows from Theorem 0.4(d) that

   (iii) $\begin{cases} \bar{A}_0,\ \bar{A}_1,\ \bar{B}_0 \text{ and } \bar{B}_1 \text{ are malnormal in } A \text{ and } B, \text{ respectively. Moreover,} \\ \text{no elements of } \bar{A}_0 \text{ are conjugate to elements of } \bar{A}_1 \text{ in } A \text{ and no elements} \\ \text{of } \bar{B}_0 \text{ are conjugate to elements of } \bar{B}_1 \text{ in } B. \end{cases}$

         Also, since $A_0,\ A_1,\ B_0$ and B_1 are finite of known order, it follows from
         (i) that
   (iv) $\bar{A}_0, \bar{A}_1, \bar{B}_0$ and $\bar{B}_1$ have solvable double coset problems.
         Let now $C = \langle u, v | u^7, v^7, uvu^{-1}v^{-1} \rangle$, let $\alpha : C \to A_1$ and let $\beta : C \to B_1$
         be defined by $\alpha(u) = a$, $\alpha(v) = b$, $\beta(u) = cd^2$ and $\beta(v) = c^3 d^5$. Then
         α and β are recursive morphisms with recursive inverses and $A \underset{C}{*} B$ is a
         well-defined amalgamated free product. Let $R_3 = x_1 b y_2 c^2 d x_2^{-1} a^{-2} y_1 d^{-3}$
         and let $\mathcal{R}_3$ be the symmetric closure of the image of R_3 in $A \underset{C}{*} B$. Then
(2) $\mathcal{R}_3$ satisfies the condition $C(4)\&T(4)$.
   Let $G_1 = \langle A \underset{C}{*} B | \mathcal{R}_3 \rangle$. Then it follows from (i)–(iv) that the conditions of
   Theorems 0.2 and 0.4 are satisfied, hence by Theorem 0.2 and Theorem 0.4
   we get
     (i$'$) G_1 has solvable word and conjugacy problems and
     (ii$'$) A and B embed into G_1.
         Denote by $\tilde{A}$ and $\tilde{B}$ the images of A and B in G_1, respectively and
         denote by $\tilde{A}_0, \tilde{A}_1, \tilde{B}_0$ and $\tilde{B}_1$ the images of A_0, A_1, B_0 and B_1 in G_1,
         respectively. Then

   (iii$'$) $\begin{cases} \tilde{A},\ \tilde{B},\ \tilde{A}_0,\ \tilde{A}_1,\ \tilde{B}_0 \text{ and } \tilde{B}_1 \text{ are malnormal in } G_1. \text{ Moreover, no non-} \\ \text{trivial elements of one of these groups is conjugate to non-trivial ele-} \\ \text{ments of any one of the others.} \end{cases}$

     (iv$'$) $\tilde{A}, \tilde{B}, \tilde{A}_0, \tilde{B}_0$ and $\tilde{B}_1$ have solvable double coset problems in G_1.

         Finally, let $G_2 = \langle G_1, t | t^{-1} \tilde{A}_0 t = \tilde{B}_0 \rangle$ with $t^{-1} \tilde{x}_1 t = \tilde{y}_1^{-1}$ and $t^{-1} \tilde{x}_2 t =$
         $\tilde{y}_2$. It follows from (ii) and (ii$'$) that G_2 is a well-defined HNN-extension
         of G_1 with stable letter t. Let $R_4 = \tilde{x}_2^3 \tilde{y}_2^4 t^2 \tilde{x}_1^2 \tilde{x}_2^{-2} t^{-3} \tilde{y}_1^3 t$, let $\mathcal{R}_4$ be the
         normal closure of R_4 in F and let $G_3 = \langle G_2 | \mathcal{R}_4 \rangle$. Then
(3) $\mathcal{R}_4$ satisfies the condition $C(6)$.
   Consequently it follow from (i$'$)–(iv$'$) and Theorems 0.3 and 0.4 that the
   obvious versions of (i)–(iv) hold true in G_3.

Example 2. Let $F = A * B$ and let $R = W^n$, where W is a cyclically reduced
word of length at least two which is not a proper power and $n \geq 2$. Let N be

the normal closure of R in F and let $G = F/N$. It is shown in [4, Th. 3.16], among other results, that if $n \geq 4$ and no letter in R has order two or if $n \geq 2$ and A and B are locally indicable then the conjugacy problem for G is solvable. Thus, if we do not assume that A and B are locally indicable, then Theorem 3.16 of [4] provides solution to the conjugacy problem, except for $n = 2$ and $n = 3$. Theorem 0.1 above applies in this case, since the presentation $\langle F \mid R \rangle$ satisfies the condition $C(6)$, by [5], provided that W contains no elements of order two, hence G has solvable conjugacy problem.

Let now $F = A \underset{C}{*} B$ where C is malnormal in both A and B, A and B nontrivial, $C \neq A$ and $C \neq B$. Let $R = W^n$, where W is cyclically reduced having length at least two, W does not contain elements of order two and W is not a C-proper power in the following sense: W cannot be written in the form $U_1 U_2 \ldots U_k$, $k \geq 2$ reduced as written in F, where $U_i = a_1^{(i)} b_1^{(i)} \ldots a_s^{(i)} b_s^{(i)}$, $a_j^{(i)} \in A$ and $b_j^{(i)} \in B$, $i = 1, \ldots, k$ $j = 1, \ldots, s$ and $C a_j^{(i)} C = C a_j^{(1)} C$ for every i and j and similarly, $C b_j^{(1)} C = C b_j^{(i)} C$. Let N be the normal closure of R in F and let $G = F/N$. Relying on the result of [5] it is shown in [15] that if $n \geq 3$ then G satisfies the condition $C(2n)$. As a consequence, it was shown that for $n \geq 4$ the conjugacy problem is solvable. The proof did not work for $n = 3$. Theorem 0.2 above shows, due to the condition $C(6)$ which is satisfied for $n = 3$, that in this case G has solvable conjugacy problem.

A similar improvement is provided by Theorem 0.3 above to a result in [16] where F is an HNN-extension and $R = W^n$. We do not go into details.

Basic to the proofs of Theorems 0.1–0.4 is the following:

Main Theorem. *Let M be an annular map with connected interior which satisfies the condition $V(6)$. Then M decomposes to annular one-layer diagrams $\mathcal{L}_{-k}, \mathcal{L}_{-k+1}, \ldots, \mathcal{L}_0, \mathcal{L}_1, \mathcal{L}_2, \ldots, \mathcal{L}_\ell$ with boundaries homotopic in M to the boundaries of M such that each of the following holds:*

(i) *$M = \cup_{i=-k}^{\ell} \mathcal{L}_i$ and $\mathrm{Int}(\mathcal{L}_i) \cap \mathrm{Int}(\mathcal{L}_j) = \emptyset$, $i \neq j$, where $\mathrm{Int}(\mathcal{L})$ denotes the interior of $\mathcal{L}$ (see Fig. 1).*

(ii) *For a boundary path μ of M, denote by $|\mu|$ the number of edges of μ. (As usual, the endpoints of an edge of μ may have valency two in M.) Denote by ω the outer boundary of M and denote by τ the inner boundary of M. Similarly, denote by $\omega(\mathcal{L}_i)$ the outer boundary of $\mathcal{L}_i$ and denote by $\tau(\mathcal{L}_i)$ the inner boundary of $\mathcal{L}_i$. Then $|\omega| \geq |\omega(\mathcal{L}_i)| \geq |\omega(\mathcal{L}_{i-1})|$ for $i = 1, 2, \ldots, \ell$ and $|\tau| \geq |\tau(\mathcal{L}_j)| \geq |\tau(\mathcal{L}_{j+1})|$ for $j = -k, -k+1, \ldots, -1$.*

(iii) *If M contains a layer with more than one region, then $|\omega| + |\tau| \geq 4$.*

(iv) *If $|\partial D \cap \omega(M)| \leq 1$ and $|\partial D \cap \tau(M)| \leq 1$ for every boundary region D of M then for every region E of M we have $|\partial E| = 4$ and $d_M(v) = 4$ for every inner vertex v of M and $d_M(w) = 3$ for every boundary vertex w of M. Moreover, every $\mathcal{L}_i$ has annular interior.*

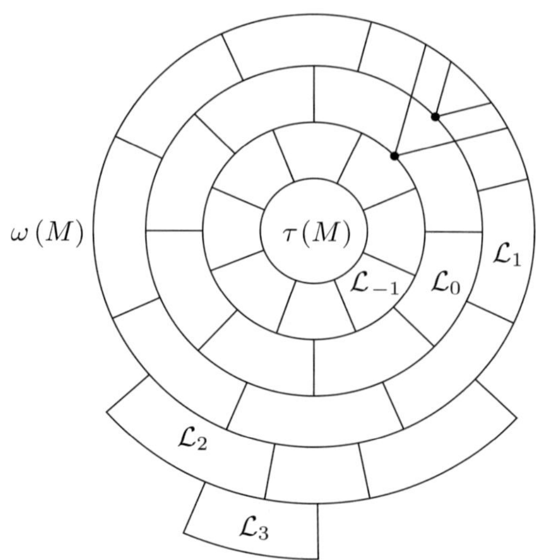

FIGURE 1

The work is organised as follows:

In Section 1 we introduce the notation and the main notions.

In Section 2 we prove the Main Theorem. Briefly, we start with the construction of the 0th layer. This is done by choosing a shortest curve going along the annulus. We continue by constructing the other layers, relying heavily on the $V(6)$-condition, which is a local combinatorial non-positive curvature condition, via a multi-argument induction. It turns out that the boundary lengths of the layers increase (does not decrease) towards the boundary components of M.

In Section 3 we prove Theorems 0.1–0.4 and the statements of Example 1.

1. Preliminaries on maps

1.1. Notation and basic notions

We follow [20, Ch. V], in general.

(a) Let M be a map. Denote by $\mathrm{Reg}(M)$ the set of all the regions of M.
(b) For a path μ of M let $o(\mu)$ denote its initial vertex and let $t(\mu)$ denote its terminal vertex.
(c) Two regions D_1 and D_2 are *neighbors* if $\partial D_1 \cap \partial D_2 \neq \emptyset$. Regions D_1 and D_2 are *proper neighbors* if $\partial D_1 \cap \partial D_2$ contains a non-empty edge.
(d) For a submap S of a map M and region D in S denote by $d_S(D)$ the number of proper neighbors of D in S. If S is clear from the context, we shall write $d(D)$ for $d_S(D)$.
(e) M is called *regular* if every edge is on the boundary of a region.

Remark. In this work we assume that all the inner vertices of a map have valency at least three. However, if N is a submap of M, inner vertices of M on the boundary of N may have valency two in N. For example, in Fig. 2, $d_N(v) = 2$ while $d_M(v) = 3$.

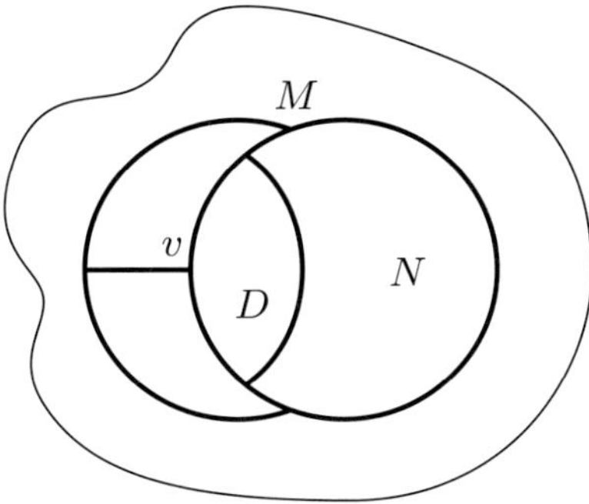

FIGURE 2

One-layer submaps play an important role in what follows, therefore the next subsection is devoted to them.

1.2. One-layer maps

Definition 1.1. Let M be a connected map which is either simply connected or annular, with hole H.

(a) Let $\mathcal{L}$ be a connected submap of M with at most one hole which contains H, if M is annular and let $\mathrm{Reg}(\mathcal{L}) = \{D_0, \ldots, D_t\}$. $\mathcal{L}$ is a *one-layer map*, if each region has at most two neighbors and the D_i are arranged such that $\partial D_i \cap \partial D_{i+1} \neq \emptyset$ and $\partial D_i \cap \partial D_j$ contains a non-empty edge only if $|i - j| \equiv 1 \ \mathrm{mod}(t+1)$.

Note that this definition differs from the definition given in [14, p. 60] in that here we allow neighboring regions to have only a vertex in common (i.e., the interior of $\mathcal{L}$ may be not connected).

(b) Let $\mathcal{L}$ be a connected simply connected one-layer submap of M with connected interior. Let $\mathrm{Reg}(\mathcal{L}) = \{D_0, \ldots, D_t\}$. If the regions D_i, $i = 0, \ldots, t$ are arranged such that $\partial D_i \cap \partial D_{i+1} \neq \emptyset$, then we shall write $\mathcal{L} = \langle D_0, \ldots, D_t \rangle$. Call D_0 the *initial* and D_t the *terminal* region of $\mathcal{L}$. Denote $h(\mathcal{L}) = D_0$, $t(\mathcal{L}) = D_t$.

(c) Denote by $|\mathcal{L}|$ the number of regions in $\mathcal{L}$. Denote by $v(\mathcal{L})$ the number of vertices with valency at least four in $\mathcal{L}$ and set $\|\mathcal{L}\| := |\mathcal{L}| + \frac{1}{2}v(\mathcal{L})$.

(d) Let $\mathcal{L}$ be an annular one-layer submap of M, the hole of which contains H. Let $\mathcal{L}_0$ be a connected simply connected one-layer submap of $\mathcal{L}$ and let $\mathcal{L}'_0$ be a connected simply connected one-layer submap of M with $h(\mathcal{L}_0) = h(\mathcal{L}'_0)$ and $t(\mathcal{L}_0) = t(\mathcal{L}'_0)$ such that the submap $\mathcal{L}'$, obtained by deleting $\mathcal{L}_0$ from $\mathcal{L}$

and joining $\mathcal{L}'_0$ to the resulting map, is an annular one-layer map. We shall say that $\mathcal{L}'$ *shortens* $\mathcal{L}$ if $|\mathcal{L}'| < |\mathcal{L}|$ or $|\mathcal{L}'| = |\mathcal{L}|$ and $\|\mathcal{L}'\| < \|\mathcal{L}\|$.

The main result of Section 1 is that every annular map which satisfies the condition $V(6)$ can be decomposed to the disjoint union of one-layer annular submaps. The idea of the proof is to notice first that every annular map has a natural decomposition into annular submaps, defined via adjacency, and then to show that under the condition $V(6)$ this decomposition is a decomposition into one-layer submaps. In the next subsection we describe this natural decomposition to annular submaps.

1.3. Layer structure in annular maps

Definition 1.2.

(a) Let M be an annular map. For any submap A denote by $\mathrm{Int}(A)$ the interior of A and by $\overline{\mathrm{Int}(A)}$ the closure of $\mathrm{Int}(A)$. Denote by $\omega(M)$ the outer boundary of M and denote by $\tau(M)$ the inner boundary of M. Let A be a regular annular submap of M with boundaries homotopic to ∂M. Denote $\mathcal{L}_0(A) = A$. Denote by $\mathcal{L}_1(A)$ the subdiagram of M consisting of $\omega(A)$ and the set of all the regions of $M \backslash A$, the boundary of which has a common vertex or a common edge with $\omega(A)$ and is to the left of $\omega(A)$ when traversing along $\omega(A)$, in clockwise direction. Similarly, denote by $\mathcal{L}_{-1}(A)$ the subdiagram of M consisting of $\tau(A)$ and the set of all the regions of M, the boundary of which has a common vertex or a common edge with $\tau(A)$ and is to the right of $\tau(A)$. For $i \geq 2$ define inductively $\mathcal{L}_i(A)$ to be the subdiagram of M consisting of $\omega(\mathcal{L}_{i-1}(A))$ and the set of regions $\{D \in M \backslash \mathcal{L}_{i-1}(A) \,|\, \partial D \cap \omega(\mathcal{L}_{i-1}(A)) \neq \emptyset\}$ and for $i \leq -2$ define $\mathcal{L}_i(A)$ to be the subdiagram of M consisting of $\tau(\mathcal{L}_{i+1}(A))$ and the set of regions $\{D \in M \backslash \mathcal{L}_{i+1}(A) \,|\, \partial D \cap \tau(\mathcal{L}_{i+1}(A)) \neq \emptyset\}$. See Fig. 1, where $\mathcal{L}_{-1}$, $\mathcal{L}_0$ and $\mathcal{L}_1$ have annular interior, while $\mathcal{L}_2$ and $\mathcal{L}_3$ are annular diagrams with simply connected interiors. Also define $\mathrm{Oust}^i_M(A) = \bigcup_{j=0}^{i} \mathcal{L}_j(A)$ for $i \geq 0$, and define $\mathrm{Inst}^i_M(A) = \bigcup_{j=i}^{0} \mathcal{L}_j(A)$ for $i \leq 0$. Denote $\omega\big(\mathrm{Oust}^i_M(A)\big)$ by ω_i and similarly, denote $\tau\big(\mathrm{Inst}^i_M(A)\big) = \tau_i$.

(b) Let M be an annular map with a decomposition $M = \mathcal{L}_{-k} \cup \mathcal{L}_{-(k-1)} \cdots \cup \mathcal{L}_0 \cup \mathcal{L}_1 \cup \mathcal{L}_2 \cdots \cup \mathcal{L}_\ell$, where $\mathcal{L}_0(= A)$ is a one-layer annular subdiagram with boundaries homotopic to ∂M and $\mathcal{L}_i$ is defined for $i \neq 0$ as above and $\mathrm{Int}(\mathcal{L}_i) \cap \mathrm{Int}(\mathcal{L}_j) = \emptyset$ for $i \neq j$. Then for every region D in $\mathcal{L}_i$, $0 < i \leq \ell - 1$, we may define $\alpha(D) = \{E \in \mathcal{L}_{i-1} \,|\, \partial E \cap \partial D \neq \emptyset\}$, $\beta(D) = \{K \in \mathcal{L}_i \,|\, \partial D \cap \partial K$ contains a non-empty edge$\}$ and $\gamma(D) = \{F \in \mathcal{L}_{i+1} \,|\, \partial F \cap \partial D \neq \emptyset\}$. Similarly, if $-k < i < 0$, define $\alpha(D) = \{E \in \mathcal{L}_{i+1} \,|\, \partial E \cap \partial D \neq \emptyset\}$, $\beta(D) = \{E \in \mathcal{L}_i \,|\, \partial D \cap \partial E$ contains a non-empty edge$\}$ and $\gamma(D) = \{E \in \mathcal{L}_{i-1} \,|\, \partial E \cap \partial D \neq \emptyset\}$. (See Fig. 3(b), where $\alpha(D) = \{E_1, E_2, E_3\}$, $\beta(D) = \{K_1, K_2\}$ and $\gamma(D) = \{F_1, F_2, F_3, F_4\}$.)

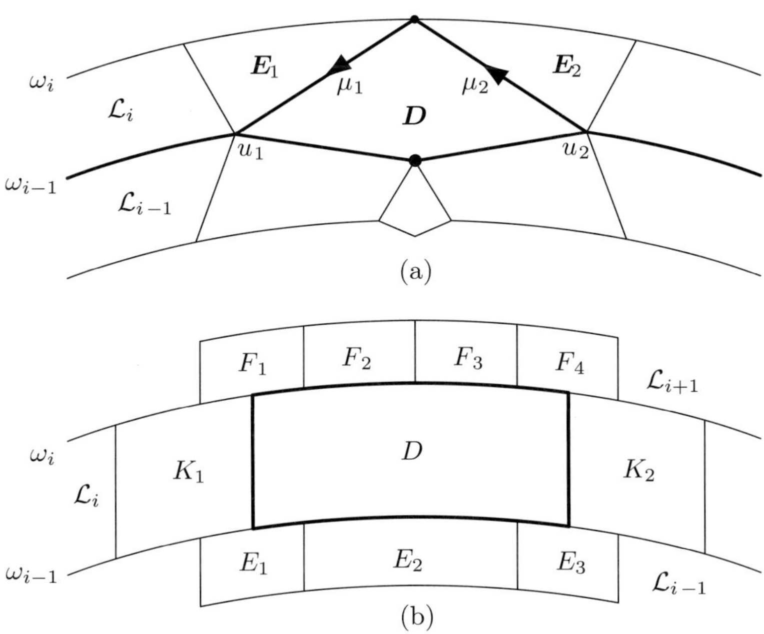

FIGURE 3

(c) Let D be a region of $\mathcal{L}_i$ for $i \geq 1$. D is *2-exceptional in* $\mathcal{L}_i$, if each of the following holds:

    1) $d_M(D) = 4$;

    2) $|\omega_{i-1} \cap \partial D| = 2$;

    3) $\omega_i \cap \partial D = \{v\}$, v a vertex;

    4) D has two proper neighbors E_1 and E_2 in $\mathcal{L}_i$, $E_1 \neq E_2$. They satisfy

$$\partial E_j \cap \partial D \cap \omega_{i-1} \neq \emptyset, \text{ for } j = 1, 2.$$

Let $\mu_1 = \partial E_1 \cap \partial D$ and let $\mu_2 = \partial E_2 \cap \partial D$. Then μ_1 and μ_2 are connected with v as the initial vertex of μ_1 and as a terminal vertex of μ_2. (See Fig. 3(a).)

If $i \leq -1$, we make the analogous definition.

(d) Let D be a region of $\mathcal{L}_i$ for $i \geq 1$. D is *3-exceptional in* $\mathcal{L}_i$, if each of the following holds (see Fig. 3(b)):

    1) $d_M(D) \geq 6$;

    2) $|\omega_{i-1} \cap \partial D| = 3$.

If $i \leq -1$, we make the analogous definition.

(e) Let D be a region of $\mathcal{L}_i$, $i \geq 1$. D is a *2-corner region in* $\mathcal{L}_i$ if $\omega_{i-1} \cap \partial D = \{v\}$, v a vertex. D is a *3-corner region in* $\mathcal{L}_i$ if $d_M(D) \geq 6$ and $|\omega_{i-1} \cap \partial D| = 1$.

From now on M will denote an annular map with connected interior and with hole H.

2. The layer structure of M

2.1. Regions in annular $V(6)$ maps

Lemma 2.1. *Let D be a region in M.*

(a) *If the boundary of D is not a simple closed curve then $\mathbb{E}^2 \setminus \{D\}$ has a unique bounded connected component H_0 and H_0 contains H. (See Fig. 4.)*

(b) *Either $\mathrm{Int}(\bar{D}) = D$ in which case $\bar{D}$ is simply connected or $\mathrm{Int}(\bar{D}) \neq D$ and $\bar{D}$ is an annular submap of M with boundaries homotopic to the boundaries of M.*

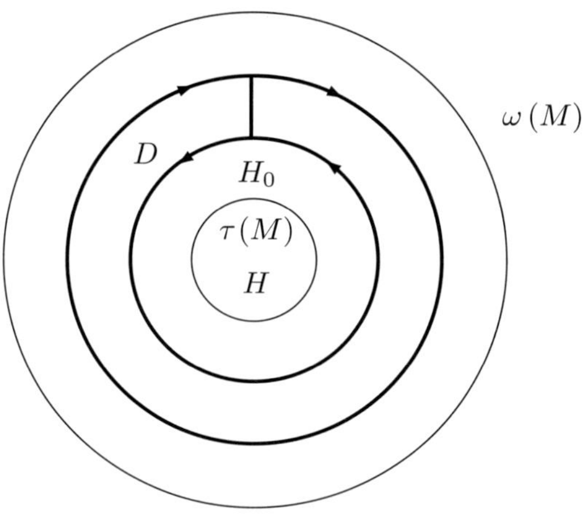

FIGURE 4

Proof. (a) Due to the planarity of M, $\mathbb{E}^2 \setminus \{D\}$ has at most one bounded connected component which contains H. By [14, p. 63] $\mathbb{E}^2 \setminus \{D\}$ may have no connected bounded components in $\mathbb{E}^2 \setminus H$. Consequently, $\mathbb{E}^2 \setminus \{D\}$ may have at most one bounded connected component H_0 and it contains H. If ∂D is not a simple closed curve then $\mathbb{E}^2 \setminus \{D\}$ necessarily contains a bounded connected component. The result follows.

(b) If $\bar{D}$ is not simply connected then the boundary of D is not a simple closed curve, hence by part (a) $\mathbb{E}^2 \setminus \{D\}$ has a unique bounded connected component and it contains H. Consequently, $\mathrm{Int}(\bar{D})$ is an annular submap of M. If on the other hand ∂D is a simple closed curve then $\mathrm{Int}(\bar{D}) = D$.

The lemma is proved. $\qquad \square$

2.2. The first layer

Lemma 2.2. *For each simple closed curve γ in M, homotopic in M to the boundary components of M, let L_γ be a regular submap of M with minimal set of regions such that $\gamma \subseteq L_\gamma$. Let Λ_0 be the collection of those L_γ for which $|L_\gamma|$ is minimal possible over all γ's and let $\mathcal{L}_0$ be an element of Λ_0 with $\|\mathcal{L}_0\|$ minimal possible. Then*

(a) *$\mathcal{L}_0$ is an annular one-layer subdiagram of M, which cannot be shortened, in the sense of Definition 1.1(d).*

(b) *$\omega(\mathcal{L}_0)$ and $\tau(\mathcal{L}_0)$ are simple closed curves and $\partial D \cap \omega(\mathcal{L}_0) \neq \emptyset$ and $\partial D \cap \tau(\mathcal{L}_0) \neq \emptyset$ for every region D in $\mathcal{L}_0$.*

Remark. We shall refer to the minimality of $|\mathcal{L}_0|$ and $\|\mathcal{L}_0\|$, like in Lemma 2.2, as "the minimality of $\mathcal{L}_0$".

Proof. (a) Consider two cases, according to which of the cases of Lemma 2.1 (b) occurs. If M contains a region D such that ∂D is not simple closed then by Lemma 2.1 (b) $\mathrm{Int}(\bar{D})$ is an annular submap with boundaries homotopic to the boundaries of M. Clearly, $\mathcal{L}_0 = \{D\}$ satisfies the requirements of the lemma. Suppose therefore that

$$\partial D \text{ is simple closed, for every region } D \text{ in } M \tag{1}$$

Let γ be a simple closed curve such that $\mathcal{L}_0 = L_\gamma$. If D is a region of $\mathcal{L}_0$ and $\partial D \cap \gamma = \emptyset$, then we can delete D from $\mathcal{L}_0$ and get a regular diagram $\mathcal{L}_0'$ with $|\mathcal{L}_0'| < |\mathcal{L}_0|$, contradicting the minimality of $\mathcal{L}_0$. Hence

$$\partial D \cap \gamma \neq \emptyset, \text{ for every } D \in \mathcal{Reg}(\mathcal{L}_0) \tag{2}$$

We claim that

$$\partial D \cap \gamma \text{ is connected for every } D \in \mathcal{Reg}(\mathcal{L}_0) \tag{3}$$

For suppose not. Then ∂D has non-consecutive subpaths $\mu_1, \ldots, \mu_r$, $r \geq 2$, such that $\partial D \cap \gamma = \bigcup_i \{\mu_i \mid i = 1, \ldots, r\}$. (See Fig. 5(a) with $r = 2$.) Let μ_1 and μ_2 be adjacent connected components of $\partial D \cap \gamma$, when traversing along γ clockwise. Since $\partial D \cap \gamma$ is not connected, it follows from Lemma 2.1 that $\mathbb{E}^2 \setminus (\{D\} \cup \gamma)$ has a connected bounded component U with $U \cap H = \emptyset$ (which contains at least one region). Now, U has a boundary cycle $v \alpha w \beta^{-1}$, where v is the terminal vertex of μ_1, w is the initial vertex of μ_2, $\alpha = \partial D \cap \partial U$ and $\beta = \partial U \cap \gamma$. (See Fig. 5(a).)

Define $\gamma' = (\gamma \setminus \beta) \cup \alpha$. Then $|L_{\gamma'}| < |L_\gamma|$ and $\|L_{\gamma'}\| \leq \|L_\gamma\|$, since $\mu_1 \beta \mu_2$ is on the boundary of at least two regions, one of which is D and at least one region of U, while α is on the boundary of one region D, again violating the minimality of L_γ. Thus $\partial D \cap \gamma$ is connected for every $D \in \mathcal{L}_0$. A similar argument shows that

$$\text{if } D_1 \text{ and } D_2 \text{ are regions in } \mathcal{L}_0 \text{ then } \partial\left(\bar{D}_1 \cup \gamma\right) \cap \partial D_2 \text{ is connected} \tag{4}$$

(See Fig. 5(b) for a configuration where this does not holds.) Consequently, all the neighbors of a region D are adjacent to D along γ in the sense that $\partial E \cap \partial D \neq \emptyset \Rightarrow \partial E \cap \partial D \cap \gamma \neq \emptyset$.

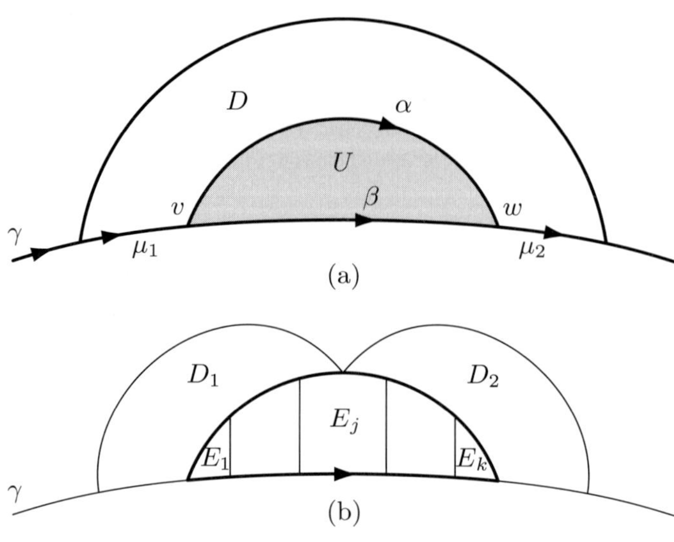

FIGURE 5

We show $\mathcal{L}_0$ is annular. Suppose not. Then $M \setminus \bar{\mathcal{L}}_0$ has a simply connected, bounded component U such that $\bar{U}$ is homeomorphic to a closed disc and in particular ∂U is a simple closed curve. (See Fig. 6.)

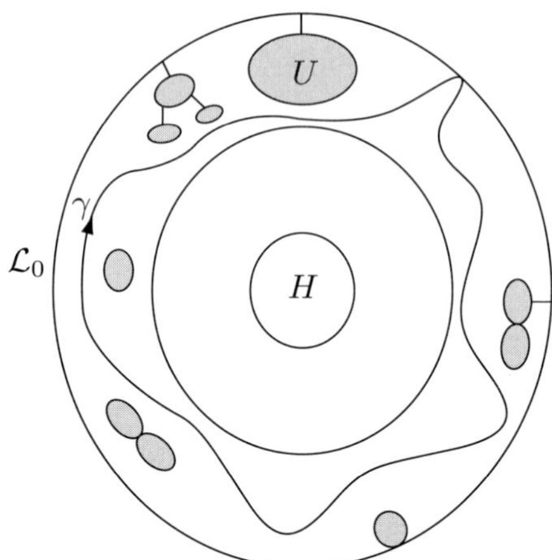

FIGURE 6

Since $\partial U \subseteq \partial \mathcal{L}_0$, there are regions $D_1, \ldots, D_m$ in $\mathcal{L}_0$ and connected components $\mu_1, \ldots, \mu_k$ of $\partial D_i \cap \partial U$, $i = 1, \ldots, k$ respectively, such that $\mu_1 \cdots \mu_k$ is a boundary cycle of U. (See Fig. 7(a).) Due to the $CN(3)$ property (see [14, p. 63]) we have

$$m \geq 4. \tag{5}$$

Since ∂U is a simple closed curve, the above proof of the statement that $\partial D_i \cap \gamma$ is a connected subpath of ∂D_i and a connected subpath of γ, applies equally well to $\partial D_i \cap \partial U$, so that $\partial D_i \cap \partial U$ is a connected subpath both of ∂D_i and of ∂U, for every $i = 1, \ldots, m$. In particular, $k = m$. Combining with (5) we get

$$k = m \geq 4\,. \tag{6}$$

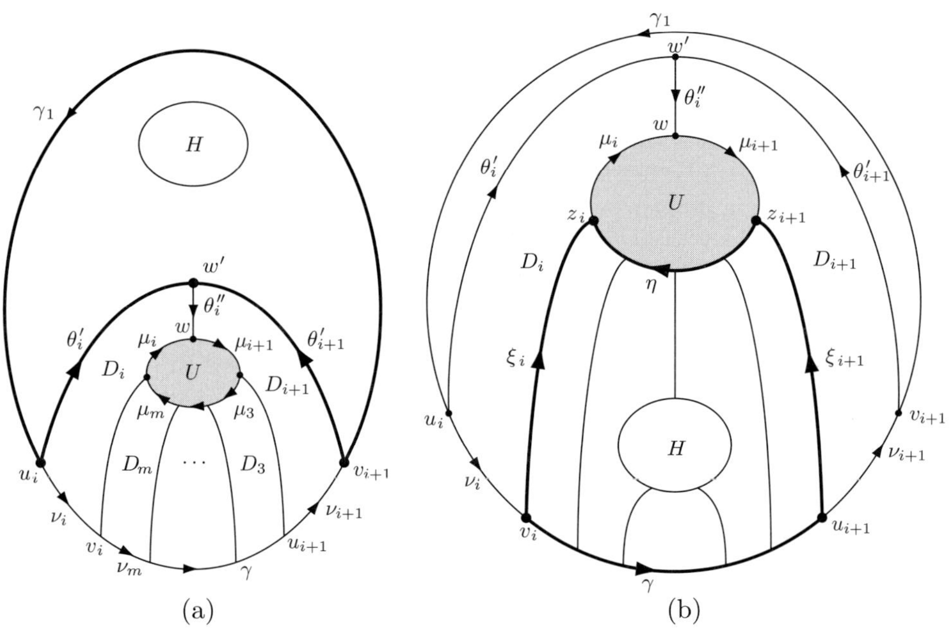

FIGURE 7

Let $\nu_i = \partial D_i \cap \gamma$ for $i = 1, \ldots, m$. Then ν_i is a connected subpath of γ and a connected subpath of ∂D_i. We propose to show that ν_i and ν_{i+1} are adjacent subpaths of γ. Suppose not and let D_i and D_{i+1} be regions such that ν_i and ν_{i+1} are not adjacent subpaths of γ. (See Fig. 7(b).) Let $\bar{\nu}_i = u_i \nu_i v_i$, let $\bar{\nu}_{i+1} = u_{i+1} \nu_{i+1} v_{i+1}$ and let $w = \partial D_i \cap \partial D_{i+1} \cap \partial U$. (See Fig. 7(a).) Then by our last assumption $v_i \neq u_{i+1}$. Let θ_i be the subpath of ∂D_i which starts at u_i, ends at w and does not contain v_i and similarly let θ_{i+1} be the subpath of ∂D_{i+1} which starts at v_{i+1}, ends at w and does not contain u_{i+1}. Let γ_1 be the subpath of γ which starts at v_{i+1}, ends at u_i and does not contain v_i and u_{i+1}. Let γ_2 be its complement in γ so that $\gamma = v_{i+1} \gamma_1 u_i \gamma_2$. Let $\theta_i = \theta_i' w' \theta_i''$ and let $\theta_{i+1} = \theta_{i+1}' w' \theta_{i+1}''$ such that $\overline{\partial D_i \cap \partial D_{i+1}} = w' \theta_i'' w = w' \theta_{i+1}'' w$. Let $\gamma' = u_i \theta_i' w' \theta_{i+1}'^{-1} v_{i+1} \gamma_1$. Then γ' is a simple closed curve. Consider two cases, according as H is contained in the bounded component of $\mathbb{E}^2 \setminus \gamma'$ or in the unbounded component of it, respectively. (See Fig. 7(a) and Fig. 7(b), respectively.) In the first case γ' is homotopic to ∂H

and $\gamma' \cap \gamma = v_{i+1}\gamma_1 u_i$. In particular, $\gamma' \neq \gamma$. Moreover, since $v_i \neq u_{i+1}$ due to the assumption that ν_i and ν_{i+1} are not adjacent, hence $|L_{\gamma'}| < |L_\gamma|$: it follows from the regularity of L_γ that the submap L_{γ_2} of L_γ, consisting of those regions the boundary of which has non-empty intersection with γ_2, contains at least four regions, by (6). On the other hand if $\gamma'' = \theta_i' w' \theta_{i+1}'^{-1}$ then $L_{\gamma''} = |\langle D_i, D_{i+1}\rangle| = 2$. Therefore,

$$|L_{\gamma'}| = |L_{\gamma_1}| + |\langle D_i, D_{i+1}\rangle| = |L_{\gamma_1}| + 2 < |L_{\gamma_1}| + 4 \leq |L_{\gamma_1}| + |L_{\gamma_2}| = |L_\gamma|,$$

i.e., $|L_{\gamma'}| < |L_\gamma|$, a contradiction.

Consider the second case. Let ξ_i be the complement of $\mu_i w \theta_i^{-1} u_i \nu_i v_i$ on ∂D_i and let ξ_{i+1} be the complement of $\mu_{i+1}^{-1} w \theta_{i+1}^{-1} v_{i+1} \nu_{i+1}^{-1} u_{i+1}$ on ∂D_{i+1} such that $\mu_i w \theta_i^{-1} u_i \nu_i v_i \xi_i z_i$ and $\nu_{i+1} v_{i+1} \theta_{i+1} w \mu_{i+1} z_{i+1} \xi_{i+1} u_{i+1}$ are boundary cycles of D_i and D_{i+1}, respectively, where z_i and z_{i+1} vertices on ∂D_i and ∂D_{i+1}, respectively. (See Fig. 7(b).) Let η be the complement of $\mu_i w \mu_{i+1}$ on ∂U, such that $z_i \mu_i w \mu_{i+1} z_{i+1} \eta$ is a boundary cycle of U.

Finally, define $\gamma'' = v_i \xi_i z_i \eta^{-1} z_{i+1} \xi_{i+1}^{-1} u_{i+1} \gamma_3$, where γ_3 is the subpath of γ_2 which starts at v_i and ends at u_{i+1}. (See Fig. 7(b).) Then $\gamma_3 \neq \emptyset$, since $v_i \neq u_{i+1}$ and γ'' is a simple closed curve homotopic to ∂H. Moreover, $L_{\gamma''}$ is a submap of L_γ which does not contain D_i and D_{i+1} and is regular. However, this violates the minimality of $|L_\gamma|$. Consequently,

$$\nu_i \text{ and } \nu_{i+1} \text{ are adjacent subpaths of } \gamma \tag{7}$$

An immediate consequence of (5), (6) and (7) and the fact that γ is a closed curve is that the ν_i exhaust the whole of γ, i.e.,

$$\mathcal{L}_0 = \{D_1, \ldots, D_m\} \tag{8}$$

Now, to show that $\mathcal{L}_0$ is annular we consider two cases according to whether U is in the bounded or unbounded component of $\mathbb{E}^2 \setminus \gamma$, respectively. Suppose first that U lies in the unbounded component of $\mathbb{E}^2 \setminus \gamma$. (See Fig. 8(a).)

Then γ is not homotopic to ∂U in $\mathbb{E}^2$. Since ∂U is a simple closed curve and since $M \cup H$ is an orientable compact surface, it follows from the Hairy Ball Theorem (see, e.g., [7, p. 124]) and the fact that the boundary of every region in $\mathcal{L}_0$ intersects γ non-trivially, that there exists a pair of regions D_i and D_{i+1} such that $\partial D_i \cap \partial U$ and $\partial D_{i+1} \cap \partial U$ are adjacent subpaths of ∂U, while $\partial D_i \cap \gamma$ and $\partial D_{i+1} \cap \gamma$ are not adjacent. (See Fig. 8(a).) Notice that if $u_i = v_{i+1}$ then $\langle D_i, D_{i+1}\rangle$ is shorter then L_γ. But this contradicts (7). Hence U lies in the bounded component of $\mathbb{E}^2 \setminus \gamma$. (See Fig. 8(b).) Notice that if we denote by V the bounded connected component of $\mathbb{E}^2 \setminus \mathcal{L}_0$ which contains H then $\partial V \subseteq \partial \mathcal{L}_0$. Therefore, it follows from (8) that there are regions $\{D_{i_1}, \ldots, D_{i_k}\}$ in $\mathcal{L}_0$, $1 \leq k \leq m$, such that $\partial V = \bigcup_{j=1}^{k} (\partial D_{i_j} \cap \partial V)$.

We claim $k \geq 2$. Suppose by way of contradiction that $k = 1$. Then $\partial V \subseteq \partial D_{i_1}$, hence ∂D_{i_1} cannot be a simple closed curve, (it has a double point z) violating our assumption (1), because it contains properly the simple closed curve ∂V. (See Fig. 8(b).) Alternatively, U would violate condition $CN(1)$. Hence $k \geq 2$. We propose to show that μ_{i_j} and μ_{i_j+1} are consecutive subpaths of ∂U. Suppose not. Then

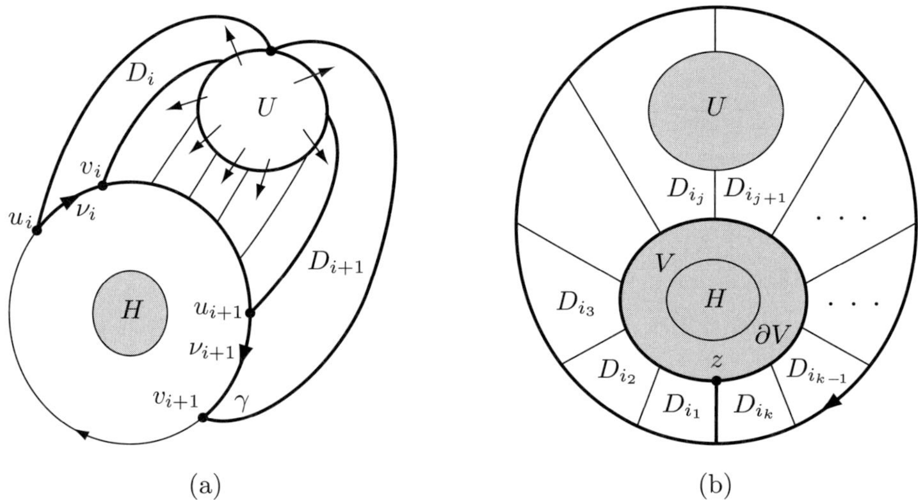

(a)   (b)

Figure 8

either ∂D_{i_j+1} or ∂D_{i_j-1} can have no non-trivial intersection with γ, due to the
planarity of M and because U and V are on the same side of γ in M. (See Fig. 9(a).)

This however, violates the definition of $\mathcal{L}_0$. Hence μ_{i_j} and $\mu_{i_{j+1}}$ are consecutive subpaths of ∂U. Now the same argument can be applied to γ (instead of
∂U), due to (7) and (8), to show that ν_{i_j} and $\nu_{i_{j+1}}$ are consecutive subpaths of
γ. (See Fig. 9(b), where we chose $j = 1$.) This argument also implies that $k = 2$,
for if $k \geq 3$ then μ_{i_1}, μ_{i_2} and μ_{i_3} are consecutive subpaths of ∂U and ν_{i_1}, ν_{i_2} and
ν_{i_3} are consecutive subpaths of γ, hence due to the property $CN(2)$ and due to
planarity $\partial D_{i_2} \subseteq \partial D_{i_1} \cup \partial D_{i_3} \cup \partial V \cup \partial U$. (See Fig. 9(c).) But then $\partial D_{i_2} \cap \gamma = \emptyset$,
contradicting our basic assumption (2) that $\partial D \cap \gamma \neq \emptyset$, for every D in $\mathcal{L}_0$. Thus
$k = 2$. Consider $L := \langle D_{i_1}, D_{i_2} \rangle$. Since L wraps V, but it cannot wrap U or
any other simply connected bounded connected component U_i of $M \setminus \mathcal{L}_0$ (because
then U_i violates the condition $CN(2)$), hence L is an annular submap of $\mathcal{L}_0$ with
outer boundary $v_0\nu_{i_1}v_1\nu_{i_2}v_2\theta_2'^{-1}w_2\mu_{i_2}^{-1}w_1\mu_{i_1}^{-1}w_0\theta_1'$, which is homotopic in M to
∂H. (See Fig. 9(d).) Hence if $|L_\gamma| \geq 3$ then L violates the minimality of L_γ, since
$|L| = 2 < 3 \leq |L_\gamma|$, i.e., $|L| < |L_\gamma|$. We show below that $|L_\gamma| = 2$ is impossible,
hence our basic assumption (by way of contradiction) that $\mathcal{L}_0$ is not annular is
false, proving $\mathcal{L}_0$ annular.

Thus, suppose $|L_\gamma| = 2$. Then $i_1 = 1, i_2 = 2$ and $\partial U \subseteq \partial D_1 \cup \partial D_2$. Since U
is simply connected, this contradicts the $CN(2)$-condition for $U \cup D_1 \cup D_2$. Thus
$\mathcal{L}_0$ is annular.

Now we prove that $\mathcal{L}_0$ is a one-layer map. Every region of $\mathcal{L}_0$ is either to the
right or to the left of γ, when traversing clockwise along γ. If D has two neighbors
E_1 and E_2 from its left, say, then due to (4) E_1 is to the left of γ and E_2 to the right
of γ. But then we can remove one of E_1 and E_2 without destroying the regularity

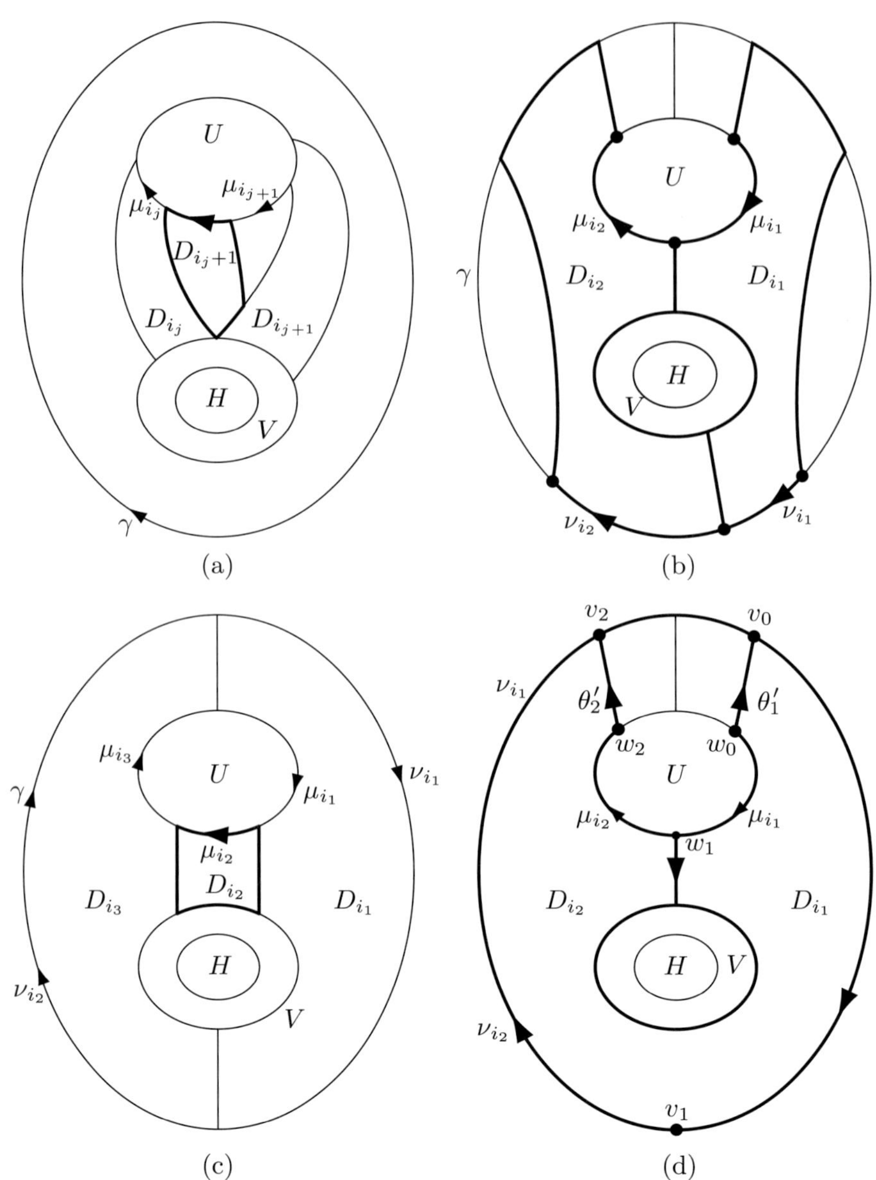

$$\textsc{Figure 9}$$

of $\mathcal{L}_0$. This however contradicts the minimality of $\mathcal{L}_0$, hence each region in $\mathcal{L}_i$ has one neighbor from its right and one neighbor from its left. But then every region in $\mathcal{L}_0$ has exactly two neighbors, hence $\mathcal{L}_0$ is an annular one-layer submap by Definition 1.1(a). The rest of part (a) of the lemma is clear from the construction of $\mathcal{L}_0$.

(b) Follows from the fact that $\mathcal{L}_0$ is an annular one-layer map, with $|\mathcal{L}_0|$ shortest possible among the annular submaps of M with boundaries homotopic to ∂M.

The lemma is proved. $\qquad\qquad\qquad\qquad\qquad\qquad\qquad\qquad\qquad\qquad\square$

From now on we shall denote by $\mathcal{L}_0$ a fixed annular one-layer submap of Lemma 2.2 and denote by $\mathcal{L}_i$ what we defined in 1.3(a) by $\mathcal{L}_i(\mathcal{L}_0)$.
We shall first prove the Main Theorem under hypothesis $\mathcal{H}$ below.
From now on we shall assume that M satisfies the condition $V(6)$.

2.3. The Hypothesis $\mathcal{H}$ (Annular layer structure)

Let M be an annular map with at least one region, simple closed boundary cycles and hole H. Hypothesis $\mathcal{H}$ requires the following:

(1) For every region D in $\mathcal{L}_i$, $i \geq 1$, $\partial D \cap \omega_{i-1}$ is connected.

($1'$) For every region D in $\mathcal{L}_i$, $i \leq -1$, $\partial D \cap \tau_{i+1}$ is connected.

(2) Let $1 \leq i \leq \ell$ and let $D_1, D_2 \in \mathcal{R}eg(\mathcal{L}_i)$ (not necessarily distinct) with $\partial D_1 \cap \partial D_2 \neq \emptyset$. Let $U = \mathbb{E}^2 \setminus \left(\omega_{i-1} \cup \{\bar{D}_1, \bar{D}_2\}\right)$. Then every (non-empty) connected bounded component of U which does not contain H, consists of a single region D, which is a 2-exceptional region of $\mathcal{L}_i$.

($2'$) The same as (3) with $-k \leq i \leq -1$ and ω_i replaced by τ_i.

Remark. Observe that $\mathcal{L}_0$ in Lemma 2.2 satisfies hypothesis $\mathcal{H}$.

Corollary to hypothesis $\mathcal{H}$. *Assume hypothesis $\mathcal{H}$. Then*

(a) *for every $i \neq 0$ either $\mathrm{Int}(\mathcal{L}_i)$ is simply connected or $\mathrm{Int}(\mathcal{L}_i)$ is annular with boundaries homotopic to ∂H.*

(b) *If $\mathcal{L}_i$ contains a region D such that $\bar{D}$ is not simply connected then $\mathcal{L}_i$ is annular, with boundaries homotopic to ∂H, $\bar{D}$ is annular with boundaries homotopic to ∂H and $|\mathcal{L}_i| \leq 2$.*

(c) *If $|\mathcal{L}_i| = 2$ and $\mathcal{L}_i$ contains a region D with $\bar{D}$ not simply connected then $\mathcal{L}_i = \{D, E\}$ for a 2-exceptional region E of $\mathcal{L}_i$.*

Proof. (a) Let i with $i \geq 1$ be the smallest positive number such that $\mathrm{Int}(\mathcal{L}_i)$ is neither annular with boundaries homotopic to ∂H, nor simply connected, while $\mathrm{Int}(\mathcal{L}_{i-1})$ is either simply connected or $\mathrm{Int}(\mathcal{L}_{i-1})$ is annular with boundaries homotopic to ∂H. Then $M \setminus \mathcal{L}_i$ contains a connected bounded simply connected submap U with $U \cap H = \emptyset$ such that $\bar{U}$ is homeomorphic to a closed disc and in particular, ∂U is a simple closed curve. Let ω_{i-1} he the outer boundary of Oust^{i-1}. Since $\bar{U}$ is simply connected and $U \cap H = \emptyset$, hence ∂U cannot be homotopic to ω_{i-1}. Therefore, it follows from the Hairy Ball Theorem, like in the proof of Lemma 2.1, that $\mathcal{L}_i$ contains regions D_1 and D_2 with $\partial D_1 \cap \partial U$ and $\partial D_2 \cap \partial U$ non-trivial, which violate assumption $\mathcal{H}(2)$, since by construction U is not in $\mathcal{L}_i$. Consequently, $U = \emptyset$ and either $\mathrm{Int}(\mathcal{L}_i)$ is annular with boundaries homotopic to ∂H, or $\mathrm{Int}(\mathcal{L}_i)$ is simply connected.

140   A. Juhász

(b) and (c) If $\mathcal{L}_i$, for $i \geq 1$, contains a region D such that $\bar{D}$ is not simply connected then by Lemma 2.1 $\bar{D}$ is annular with boundaries homotopic in M to ∂H. Hence, it follows from part (a) that $\mathcal{L}_i$ is annular with boundaries homotopic to ∂H. Suppose $|\mathcal{L}_i| \geq 2$. Since by the definition of $\mathcal{L}_i$ the boundary of every region of $\mathcal{L}_i$ intersects ω_{i-1} non-trivially, the planarity of M implies that the outer boundary of $\mathcal{L}_i$ coincides with the outer boundary of D: if $E \neq D$ is any region of $\mathcal{L}_i$ with $\partial E \cap \omega_i$ a non-trivial edge, then E should "cross" D in order that ∂E intersects ω_{i-1} non-trivially. Hence every region $E \neq D$ of $\mathcal{L}_i$ is contained in a bounded connected component of $\bar{\mathcal{L}}_i \setminus \{\omega_{i-1} \cup \bar{D}\}$. Therefore, to show $|\mathcal{L}_i| \leq 2$ it is enough to show that $\bar{\mathcal{L}}_i \setminus \{\omega_{i-1} \cup \bar{D}\}$ consists of a simple bounded component U, which consists of a 2-exceptional region of $\mathcal{L}_i$. Thus, consider $\bar{\mathcal{L}}_i \setminus \{\omega_{i-1} \cup \bar{D}\}$. It has a bounded, connected and simply connected component U (which does not intersect H). (See Fig. 10(b).) By assumption $\mathcal{H}(2)$ for U, in $\mathcal{L}_i$ we have $U = \{E\}$, for some 2-exceptional region E of $\mathcal{L}_i$.

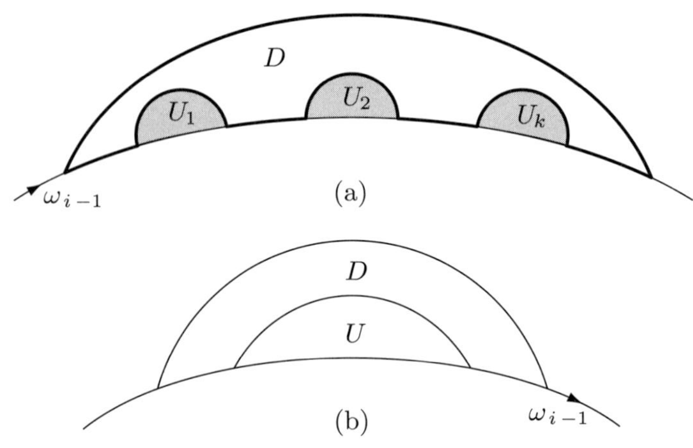

Figure 10

Finally, if $\bar{\mathcal{L}}_i \setminus \{\omega_{i-1} \cup \bar{D}\}$ has more than one bounded component, say $U_1, \ldots, U_k$, $k \geq 2$, then $\partial D \cap \omega_{i-1}$ is not connected, violating assumption $\mathcal{H}(1)$. (See Fig. 10(a).) Consequently, $\bar{\mathcal{L}}_i \setminus \{\omega_{i-1} \cup \bar{D}\} = \{E\}$ and as explained above, $\mathcal{L}_i = \{D, E\}$, as required.

The corollary is proved. $\qquad\qquad\square$

2.4. Regions in layers, under hypothesis $\mathcal{H}$

For convenience, for a region D in $\mathcal{L}_i$, we shall denote $a(D) = |\alpha(D)|$, $b(D) = |\beta(D)|$ and $c(D) = |\gamma(D)|$.

Proposition 2.1. *Let A be an annular map with connected interior which satisfies condition $\mathcal{H}$. Let $D \in \mathcal{L}_i$ be a region, for some i, for $i \geq 1$. Then each of the following holds:*

(1) $\partial D \cap \omega_i \neq \emptyset$ and is connected.

(2) $b(D) \leq 2$. Let $E \in \beta(D)$. If $\mathcal{L}_i$ is simply connected or $\mathcal{L}_i$ is annular with $|\mathcal{L}_i| \geq 3$, then $\partial D \cap \partial E$ is connected. (See Fig. 11(a).)

If $\mathcal{L}_i = \{D, E\}$ and $\mathcal{L}_i$ is annular, then $\partial D \cap \partial E = \mu_1 \cup \mu_2$, where μ_1 and μ_2 are (connected) boundary paths of ∂E. (See Fig. 11(b).)

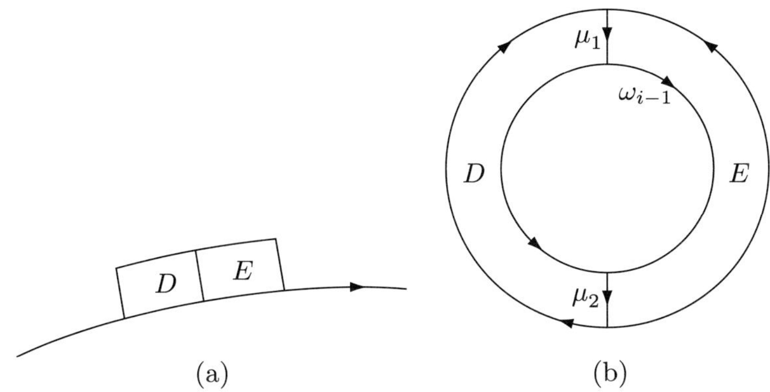

FIGURE 11

(3) $a(D) \leq 3$. Suppose $a(D) = 3$. Then D is 3-exceptional in $\mathcal{L}_i$. Let L be the closure of the connected component of $\mathrm{Int}(\mathcal{L}_i)$ which contains D and let $D = D_1, \ldots, D_k$, $k \geq 1$ be all the 3-exceptional regions of L, occurring in this order from left to right. (See Fig. 12(d).) If $k \geq 2$ then for every j with $j \leq k - 1$, a 3-corner region or a 2-corner region which is not a proper neighbor of a 2-exceptional region, occurs between D_i and D_{i+1}, which is to the right of D_i and to the left of D_{i+1} (See Fig. 12(a), where $k = 2$ and where 2-corner region E_2 occurs between D_1 and D_2, and see Fig. 12(d), where 3-corner region E occurs between D and D_1.)

(4) Suppose $d(D) \leq 5$. Then $a(D) \leq 2$. If $a(D) = 2$ then D is 2-exceptional and if $E \in \beta(D)$ then E is a 2-corner region. (Fig. 3(a).)

(5) Let w be a vertex of ω_i. Then $d_{\mathcal{L}_i}(w) \leq 4$. If $d_{\mathcal{L}_i}(w) = 4$ and D_1, D_2 and D_3 are the regions of $\mathcal{L}_i$ which contain w on their boundary, then D_1 and D_3 are 2-corner regions of $\mathcal{L}_i$ and D_2 is 2-exceptional in $\mathcal{L}_i$.

(6) $\mathcal{L}_i$ is a one-layer annular diagram (not necessary with connected interior) with the property that $\partial D \cap \omega_{i-1} \neq \emptyset$ and $\partial D \cap \omega_i \neq \emptyset$, for every region D in $\mathcal{L}_i$.

Remark. The corresponding statement holds true for $i \leq -1$.

Proof. Suppose the proposition is false. Then there exists a smallest index i, $i \geq 1$ such that either the proposition holds for every $\mathcal{L}_j$ with $1 \leq j < i$ and does not holds for $\mathcal{L}_i$, or $i = 1$.

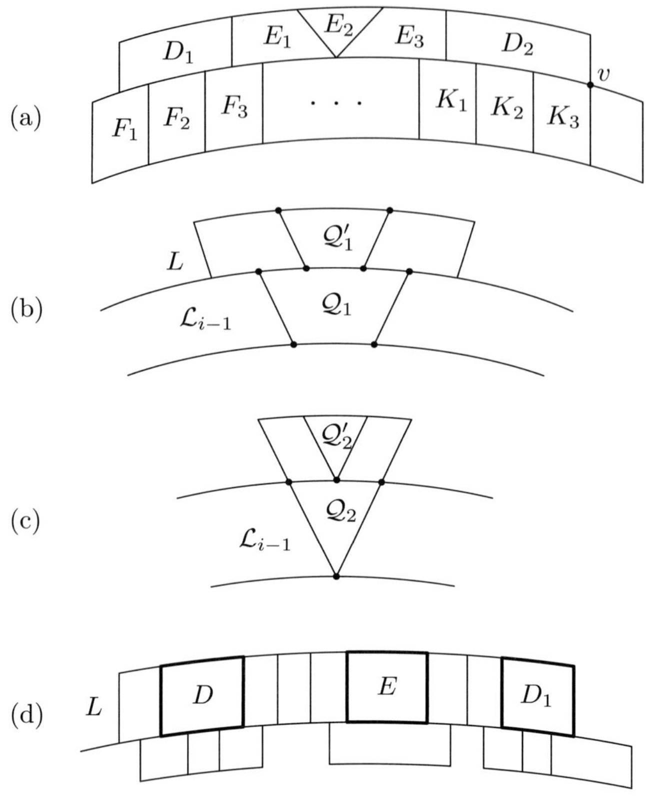

FIGURE 12

(1) Suppose $\mathcal{L}_i$ contains a region D with $\partial D \cap \omega_i = \emptyset$. Let $\mu = \partial D \cap \omega_{i-1}$. By $\mathcal{H}(1)$, μ is connected. Let $\partial D = v\mu u\nu$, v and u vertices and let $\nu = \nu_1 \cdots \nu_k$, $k \geq 1$, where $\nu_j = \partial D \cap \partial E_j$, for some regions $E_j \in \mathcal{L}_i$. (Notice that ν_j is connected, due to the $CN(2)$ property of simply connected $V(6)$ maps. See [14].) If $k \geq 2$ then $|\mathcal{L}_i| \geq 3$ hence by the corollary to hypothesis $\mathcal{H}$, $\bar{E}$ is simply connected, for every region E in $\mathcal{L}_i$. Since ∂D is clearly not homotopic to ω_{i-1}, we can use the Hairy Ball Theorem to conclude that there exists a pair of regions E_1 and E_2 such that the triple (E_1, E_2, ω_{i-1}) violates $\mathcal{H}(2)$. (See the proof of Lemma 2.2 for details.) Therefore, $k = 1$. If $\bar{E}_1$ is simply connected then $(\partial E_1 \cap \omega_{i-1}) \cup (\partial D \cap \omega_{i-1}) \neq \omega_{i-1}$ (for if equality holds then $\bar{E}_1$ is annular), This however implies that $\partial E_1 \cap \omega_{i-1}$ is not connected, violating $\mathcal{H}(1)$. (See Fig. 13(a).) Consequently $\bar{E}_1$ is not simply connected. Therefore by part (c) of the corollary to hypothesis $\mathcal{H}$, $\mathcal{L}_i = \{E_1, D\}$ and D is a 2-exceptional region of $\mathcal{L}_i$. Hence, by the definition of 2-exceptional regions $\partial D \cap \omega_i \neq \emptyset$ and is connected, a contradiction which proves part **(1)**.

(2) If $|\mathcal{L}_i| \leq 2$ then the claim is clear, hence we may assume that $|\mathcal{L}_i| \geq 3$.

Hence by the corollary to assumption $\mathcal{H}$,

$$\text{for every region } F \text{ of } \mathcal{L}_i,\ \bar{F} \text{ is simply connected} \qquad (*)$$

Now, $\partial D \cap \omega_{i-1}$ is connected, by assumption $\mathcal{H}(1)$ and $\partial D \cap \omega_i$ is connected by part (1) of the proposition. Let u_1 and v_1 be the endpoints of $\partial D \cap \omega_{i-1}$ and let u_2 and v_2 be the endpoints of $\partial D \cap \omega_i$, respectively. ($u_1 = v_1$ or $u_2 = v_2$ are not excluded.) Let $u_1\mu_1 u_2$ and $v_1\mu_2 v_2$ be the boundary paths of D which are in the closure of $\mathcal{L}_i$ and which intersect ω_{i-1} in u_1 and v_1, respectively and intersect ω_i in u_2 and v_2, respectively. (See Fig. 13(b).)

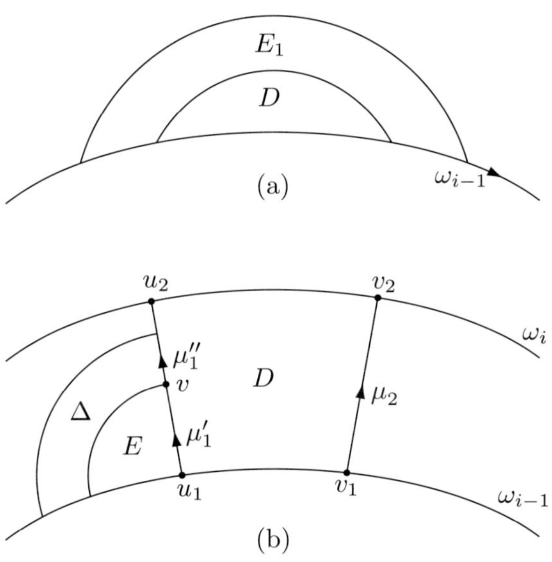

$$\text{FIGURE } 13$$

We claim that (the open path) μ_1 if non-empty, is an interior path of $\mathcal{L}_i$, because if not, then μ_1 intersects $\partial \mathcal{L}_i$ in a vertex different from u_1 and u_2. But since $\mathcal{L}_i$ is annular by the corollary to the assumption $\mathcal{H}$, $\partial \mathcal{L}_i = \omega_i \cup \omega_{i-1}$, hence, either $\mu_1 \cap \omega_{i-1} \neq \{u_1\}$ or $\mu_1 \cap \omega_i \neq \{u_2\}$. However, $\mu_1 \subseteq \partial D$, because $\partial D \cap \omega_i$ and $\partial D \cap \omega_{i-1}$ are connected. Hence μ_1 is connected and $\mu_1 \cap \omega_i = \{u_2\}$. Thus μ_1 is an interior subpath of $\mathcal{L}_i$. If $\mu_1 = \emptyset$ then μ_1 does not contribute to $b(D)$, hence may assume $\mu_1 \neq \emptyset$. Let E be a proper neighbor of D which contains u_1. Since $\mathcal{L}_i$ is annular or simply connected and μ_1 is an interior path of $\mathcal{L}_i$, such a region exists. (See Fig. 13(b).) Let $\mu_1' = \partial E \cap \partial D$ and let $\mu_1 = \mu_1' v \mu_1''$. Then $\mu_1'' \cap \omega_{i-1} \subseteq \mu_1 \cap \omega_{i-1} = \{u_1\}$, hence $\mu_1'' \cap \omega_{i-1} \subseteq \mu_1'' \cap \{u_1\}$. But since $\mu_1' \neq \emptyset$, hence $\mu_1'' \cap \{u_1\} = \emptyset$. Thus,

$$\mu_1'' \cap \omega_{i-1} = \emptyset.$$

We propose to show that $\mu_1 = \emptyset$. Suppose not. Now, μ_1' and μ_1'' are simple subpaths of ∂D. Since μ_1'' is an interior subpath of $\mathcal{L}_i$, hence $\mathcal{L}_i$ contains a region

Δ such that $v \in \partial\Delta \cap \mu_1''$. By definition of $\mathcal{L}_i$, $\partial\Delta \cap \omega_{i-1} \neq \emptyset$. But then due to the planarity of M, $\partial E \cap \omega_i = \emptyset$, violating part **(1)** of the proposition. Therefore $\mu_1'' = \emptyset$. (Alternatively, $(\Delta, D, \omega_{i-1})$ violates $\mathcal{H}(2)$ unless E is a 2-exceptional region. But then, by the definition of 2-exceptional regions, $v \in \omega_i$. Since $v \in \mu_1$ and $\mu_1 \cap \omega_i = \{u_2\}$, hence $v = u_2$ and $\mu_1'' = \emptyset$.) Consequently, $\Delta = \emptyset$, $\mu_1' = \mu_1$ and E is the only proper neighbor of D in $\mathcal{L}_i$ with common boundary on μ_1. A similar consideration for μ_2 shows that if $\mu_2 \neq \emptyset$ then D has a unique proper neighbor E' with $\partial E' \cap \partial D$ on μ_2. Consequently, $b(D) \leq 2$. If E and D are as in the proposition then by the property $CN(2)$, $\partial E \cap \partial D$ is connected.

(3) Suppose $a(D) \geq 4$ and let $\mu = \partial D \cap \omega_{i-1}$. By $\mathcal{H}(1)$ μ is connected. Let $\mu = \mu_1 v_1 \mu_2 \cdots \mu_k v_k$, v_i vertices and $k \geq 4$, such that $\mu_i = \partial D \cap \partial E_j$, for some region E_j in $\mathcal{L}_{i-1}$, $j = 1, \ldots, k$. (Notice that the μ_i are connected by part (1) of the proposition for $\mathcal{L}_{i-1}$, if $i \geq 2$ and by the definition of $\mathcal{L}_0$, if $i = 1$.) Now, since M is annular, $E_2, \ldots, E_{k-1}$ are interior regions of $\mathcal{L}_{i-1}$, hence

$$d_{\mathcal{L}_{i-1}}(v_j) = d_M(v_j) - d_{\mathcal{L}_i}(v_j) + 2 \tag{2.1}$$

Consequently, since $d_M(v_j) \geq 3$ and $d_{\mathcal{L}_i}(v_j) = 2$ hence

$$d_{\mathcal{L}_{i-1}}(v_j) = d_M(v_j) \geq 3. \tag{2.2}$$

Suppose first $i \geq 2$. If $d_{\mathcal{L}_{i-1}}(v_2) \geq 4$ then $d_{\mathcal{L}_{i-1}}(v_1) = d_{\mathcal{L}_{i-1}}(v_3) = 2$, by part **(5)** of the proposition for $\mathcal{L}_{i-1}$, a contradiction. Consequently, it follows from (2.2) that $d_{\mathcal{L}_{i-1}}(v_2) = 3$ and hence $d_{\mathcal{L}_{i-1}}(v_1) = d_{\mathcal{L}_{i-1}}(v_3) \leq 3$, again by part **(5)** of the proposition for $\mathcal{L}_{i-1}$. Hence by (2.2) $d_{\mathcal{L}_{i-1}}(v_j) = 3$, for $j = 1, 2, 3$. But then $d_M(v_j) = 3$ by (2.1), hence $d_M(E_j) \geq 6$, for $j = 1, 2, 3$, by the $V(6)$-condition.

Since $b(E_j) = 2$, by part **(2)** of the proposition, E_j are interior regions of $\mathcal{L}_{i-1}$ for $j = 2, 3$, hence it follows that E_2 and E_3 are 3-exceptional by definition, violating part **(3)** for $\mathcal{L}_{i-1}$. Thus $a(D) \leq 3$. Suppose $a(D) = 3$. Then (D) is 3-exceptional, due to condition $V(6)$.

Suppose now that $i = 1$. Define $L := \mathcal{L}_0 \setminus \{E_2, \ldots, E_{k-1}\} \cup \{D\}$ It is an annular submap of A with boundaries homotopic to ∂M such that

$$|L| = |\mathcal{L}_0| - (k - 2) + 1 = |\mathcal{L}_0| + 3 - k \leq |\mathcal{L}_0| - 1, \text{ since } k \geq 4,$$

violating the minimality of $|\mathcal{L}_0|$.

Finally, assume D_1 and D_2 are adjacent 3-exceptional regions of $\mathcal{L}_i$ in the sense that if $L := \langle D_1, E_1, \ldots, E_k, D_2 \rangle$ is a submap of $\mathcal{L}_i$ with connected interior then none of the E_i is 3-exceptional. Then $\alpha(D_1) = \langle F_1, F_2, F_3 \rangle$ and $\alpha(D_2) = \langle K_1, K_2, K_3 \rangle$. It follows that F_2 and K_2 are 3-exceptional. (See Fig. 12(a).)

Suppose first $i \geq 2$. By the induction hypothesis for $\mathcal{L}_{i-1}$, either there exists a 3-corner region Q_1 in $\alpha(L)$ or there exists a 2-corner region Q_2 which is not adjacent to a 2-exceptional region. It follows easily that $\gamma(Q_1)$ contains a 3-corner region Q_1' and $\gamma(Q_2)$ contains a 2-corner region Q_2' of L, proving our claim. (See Fig. 12(b) and (c).) If $i = 1$ then it is easy to see that if L contains no corner regions as stated in the proposition, then L shortens $\mathcal{L}_0$, a contradiction, to the minimality of $\mathcal{L}_0$.

(4) Suppose $a(D) \geq 3$ and let $\mu = \partial D \cap \omega_{i-1}$. Then μ is connected and we may decompose μ into $\mu_1 v_1 \mu_2 v_2 \cdots \mu_k$, $k \geq 3$. By the $V(6)$-condition $d_M(v_j) \geq 4$, for $j = 1, \ldots, k$ and since $d_{\mathcal{L}_i}(v_j) = 2$, it follows from (2.1) in the proof of part **(3)** of the proposition that $d_{\mathcal{L}_{i-1}}(v_j) \geq 4$ for $j = 1, 2$. But this violates part **(5)** of the proposition for $\mathcal{L}_{i-1}$, if $i \geq 2$. (See Fig. 14.) Hence $i = 1$. However, then $\{D\}$ shortens $\mathcal{L}_0$. Consequently, $a(D) \leq 2$. Suppose $a(D) = 2$ and let u_1 and u_2 be the endpoints of μ. Then $d_{\mathcal{L}_i}(v_1) = 2$ hence it follows from the $V(6)$ condition and from (2.1) that $d_{\mathcal{L}_{i-1}}(v_1) \geq 4$. Consequently, if $i \geq 2$ then u_1 and u_2 have valency two in $\mathcal{L}_{i-1}$, due to part **(5)** of the proposition for $\mathcal{L}_{i-1}$. But then again by (2.1) $d_{\mathcal{L}_i}(u_j) \geq 4$, $j = 1, 2$. In particular, D has proper neighbors E_1 and E_2 in $\mathcal{L}_i$ with $|\alpha(E_1)| = |\alpha(E_2)| = 0$. Since $b(E_1), b(E_2) \leq 2$, they are 2-corner regions, as stated. (See Fig. 3(a).) If $i = 1$ and u_1 or u_2 has valency at least three then again $\{D\}$ shortens $\mathcal{L}_0$.

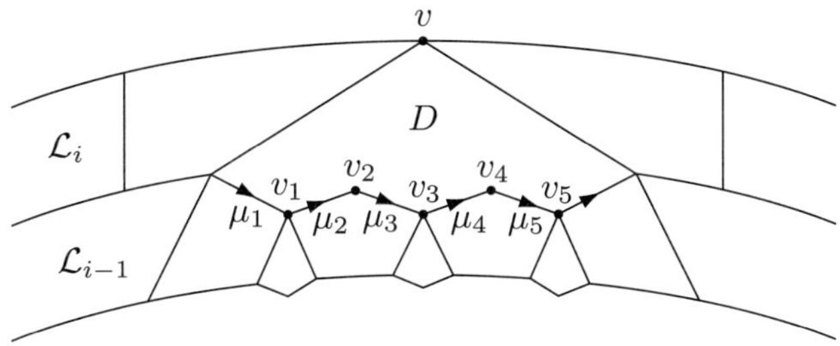

FIGURE 14

(5) Let $k = d_{\mathcal{L}_i}(w)$. Suppose $k \geq 5$ and let $D_1, D_2, \ldots, D_{k-1}$ be all the regions of $\mathcal{L}_i$ which contain w on their boundary.

We claim that $D_q \neq D_j$ for $2 \leq q < j \leq k - 2$. For suppose $D_q = D_j$. Then either H in the unbounded connected component of $\mathbb{E}^2 \setminus \{D_q\}$ or H is in a bounded connected component of $\mathbb{E}^2 \setminus \{D_q\}$. The first case is impossible due to the property $CN(2)$ (see [14]) and the second case cannot occur because then $\partial E_1 \cap \omega_{i-1} = \emptyset$, violating the definition of Oust^i. Let $L_1 = \mathbb{E}^2 \setminus (\omega_{i-1} \cup \{\bar{D}_1, \bar{D}_3\})$. By $\mathcal{H}(2)$, every connected bounded component of L_1 which does not contain H is a 2-exceptional region in $\mathcal{L}_i$. Clearly, D_2 belongs to one of these components, hence D_2 is a 2-exceptional region in $\mathcal{L}_i$. Define $L_2 = \mathbb{E}^2 \setminus (\omega_{i-1} \cup \{\bar{D}_2, \bar{D}_4\})$. By the same argument we conclude that D_3 is a 2-exceptional region in $\mathcal{L}_i$. This however violates the last statement of part **(4)**. Consequently, $k \leq 4$, D_2 is 2-exceptional and D_1 and D_3 are 2-corner regions, by part **(4)**.

(6) By the corollary to assumption $\mathcal{H}$, $\mathcal{L}_i$ is an annular map. By the definition of $\mathcal{L}_i$, $\partial D \cap \omega_{i-1} \neq \emptyset$, for every region D in $\mathcal{L}_i$ and by part **(1)** of the proposition $\partial D \cap \omega_i \neq \emptyset$, for every region D in $\mathcal{L}_i$. Finally, it follows from part **(2)** that $\mathcal{L}_i$ is a one-layer map.

The proposition is proved. $\qquad\qquad\square$

2.5. The structure of $\mathcal{L}_i$

The description of the layers of M is given in the following proposition.

Proposition 2.2. *Let M be an annular map with connected interior which satisfies the condition $V(6)$. Let $\mathcal{L}_0$ be an annular one-layer submap with boundaries homotopic to $\omega(M)$, which is shortest possible (see Lemma 2.1). Let $\mathcal{L}_{-k}$, $\mathcal{L}_{-k+1}$, ..., $\mathcal{L}_{-1}$, $\mathcal{L}_0$, $\mathcal{L}_1$, ..., $\mathcal{L}_\ell$ be a layer decomposition of M. (See Definitions 1.2.) Let $\mathcal{L}_i$ be one of the layers, $i \geq 1$. Then each of the following holds:*

(a) *M satisfies condition $(\mathcal{H})$.*

(b) *$\mathcal{L}_i$ is an annular one-layer submap.*

(c) *If $v \in \omega_i$ is a vertex then $d_{\mathcal{L}_i}(v) \leq 4$.*

Suppose $d_{\mathcal{L}_i}(v) = 4$ and let E_1, E_2, E_3 be the regions of $\mathcal{L}_i$ which contain v on their boundary. Then E_1 and E_3 are 2-corner regions of $\mathcal{L}_i$ and E_2 is a 2-exceptional region of $\mathcal{L}_i$. In particular, the vertices adjacent to v have valency two in M.

(d) *For every region D in $\mathcal{L}_i$, $b(D) \leq 2$. If D is an inner region of $\mathcal{L}_i$ then $b(D) = 2$ and if D is an extremal region of $\mathcal{L}_i$ then $b(D) \leq 1$.*

(Notice that if $\mathrm{Int}(\mathcal{L}_i)$ is annular then every region is inner in $\mathcal{L}_i$ and if $\mathrm{Int}(\mathcal{L}_i)$ is not annular then it is the disjoint union of simply connected submaps, each of which either consists of a single region D that necessarily is extremal and $b(D) = 0$ or consists more than one region in which case two of the regions are extremal and the rest are inner regions.) (See Fig. 15.)

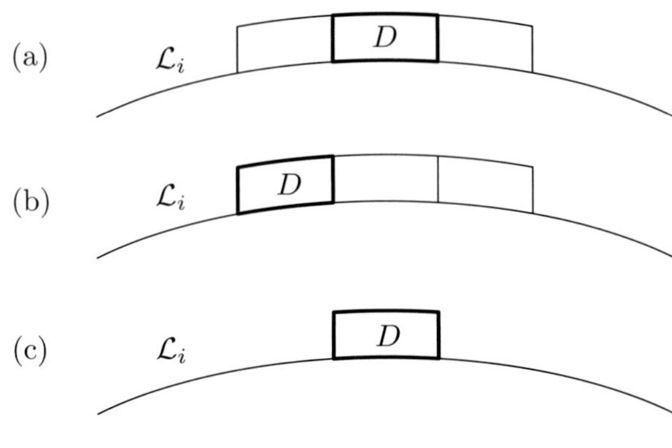

FIGURE 15

(e) *$a(D) \leq 3$ for every region D in $\mathcal{L}_i$.*

Suppose $a(D) = 3$. Then D is a 3-exceptional region and if D' is another 3-exceptional region of $\mathcal{L}_i$ which is adjacent to D as a 3-exceptional region in the connected component L of $\mathrm{Int}(\mathcal{L}_i)$ which contains D then either there is a region E in L between D and D' which is a 3-corner region of $\mathcal{L}_i$ or a 2-corner region of $\mathcal{L}_i$ which is not a proper neighbor of a 2-exceptional region of $\mathcal{L}_i$.

Proof. First observe that if part (a) of the proposition is satisfied then parts (b), (c), (d), and (e) follow by Proposition 2.1, parts (6), (5), (2) and (3), respectively. Hence, we concentrate on the proof of part (a).

By Lemma 2.2, M has an annular regular one-layer submap $\mathcal{L}_0$ with boundaries homotopic to ∂M such that $|M|$ and $\|M\|$ are minimal possible. We prove the proposition by induction on $|M|$, the case $|M| = 1$ being clear. By Remark 2.3, $\mathcal{L}_0$ satisfies condition $\mathcal{H}$. Hence, if M contains a region D in a layer $\mathcal{L}_i$ which violates condition $\mathcal{H}(1)$ or a pair of regions (D_1, D_2) in layer $\mathcal{L}_i$ which violates condition $\mathcal{H}(2)$, then either $i \geq 1$ or $i \leq -1$. We shall call such a region a *refuting region* and such pairs *refuting pairs*. We shall assume $i \geq 1$; the case $i \leq -1$ can be dealt with by the same arguments. We prove first $\mathcal{H}(2)$. Suppose the proposition holds true in layer $\mathcal{L}_{i-1}$ while $\mathcal{L}_i$ contains a refuting pair (D_1, D_2). We prove that this leads to a contradiction.

Let $S_0 = \omega_{i-1} \cup \{D_1, D_2\}$ and let U be a bounded, connected component of $\mathbb{E}^2 \setminus S_0$ which does not contain H. Then U is simply connected. Suppose that among all the bounded, connected components of $\mathbb{E}^2 \setminus S_0$ which do not contain H, U contains minimal number of regions. Let $S = \mathrm{Oust}^{i-1} \cup U$. Clearly S is a regular annular map which does not contain D_1 and D_2, hence $|S| < |M|$ and the claim of the proposition holds true for S by the induction hypothesis. Therefore, S has a layer structure $\Lambda = \{\mathcal{L}_0, \mathcal{L}_1, \ldots, \mathcal{L}_{i-1}, \mathcal{L}'_i, \ldots, \mathcal{L}'_{i+p}\}$ each $\mathcal{L}_j$ and $\mathcal{L}'_j$ being an annular one-layer map. We propose to show that $p = 0$. Suppose by way of contradiction that $p \geq 1$ and consider $\mathcal{L}'_{i+p}$. It follows from the minimality assumption that its interior is connected and simply connected. Let $\mathcal{L}'_{i+p} = \langle E_1, \ldots, E_k \rangle$, $k \geq 1$. (See Fig. 16.) Assume first $k = 1$ and let $D = E_1$. (See Fig. 16(a).) Then $c(D) \leq 2$, since $\gamma(D) = \{D_1, D_2\}$, $b(D) = 0$, since D has no neighbors in $\mathcal{L}_{i+p}$. Hence $a(D) = d(D) - b(D) - c(D) \geq d(D) - 2$. Now $a(D) \leq 3$ due to the induction hypothesis for S, since by Proposition 2.1, $a(D) = 3$ implies $d(D) \geq 6$ and hence $a(D) \geq 6 - 2 = 4$. Therefore $a(D) \leq 2$. But since $d(D) \geq 4$, hence $a(D) = 2$ and hence $d(D) = 4$ and $c(D) = 2$. Thus,

$$a(D) = 2, \quad d_M(D) = 4 \quad and \quad c(D) = 2 \tag{$*$}$$

Let $\mu = \omega_{i+p-1} \cap \partial D$. Then $\mu = v_1 \mu_1 v_2 \mu_2 v_3$, v_1, v_2 and v_3 vertices. Since v_2 is an inner vertex of S and of M, $d_S(v_2) \geq 4$ due to the $V(6)$-condition and because $d_{\mathcal{L}_{i+p}}(v_2) = 2$. Hence $d_{\mathcal{L}_{i+p-1}}(v_2) \geq 4 + 2 - 2 = 4$ by formula (2.1). By Proposition 2.1 $d_{\mathcal{L}_{i+p-1}}(v_2) \leq 4$, since $d_M(D) = 4$ by ($*$), hence $d_{\mathcal{L}_{i+p-1}}(v_2) = 4$ by the $V(6)$-condition. Consequently, if F_1, F_2, F_3 are the regions of $\mathcal{L}_{i+p-1}$ which contain v_2 on their boundary, then by the induction hypothesis for S, F_2 is 2-exceptional, by Proposition 2.1. Hence, either F_1 is a proper boundary region of S, or F_1 is a 2-corner region, by Proposition 2.1. In the first case, since $|\mathcal{L}_{i+p}| = 1$, by assumption, the regions which contain v_1 on their boundary are D_1, D and F_1, therefore $d_S(v_1) = 3$, violating condition $V(6)$. If the second case occurs $\partial F_1 \cap \partial D_1 = \{v_1\}$. But then $d_M(F_1) = b(F_1) + c(F_1) + a(F_1) \leq 2 + 1 + 0 = 3$, i.e., $d_M(F_1) \leq 3$, violating the condition $V(6)$ again. Consequently, $|\mathcal{L}_{i+p}| \geq 2$ and it contains an extremal region E ($E = E_1$ or $E = E_k$) which has proper common edge with exactly one of D_1 and D_2. (See Fig. 16(b).) Hence $c(E) = 1$. Since E is extremal, and $|\mathcal{L}_{i+p}| \geq 2$, hence $b(E) = 1$ and consequently, $4 \leq d(E) = a(E) + b(E) + c(E) = a(E) + 2$, i.e., $a(E) \geq 2$. Like in $(*)$ of the case $|\mathcal{L}_{i+p}| = 1$, this implies $d(E) = 4$. From this point on, the arguments of the case $|\mathcal{L}_{i+p}| = 1$ for D can be repeated for E and they lead to a contradiction, like in the case when $|\mathcal{L}_{i+p}| = 1$. Consequently, $p = 0$ as claimed.

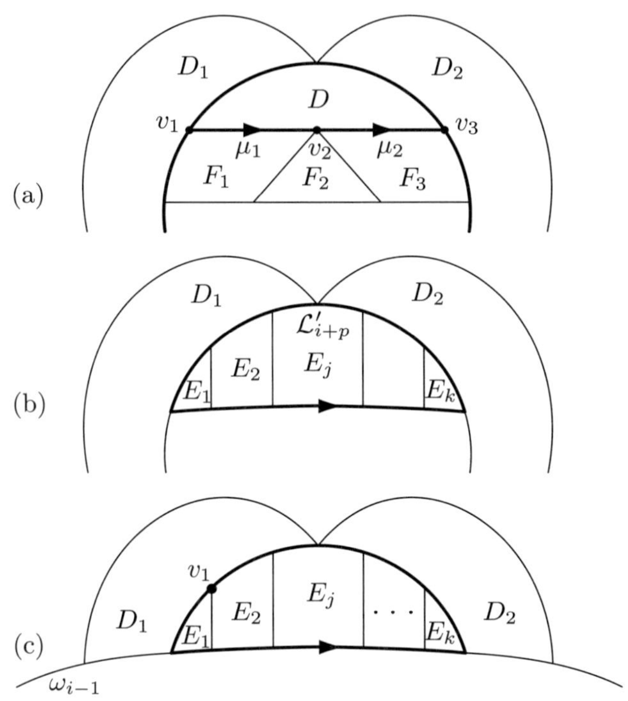

FIGURE 16

Hence $U \subseteq \mathcal{L}'_i$. (See Fig. 16(c).) Without loss of generality we may assume that (D_1, D_2) is such that $|U|$ is minimal possible, where (D_1, D_2) runs over all the refuting pairs. We claim that $|U| = 1$. For suppose $U = \langle E_1, \ldots, E_k \rangle$, $k \geq 2$. All the neighbors of E_j, $1 \leq j \leq k$, are either in $\mathcal{L}_i$ or in $\mathcal{L}_{i-1}$ and there is at most one region E_j with the property that $\partial D_1 \cap \partial E_j \neq \emptyset$ and also $\partial D_2 \cap \partial E_j \neq \emptyset$. Hence, if E_j is such a region, then no region to its right (if exists) and no region to its left (if exists) has this property. In particular, since $k \geq 2$, at least one of E_1 and E_k does not have this property and without loss of generality we may assume that E_1 does not have this property. But then (D_1, E_2) is a refuting pair in $\mathcal{L}_i$ such that a bounded connected simply connected component S_1 of $\mathbb{E} \setminus \left(\text{Oust}^{i-1} \cup \bar{D}_1 \cup \bar{E}_2 \right)$ consists of the single region E_1. Thus $|S_1| = 1 < 2 \leq k \leq |U|$, violating the minimality of $|U|$, unless E_1 is a 2-exceptional region and E_2 is a 2-corner region.

Therefore $a(E_2) = 0$ and $b(E_2) \leq 2$. Now $\gamma_M(E_2) \subseteq \{D_1, D_2\}$, hence $c(E_2) \leq 2$. Consequently, $d_M(E_2) = a(E_2) + b(E_2) + c(E_2) \leq 0 + 2 + 2 = 4$. Therefore, by the $V(6)$-condition $d_M(E_2) = 4$ and ∂E_2 contains no vertices with valency 3 in M. But the vertex $v = \partial E_1 \cap \partial E_2 \cap \partial D_1$ has valency 3 in M, because all the neighbors of E_2 in M are E_1, E_3, D_1 and D_2 and $\partial E_3 \cap \partial D_1 = \emptyset$. (Otherwise $\gamma(E_2) = \{D_1\}$ and hence $d_M(E_2) = 3$.) This is a contradiction. Consequently $k = 1$, $U = \{E_1\}$ and since $c(E_1) = 0$ and $\beta(E_1) = \{D_1, D_2\}$, we get that E_1 is a 2-exceptional region, as stated.

Finally, we prove condition $\mathcal{H}(1)$. Let D be a region of $\mathcal{L}_i$ with $\partial D \cap \omega_{i-1}$ not connected. Then $\mathbb{E}^2 \setminus (\bar{D} \cup \omega_{i-1})$ has a bounded connected component U. Let $S = \omega_{i-1} \cup U$. (See Fig. 10(b).) Then $|S| < |\mathrm{Oust}^i|$, hence the claims of the proposition hold true for S. Now, the arguments of the proof of the condition $\mathcal{H}(2)$ can be repeated, to give that $U \subseteq \mathcal{L}_i$ and $U = \{E\}$ for some E in $\mathcal{L}_i$. Thus, $c(E) = 0$, $b(E) = 1$, hence $a(E) \geq 3$. Since by induction hypothesis S satisfies the assertions of the proposition, $a(E) = 3$ by part (e) and $d(E) \geq 6$. But then $a(E) = d(E) - \big(b(E) + c(E)\big) \geq 6 - (1 + 0) = 5 > 3$ violating part (e) of the proposition for S.

The proposition is proved. $\square$

2.6. Proof of the Main Theorem

Let D be a region in $\mathcal{L}_s$, $s \geq 1$. Let $i(D)$ be the number of inner edges of E in $\mathcal{L}_s$. Thus, $i(D) = a(D) + b(D)$. Define $\sigma(D) = \big(d_M(D) - i(D)\big) - a(D)$. Then

$$\sigma(D) = |\partial D \cap \omega_{s-1}| - |\partial D \cap \omega_s| \qquad (*)$$

We shall need the following lemma for the proof of the Main Theorem.

Lemma 2.3. *For $i \geq 0$, $\sigma(\mathcal{L}_{i+1}) \geq \sigma(\mathcal{L}_i)$ and for $i \leq 0$ $\sigma(\mathcal{L}_i) \geq \sigma(\mathcal{L}_{i+1})$.*

Proof. We shall consider only the case $i > 0$. The case $i \leq 0$ is analogous. Suppose first that $\mathrm{Int}(\mathcal{L}_{i+1})$ is annular and let $D_1, D_2, \ldots, D_k$ be all the distinct 2-exceptional or 3-exceptional regions of $\mathcal{L}_{i+1}$. Then it follows from parts (c) and (e) of Proposition 2.2 that for $j = 1, \ldots, k - 1$, there are corner regions E_j between D_j and D_{j+1} and there is a corner region E_k between D_k and D_1. Hence,

$$\sigma(\mathcal{L}_{i+1}) = \sum_{j=1}^{k} \sigma(D_j) + \sum_{j=1}^{k} \sigma(E_j) + \sum_{j=1}^{s} \sigma(F_j),$$

where

$$\mathcal{L}_{i+1} = \bigcup_{j=1}^{k} \{D_j\} \cup \bigcup_{j=1}^{k} \{E_j\} \cup \bigcup_{j=1}^{s} \{F_j\},$$

where the F_j are not exceptional regions. Since F_j is not exceptional, hence it follows from parts (d) and (e) of Proposition 2.2 that $\sigma(F_j) \geq 0$:

$$\sigma(F_j) = d(F_j) - i(F_j) - a(F_j).$$

Since $i(F_j) = a(F_j) + b(F_j)$, hence

$$\sigma(F_j) = d(F_j) - \big(a(F_j) + b(F_j)\big) - a(F_j)$$
$$= d(F_j) - 2a(F_j) - b(F_j) \geq 6 - 2 \cdot 2 - 2 = 0,$$

i.e., $\sigma(F_j) \geq 0$. Also, by similar calculations we get $\sigma(D_j) \geq -2$ and $\sigma(E_j) \geq 2$, by parts (c), (d) and (e) of Proposition 2.2. Thus, $\sigma(F_j) \geq 0$, $\sigma(D_j) \geq -2$ and $\sigma(E_j) \geq 2$. Consequently, $\sigma(\mathcal{L}_{i+1}) \geq -2k + 2k + 0s = 0$, as required. If $\mathrm{Int}(\mathcal{L}_{i+1})$ is simply connected, and L is a connected component of $\mathrm{Int}(\mathcal{L}_{i+1})$, $L = \langle K_1, \ldots, K_k \rangle$, then $b(K_1) = 1$ and $b(K_k) = 1$. Assume first that K_1 is not exceptional. If $d(K_1) \geq 6$ then $i(K_1) \leq 3$ and $a(K_1) \leq 2$, hence $\sigma(K_1) \geq (6 - 3) - 2 = 1$, and if $d(K_1) \leq 5$ then $i(K_1) \leq 2$ and $a(K_1) \leq 1$, hence $\sigma(K_1) \geq 4 - 1 - 2 = 1$ and similarly, if K_k is not exceptional, then $\sigma(K_k) \geq 1$.

       Hence if K_1 is not exceptional, then $\sigma(K_1) \geq 1$. $(**)$

Hence, if $D_1, \ldots, D_r$ are all the exceptional regions of L then, by parts (c) and (e) of Proposition 2.2 there are corner regions $F_1, \ldots, F_{r-1}$ between $D_1, \ldots, D_r$, consequently, by $(*)$ and

$$\sigma(L) = \sum_{j=1}^{k} \sigma(K_j) = \sigma(K_1) + \sum_{i=2}^{k-1} \sigma(K_i) + \sigma(K_k)$$
$$\geq 1 + \sum_{j=1}^{r} \sigma(D_j) + \sum_{j=1}^{r-1} \sigma(F_j) + 1$$
$$\geq 1 + (2 \cdot (r - 1) - 2 \cdot r) + 1 = 0.$$

If E_1 is exceptional and $d(E_1) \geq 6$, then $\sigma(E_1) \geq 2 - 3 = -1$ and if $d(E_1) \leq 5$ then $\sigma(E_1) \geq 1 - 2 = -1$. Similarly, if E_k is exceptional, then $\sigma(E_k) \geq -1$. Hence, if E_1 and E_k are both exceptional, then $\sigma(L) \geq -1 + 2(r-1) - 2(r-2) + (-1) \geq 0$. Finally, if E_1 is exceptional and E_k is not, then $\sigma(L) \geq -1 + 2(r - 1) - 2(r - 1) + 1 = 0$.

    The lemma is proved.   □

    Now we turn to the proof of the Main Theorem.

Proof. (i) Let $\mathcal{L}_i$ be as defined in Definition 1.2(a). Then $\mathcal{L}_i$ is a one-layer annular map by Proposition 2.2(b).

(ii) $\sigma(\mathcal{L}_{i+1}) \geq 0$ for every $i \geq 0$ by Lemma 2.3, hence it follows from $(*)$ by summation of $\sigma(K)$, where K runs over all the regions of $\mathcal{L}_{i+1}$ that

(1) $|\omega(\mathcal{L}_{i+1})| \geq |\tau(\mathcal{L}_{i+1})|$;

(2) By the definition of the layers $\mathcal{L}_i$ we have $\tau(\mathcal{L}_{i+1}) = \omega(\mathcal{L}_i)$.

From (1) and (2) we get $|\omega(\mathcal{L}_{i+1})| \geq |\omega(\mathcal{L}_i)|$. If $i \leq 0$ then the same argument applies. Since $\omega = \omega(\mathcal{L}_\ell)$ and $\tau = \tau(\mathcal{L}_{-k})$, we get the desired weak inequality.

(iii) Assume that $|\mathcal{L}_i| \geq 2$ for some i with $i \geq 0$. Then by part (i) $|\omega(\mathcal{L}_{i+1})| \geq |\omega(\mathcal{L}_i)|$ and inductively we get that $|\omega(M)| \geq |\omega(\mathcal{L}_i)| \geq 2$. If for some j with $j < 0$ we have $|\tau(\mathcal{L}_j)| \geq 2$ or if for some i with $i \geq 0$ we have $|\omega(\mathcal{L}_i)| \geq 3$ then we are done, by the same inductive argument. Consequently, we may assume without loss of generality that $|\tau(\mathcal{L}_0)| = 1$ and $|\omega(\mathcal{L}_1)| = 2$ and M consists of $\mathcal{L}_0$ and $\mathcal{L}_1$. Let $\mathcal{L}_0 = \{D\}$ and let $\mathcal{L}_1 = \{D_1, D_2\}$. (See Fig. 17(a), Fig. 17(b) and Fig. 17(c).) Let $\mu_0 = \tau(M)$, $\mu_1 = \partial D \cap \partial D_1$ and $\mu_2 = \partial D \cap \partial D_2$. Then $w\mu_0 w\nu_1 v_0 \mu_1 v_1 \mu_2 v_0 \nu_1^{-1} w$ is a boundary cycle of D. (ν_1 may be empty.) Consequently, $|\tau(M)| = |\mu_0| \geq d(D) - 4$. Hence, if $d(D) \geq 6$ then $|\tau(M)| \geq 2$ and the result follows. Assume therefore that $d(D) \leq 5$. Then one of v_0 and v_1 on Fig. 17 is an inner vertex with valency greater than or equal to four. But then $|\partial D_2 \cap \omega(A)| \geq 2$ and again $|\omega| + |\tau| \geq 4$.

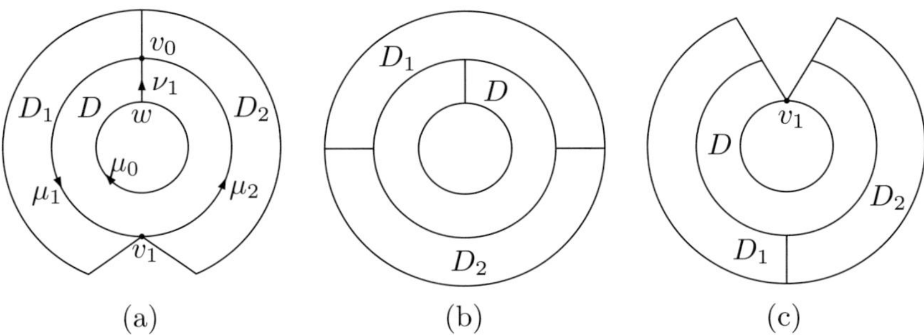

$$\textsc{Figure } 17$$

(iv) First observe that if M contains a region E in $\mathcal{L}_i$, $i \geq 0$ which is a corner region of $\mathcal{L}_i$ then it follows easily that $\mathcal{L}_\ell$ contains a corner region F of M. See Figs. 12(b) and 12(c) where Q_1 and Q_2 are 2-corner and 3-corner regions in $\mathcal{L}_i$ and Q_1' and Q_2' are "induced" corner regions in $\gamma(Q_1)$ and $\gamma(Q_2)$, respectively. But then $|\partial F \cap \omega(M)| \geq 2$. Also, if M contains an exceptional region E in $\mathcal{L}_\ell$ then by Proposition 2.2(e) $\mathcal{L}_\ell$ contains a corner region E' hence by the last observation $|\partial E' \cap \omega(M)| \geq 2$. The analogous statements are true if $i \leq 0$. Therefore

$$\text{If } E \text{ is a region of } \mathcal{L}_i \text{ then } E \text{ is neither} \atop \text{exceptional nor a corner region.} \qquad (*)$$

It follows from $(*)$ and Proposition 2.1(5) that

$$\text{If } v \in \partial \mathcal{L}_i \cap \partial \mathcal{L}_{i+1} \text{ then } d_{\mathcal{L}_i}(v),\ d_{\mathcal{L}_{i+1}}(v) \leq 3 \atop \text{and hence } d_M(v) \leq 4. \qquad (**)$$

We show that $d(E) = 4$, for every region E of M. Suppose first $d(E) = 5$. Then due to $(*)$ and the $V(6)$-condition $a(E) = 1$, $b(E) \leq 2$, hence $\gamma(E) \geq 5 - (1 + 2) = 2$ and $\partial E \cap \partial \left(\widehat{\gamma(E)} \right) = \mu_1 v \mu_2$, where μ_1 and μ_2 are pieces with $|\mu_1|, |\mu_2| \geq 1$.

Here $\widehat{\gamma(E)}$ is the subdiagram of $\mathcal{L}_{i+1}$ consisting of the regions of $\gamma(E)$. Since v is a boundary vertex of E, hence $d_M(v) \geq 4$. But $d_{\mathcal{L}_i}(v) = 2$, because v separates μ_1 and μ_2. Hence $d_{\mathcal{L}_{i+1}}(v) \geq 4$, violating $(**)$. Hence, if $d(E) \neq 4$ then $d(E) \geq 6$. Suppose $d(E) \geq 6$. If $E \in \mathcal{L}_\ell$ then $|\partial E \cap \omega(M)| \geq 6 - (\alpha(E) + \beta(E)) \geq 6 - 4 = 2$. If $E \in \mathcal{L}_\ell$, then it follows from $(*)$ and $(**)$ and the $V(6)$ condition that $\gamma(E)$

contains no region F with $d(F) \leq 5$. Hence if $d(E) \geq 6$ for some region E in some $\mathcal{L}_i$ then by repeating this argument we get a region E' in $\mathcal{L}_\ell$ with $d(E') \geq 6$. Therefore, as above $|\partial E' \cap \omega(M)| \geq 2$, a contradiction. Consequently $d(E) = 4$ for every region E of M. It follows from $(**)$ and the $V(6)$ condition that every inner vertex has valency four and every boundary vertex has valency three, as required.

The Main Theorem is proved. $\qquad\square$

3. Proofs of the main results

3.1. Preliminaries

We recall some basic notions and results from [20] concerning $\mathcal{R}$-diagrams over F, where F is either a free product, or an amalgamated free product or an HNN-extension.

3.1.1. Words in F.

Definition 3.1. Let F be a free product or an amalgamated free product. Let U and V be reduced words in F and let $W = UV$. Then UV is in *reduced form* if there is neither cancellation nor consolidation between U and V when forming the product UV; W is in *semi-reduced form* if there is no cancellation between U and V when forming the product UV. However, consolidation is allowed.

Let F be an HNN-extension, $F = \langle H, t | t^{-1} A_{-1} t = A_1 \rangle$ and let $g_0 t^{\varepsilon_1} \cdots g_{n-1} t^{\varepsilon_n} g_n$, $g_i \in H$ be an element of F. It is in *reduced form* if no subword $t^{-1} g_i t$ with $g_i \in A_{-1}$ and no subword $t g_j t^{-1}$ with $g_j \in A_1$ occurs in it.

Denote by $\|W\|$ the length of the reduced word W in F.

Definition 3.2. Let F be a free product or an amalgamated free product. Let $W = y_1 y_2 \cdots y_n$ in reduced form, where the y_i-s are letters. W is *cyclically reduced* if either $\|W\| \leq 1$ or y_1 and y_n are in different factors of F; W is *weakly cyclically reduced* if either $\|W\| \leq 1$ or $y_n \neq y_1^{-1}$. Thus there is no cancellation between y_1 and y_n although consolidation is allowed.

Let F be an HNN-extension, as above. If U and V have normal forms $U = h_0 t^{\varepsilon_1} \cdots h_n t^{\varepsilon_n} h_{n+1}$ and $V = h'_0 t^{\delta_1} \cdots h'_m t^{\delta_n} h'_{m+1}$ we say that there is cancellation in forming the product UV if either $\varepsilon_n = -1$, $h_{n+1} h'_0 \in A_{-1}$ and $\delta_1 = 1$ or if $\varepsilon_n = 1$, $h_{n+1} h'_0 \in A_1$ and $\delta_1 = -1$. We say that $W = UV$ is in *reduced form* if there is no cancellation in forming the product UV.

Definition 3.3. Let F be a free product or an amalgamated free product. A subset $\mathcal{R}$ of F is *symmetrized* if every $R \in \mathcal{R}$ is weakly cyclically reduced, and every weakly cyclically reduced conjugate of R and R^{-1} is also in $\mathcal{R}$.

Let F be an HNN-extension. A subset $\mathcal{R}$ of F is *symmetrised* if every $R \in \mathcal{R}$ is cyclically reduced and all cyclically reduced conjugates of $R^{\pm 1}$ are in $\mathcal{R}$.

Definition 3.4. A word U is called a *piece* if $\mathcal{R}$ contains distinct elements R_1 and R_2 with semireduced forms $R_1 = UC_1$ and $R_2 = UC_2$.

3.1.2. Diagrams over F. The discussion in [20] on pp. 276–278 for free products, on pp. 286–287 for amalgamated free products and on pp. 291–292 for HNN-extensions, together with the property $CN(2)$ yield the following:

Lemma 3.1. ([20]) *Let $F = *X_i$ be the free product of groups X_i or an amalgamated free product of groups or an HNN-extension $\langle H, t | t^{-1} A_{-1} t = A_1 \rangle$. Let $\mathcal{R}$ be a symmetrized subset of F which satisfies the condition $V(6)$ and let $P_1, \ldots, P_n$ be a sequence of conjugates of elements of $\mathcal{R}$ (an $\mathcal{R}$-sequence). Let $W = P_1 \cdots P_n$. Then there exists a connected and simply connected $\mathcal{R}$-diagram M, with a distinguished vertex 0 on ∂M and a labeling function $\Phi : M \to F$, such that each of the following holds:*

    (1) *There is a boundary cycle $\sigma_1 \cdots \sigma_t$ of M beginning at 0 such that $\Phi(\sigma_1) \cdots \Phi(\sigma_t)$ is a semi-reduced form of W.*

    (2) *If D is any region of M and $\alpha_1, \ldots, \alpha_s$ are the edges of a boundary cycle of D, then $\Phi(\alpha_1) \cdots \Phi(\alpha_s)$ is in semireduced form and is a weakly cyclically reduced conjugate of one of the P_i. Moreover, there are boundary cycles ω with $\Phi(\omega)$ in reduced and cyclically reduced form.*

Remark. For amalgamated free products and HNN-extensions in [20] the condition $C'(1/6)$ was assumed in place of condition $V(6)$. Condition $C'(1/6)$ was used in order to guarantee that no loops with boundary of length 1 occur in the process of the diagram construction (see [20, p. 288, l.12–18]). But this is an example of the $CN(1)$-property, proved for simply connected $W(6)$-diagrams in [14]. Hence we can carry out the construction of the diagram under the $V(6)$ condition.

We also have the analogue of Lemma 9.2 in [20].

Lemma 3.2. *Let F be a free product or an amalgamated free product or an HNN-extension and let $\mathcal{R}$ be a symmetrized subset of F which satisfies the condition $V(6)$. Then each of the following holds:*

    (1) *If M is a reduced $\mathcal{R}$-diagram, then the label on an interior edge is a piece.*

    (2) *The diagram of a minimal $\mathcal{R}$-sequence is reduced.*

Proof. (1) follows as in the free presentation case.

(2) Following the proof of Lemma 9.2(2) in [20, p. 277], we have to prove that if D is a region of M, then no two distinct edges of D may be identified. But this follows from [14], since a simply connected $W(6)$ diagram, and hence a simply connected $V(6)$ diagram, has this property by [14, Th. 2.3 p. 61].

    The lemma is proved.   □

As in [20, p. 291] we shall consider several reduced forms for the same element. We shall not allow edges with label 1, but half edges may have 1 as a a label. We have the following lemma.

Lemma 3.3.

    (a) *Let $F = \langle H, t | t^{-1} A_{-1} t = A_1 \rangle$ with corresponding isomorphism $\psi : A_{-1} \to A_1$. Let $U = g_0 t^{\varepsilon_1} g_1 \ldots t^{\varepsilon_n} g_n$ and $V = g_0' t^{\delta_1} \ldots t^{\delta_m} g_m'$, ε_i, $\delta_i \in \{1, -1\}$, $g_i \in$*

H, $g_i' \in H$ be reduced words in F_1. Then $U = V$ if and only if each of the following hold:

(i) $m = n$;

(ii) $\varepsilon_i = \delta_i$, $i = 1, \ldots, n$;

(iii) there exist elements $d_0, \ldots, d_{n-1}$ and $d_0', \ldots, d_{n-1}'$ in $A_{-1} \cup A_1$ which satisfy the following conditions:

$$d_0 = g_0^{-1} g_0', \qquad d_0' = \psi^{\varepsilon_1}(d_0), \qquad d_0 \in A_{-\varepsilon_1}, \qquad d_0' \in A_{-\varepsilon_1}$$

$$\vdots$$

$$d_i = g_i^{-1} d_{i-1}' g_i', \qquad d_i' = \psi^{\varepsilon_{i+1}}(d_i), \qquad d_i \in A_{\varepsilon_{i+1}}, \qquad d_i' \in A_{-\varepsilon_{i+1}}$$

$$\vdots$$

$$d_{n-1}' = g_n g_n'^{-1} \qquad g_i' = d_{i-1}' \qquad g_i d_i \in A_{-\varepsilon_i} g_i A_{\varepsilon_{i+1}}$$

(b) Let $F = A \underset{C}{*} B$, and let $W_1 = a_1 b_1 \ldots a_n b_n$ and $W_2 = a_1' b_1' \ldots a_m' b_m'$ reduced words in F. $W_1 = W_2$ in F if and only if $m = n$ and there exist elements $c_1, \ldots, c_{n-1}$ in C such that

$$a_1' c_1 = a_1, \qquad b_1' c_2 = c_1 b_1, \qquad b_i' c_{i+1} = c_i b_i, \qquad c_{n-1} b_n = b_n'.$$

The proofs are immediate, we omit them.

Finally, we shall need the following results for the solution of the conjugacy problem.

Recall from [8, p. 80] the notion of isodiametric function. Let $\mathcal{P}$ be a group presentation and let M be a van Kampen diagram for $\mathcal{P}$ with (reduced) boundary label W and base point v_0, where W starts. Define $Diam_{v_0}(M)$ to be the length of the longest simple path in M which starts at v_0. A function $f : \mathbb{N} \to \mathbb{N}$ is called an *isodiametric function for* $\mathcal{P}$ if for all $n \in \mathbb{N}$ and all null homotopic reduced words U of length at most n there exists a van Kampen diagram M for U with a basepoint v_0 on the boundary of M, where U starts, such that $Diam_{v_0}(M) \leq f(n)$. Isodiametric function of isomorphic groups are equivalent.

Lemma 3.4. *Let H be a group with a finite free-presentation $\langle Z | \mathcal{S} \rangle$, which has solvable word problem, with isodiametric function f. (Thus f is recursive.) Let $a_1, \ldots, a_n$, $n \geq 2$, be elements of H represented by freely reduced words $A_1, \ldots, A_n$ respectively, in the free group $F(Z)$, freely generated by Z, such that $a_1 \cdot a_2 \cdots a_n = 1$ in H. Denote by $|\cdot|$ the length of a reduced word in $F(Z)$. There are reduced words $X_1, \ldots, X_n$ in $F(Z)$ with $|X_i| \leq f\left(\Sigma_{i=1}^n |A_i|\right)$, such that if $x_1, \ldots, x_n$ are the images of $X_1, \ldots, X_n$, respectively by the natural projection on H, then $x_1 x_2^{-1} = a_1$, $x_2 x_3^{-1} = a_2, \ldots, x_n x_1^{-1} = a_n$.*

Proof. Since $a_1 \cdots a_n = 1$ in H, hence $A_1 \cdot A_2 \cdots A_n \in N$, where N is the normal closure of $\mathcal{S}$ in $F(Z)$. Since $\langle Z | \mathcal{S} \rangle$ has solvable word problem, it follows from [8, Th. 2.1] that it has recursive isodiametric function f. Consequently, there exists a reduced van Kampen diagram M with labeling function Φ, a (not-necessarily reduced) boundary cycle $\alpha_1 v_1 \alpha_2 v_2 \cdots \alpha_n v_n$, v_i vertices, α_i boundary paths with

$\Phi(\alpha_i) = A_i$, such that if v is any vertex of M and ξ_i is a shortest path in M with initial vertex v_i and terminal vertex v, then $|\xi_i| \le f_0\left(\Sigma_{i=1}^n |A_i|\right)$ (see Fig. 18). Now, $\xi_1\xi_2^{-1}\alpha_1^{-1}$ is a closed curve in M, hence $\Phi(\xi_1\xi_2^{-1}\alpha_2^{-1}) \in N$ and similarly, $\Phi(\xi_i\xi_{i+1}^{-1}\alpha_{i+1}^{-1}) \in N$ for $i = 2,\dots,k$. (We count $\mod n$. Thus $n+1 = 1$.) Define $X_i = \Phi(\xi_i)$, $i = 1,\dots,n$. Then $X_iX_{i+1}^{-1}A_i^{-1} \in N$, hence $x_ix_{i+1}^{-1} = a_i$ in H, as required.

The lemma is proved. $\square$

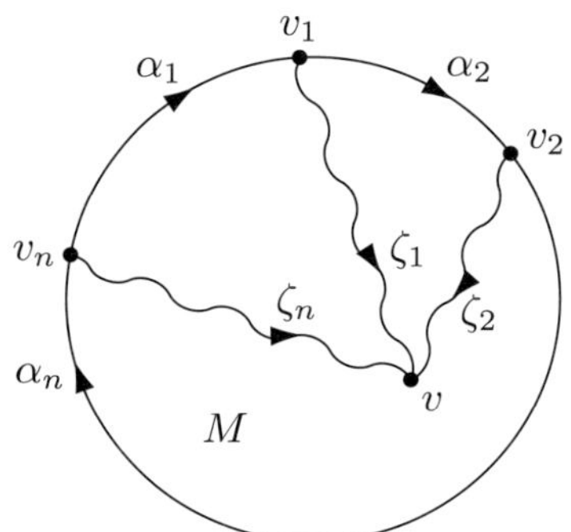

FIGURE 18

3.2. The Proof of Theorem 0.1

We start the proof of Theorem 0.1 with the word problem.

Let $\mathcal{P}_1 = \langle X_1 | \mathcal{R}_1 \rangle$ and let $\mathcal{P}_2 = \langle X_2 | \mathcal{R}_2 \rangle$ be free finite presentations of A and B respectively. Then $\mathcal{P}_0 = \langle X_1 \cup X_2 | \mathcal{R}_1 \cup \mathcal{R}_2 \rangle$ is a free presentation of $A * B$. Let $\pi_1 : F(X_1) \to A$ and $\pi_2 : F(X_2) \to B$ be the natural projections, and let $\pi_0 : F(X_1 \cup X_2) \to A * B$ be the natural projection on $A * B$. Let $\pi : A * B \to G$ be the natural projection and let N be the normal closure $\mathcal{R}$ in F. Since A and B have solvable word problems, it follows from [8, Th. 2.1] that $\mathcal{P}_1$ and $\mathcal{P}_2$, and hence also $\mathcal{P}_0$, have recursive isoperimetric functions f_1, f_2 and f_0, respectively.

Let $R \in \mathcal{R}$ be given by $a_1b_1a_2b_2 \cdots a_nb_n$, $1 \ne a_i \in A$, $1 \ne b_i \in B, n \ge 2$. Fix freely reduced words $A_1,\dots,A_n$ in $F(X_1)$ with $\pi_1(A_i) = a_i$ and fix freely reduced words $B_1,\dots,B_n$ in $F(X_2)$, with $\pi_2(B_i) = b_i$, $i = 1,\dots,n$. Define $\bar{R} = A_1B_1 \cdots A_nB_n$. Then $\bar{R}$ is cyclically reduced in $F(X_1 \cup X_2)$. Let $\bar{\mathcal{R}} = \{\bar{R} \mid R \in \mathcal{R}\}$. Define $\mathrm{Supp}(\bar{R}) = \cup\{A_i\}\bigcup\cup\{B_i\}$, $i = 1,\dots,n$. Now let $R' \in \mathcal{R}$ be another element with $\bar{R}' = A_1'B_1'A_2'B_2' \cdots A_m'B_m'$. We follow the convention that if $\pi_1(A_i) = \pi_1(A_j')$ for some i and j, then we choose $A_j' = A_i$. Denote $\mathrm{Supp}(\bar{\mathcal{R}}) = \cup_{R \in \mathcal{R}}\mathrm{Supp}(\bar{R})$. Thus $\mathrm{Supp}(\bar{\mathcal{R}})$ is a finite set of freely reduced words in $F(X_1 \cup X_2)$. Denote by N_1 the normal closure of $\mathcal{R}_1$ in $F(X_1)$ and denote by N_2 the normal closure of $\mathcal{R}_2$ in $F(X_2)$. Finally, denote by N_0 the normal closure of $\mathcal{R}_1 \cup \mathcal{R}_2 \cup \bar{\mathcal{R}}$ in $F(X_1 \cup X_2)$.

Let $\mathcal{P} = \langle X_1 \cup X_2 | \mathcal{R}_1 \cup \mathcal{R}_2 \cup \bar{\mathcal{R}} \rangle$. We propose to show that $\mathcal{P}$ has a recursive isoperimetric function f, depending on f_1, f_2 and f_0. To this end, for every cyclically reduced word W_0 in N_0 we construct a van Kampen diagram M_0 over $F(X_1 \cup X_2)$ with boundary label W_0, and show that the number of regions in M_0 is bounded by $f(|\partial M_0|)$, where f is a function $f : \mathbb{N} \to \mathbb{N}$ composed from f_1, f_2 and f_0 in such a way that if f_0, f_1 and f_2 are recursive, then f is recursive as well.

Let W_0 be a cyclically reduced word in N_0 and let $w = \pi_0(W_0)$. We may assume that $\|w\| \geq 2$. Without loss of generality, we may assume that w is cyclically reduced, $w = h_1 k_1 \ldots h_g k_g, h_i \in A, k_i \in B$ and $W_0 = H_1 K_1 \ldots H_g K_g$. (If not, then take an appropriate cyclic conjugate.) Therefore, by Lemma 3.2, there exists a connected simply connected $\mathcal{R}$-diagram M with w as a boundary label. We construct a diagram M_0 out of M, having W_0 as a boundary label as follows: Replace the boundary labels $R \in \mathcal{R}$ on the regions of M by the corresponding $\bar{R} \in \bar{\mathcal{R}}$, and replace every secondary vertex v of M (see [20]) together with the half-edges emanating from v by a connected and simply connected diagram M_v over $F(X_1)$ or over $F(X_2)$, depending on whether the labels of the half edges emanating from v are in A or are in B . More precisely, let v be a secondary vertex of M. Then either all the half edges $\xi_1, \ldots, \xi_n$ emanating from v are labeled by elements of $F(X_1)$, or all the half edges are labeled by elements of $F(X_2)$ (see [20]). Suppose the first, and let $D_1, \ldots, D_n$ be the regions of M which contain v on their boundary. Let $d(v) = n$ and let $\mu_1, \ldots, \mu_n$ be the edges of $D_1, \ldots, D_n$ respectively, which contain v, and let $u_1, \ldots, u_n$ be their label in A, respectively, occurring in this order, anticlockwise. Let $U_1, \ldots, U_n$ be the words in $\mathrm{Supp}(\mathcal{R})$ with $\pi_1(U_i) = u_i, i = 1, \ldots, n$. We claim that $U_1 \cdots U_n \in N_1$. We have $\mu_1 = \xi_1^{-1}\xi_2, \mu_2 = \xi_2^{-1}\xi_3, \ldots, \mu_n = \xi_n^{-1}\xi_1$. Consequently, $u_1 \cdots u_n = \Phi(\mu_1 \cdots \mu_n) = \Phi(\xi_1^{-1}\xi_2 \cdot \xi_2^{-1}\xi_3 \cdot \xi_n^{-1}\xi_1) = \Phi(v) = 1$, i.e., $u_1 \cdots u_n = 1$ in A. Therefore $U_1 \cdots U_n \in N_1$ as claimed. Hence, there exists a reduced simply connected and connected $\mathcal{R}_1$-diagram M_v over $F(X_1)$ (with not necessarily reduced boundary) having boundary label $U_1 \cdots U_n$. Proceed similarly with the vertices having emanating edges with labels in $F(X_2)$. Denote $F_0 = F(X_1 \cup X_2)$ and let $\mathcal{R}_0 = \mathcal{R}_1 \cup \mathcal{R}_2 \cup \bar{\mathcal{R}}$. Then we get a connected and simply connected $\mathcal{R}_0$-diagram M_0 over F_0 with boundary label W_0. We estimate $V(M_0)$ – the number of regions in M_0. Denote by $\|w\|$ the free product length of $w \in A * B$ and denote the length of W in F_0 by $|W|_{F_0}$. First of all, we have

$$V(M_0) = V(M) + \Sigma_v V(M_v), \tag{1}$$

where v runs over the vertices of M.

Now, by [14, Th. 4, p. 76],

$$V(M) \leq c\|w\|^2 \leq c|W_0|^2_{F_0}, \text{ where } c \leq \tfrac{1}{6}, \text{ i.e.,}$$

$$V(M) \leq c|W_0|^2_{F_0}. \tag{2}$$

If $d_M(v) = d$ and $L = \max\left\{|U|_{F_0} \mid U \in \mathrm{Supp}(\mathcal{R})\right\}$, then $|\partial M_v|_{F_0} \leq dL$, hence

$$V(M_v) \leq f_0(dL) \tag{3}$$

(we assumed that f_0 is monotone).

Thus, in order to estimate $V(M_v)$, we have to estimate d. Let $r = \max\{\|R\|\}$, $R \in \mathcal{R}$. Now, there are d edges emanating from v, each having length one in F_0. Consequently,

$$1 \cdot d(v) \leq \sum_{D \in M} \|\Phi(\partial D)\| \leq V(M) \cdot r \overset{\text{by (2)}}{\leq} c|W_0|^2_{F_0} \cdot r,$$

hence by (3),

$$V(M_v) \leq f_0\left(cr|W_0|^2_{F_0}L\right) \tag{4}$$

Now, the number of vertices in M less than or equal to $2rV(M) \overset{\text{by (2)}}{\leq} 2rc|W_0|^2_{F_0}$, i.e.,

$$\text{the number of vertices in } M \text{ is less than or equal to } 2rc|W_0|^2_{F_0}. \tag{5}$$

Combining (1) and (2) with (4) and (5), we get

$$V(M_0) = V(M) + \sum_v (M_v) \leq c|W_0|^2_{F_0} + 2rc|W_0|^2_{F_0} \cdot f_0(cr|W_0|^2_1 L).$$

Define $f(x) = cx^2 + 2rcx^2 f_0(rcx^2 L) = cx^2(1 + 2rf_0(rcx^2 L))$. Then $f(x)$ is an isoperimetric function for M_0 and is recursive. Consequently, G has solvable word problem, by [8, Th. 2.1].

We now turn to the solution of the conjugacy problem. First we prove below two results concerning annular diagrams which provide a bound on the length of their paths and on the valency of vertices in terms of the length of their boundaries.

Lemma 3.5. *Let $F = A * B$ and let M be an annular $\mathcal{R}$-diagram over F with connected interior. Let notation be as above. Let δ be a positive integer and assume that $d_M(v) \leq \delta$ for every vertex v of M. Let M_1 be the diagram constructed above from M. Let μ be a path in M with label $u_1 \cdots u_n$ such that $u_i = \pi_1(U_i)$ if $u_i \in A$ and $u_i = \pi_2(U_i)$ if $u_i \in B$, for suitable $U_i \in \operatorname{Supp}(\mathcal{R})$. Denote by $|\mu|_{F_0}$ the length of $U_1 \cdots U_n$ in $F_0 := F(X_1 \cup X_2)$. Let f_0 be an isodiametric function for F over F_0.*

(i) *If μ is a common boundary of two regions in M, then $|\mu|_{F_0} \leq rL + 2f_0(\delta L)$.*
(ii) *If ω is a simple closed curve in M with $\|\omega\| = k$, then $|\omega|_{F_0} \leq k(rL + 2f_0(\delta L))$.*

Proof. Part (ii) is an immediate consequence of (i), hence we show (i). Let v_1 and v_2 be the endpoints of μ. Then we may decompose μ into $\mu = v_1 \xi_1 w_1 \hat\mu w_2 \xi_2 v_2$, where w_1 and w_2 are primary vertices, v_1 and v_2 may be primary or secondary vertices, and ξ_1 and ξ_2 are half edges or are empty, according as v_1 and v_2 are secondary or primary vertices. We have

$$|\mu|_{F_0} = |\xi_1|_{F_0} + |\hat\mu|_{F_0} + |\xi_2|_{F_0}. \tag{1}$$

Since every half edge emanates from a vertex of M and every vertex has valency $\leq \delta$, it follows from Lemma 3.4 that we may choose the ξ_i such that

$$|\xi_1|_{F_0}, |\xi_2|_{F_0} \leq f_0(\delta L), \tag{2}$$

where f_0 is an isodiametric function for $\mathcal{P}$ over F_0.

Clearly,

$$|\hat{\mu}|_{F_0} \leq rL \qquad (3)$$

Hence (1), (2) and (3) yield $|\mu|_{F_0} \leq rL + 2f_0(\delta L)$.

The lemma is proved. $\qquad\square$

Lemma 3.6. *Let M be an annular $V(6)$-map with boundary components ω and τ and assume $|\omega| + |\tau| \geq 5$. Let v be any vertex of M and let $d = d_M(v)$. Then $d \leq |\omega| + |\tau| + 3$.*

Proof. We prove the lemma by induction on n – the number of layers of M. We prove the cases $n = 1$ and $n = 2$ separately. Thus assume $n = 1$ and assume $v \in \omega$. Let $D_1, \ldots, D_q$ be the regions of M which contain v on their boundary. Then $q \in \{d-1, d-2\}$ and the D_is contribute at least one to $|\omega|$ and at least $2(d-3)$ to $|\tau|$, due to the $C(4)$ condition and Proposition 2.2. Thus $1 + 2(d - 3) \leq |\omega| + |\tau|$, hence $d \leq \frac{1}{2}(|\omega| + |\tau| + 5) \leq |\omega| + |\tau| + 3$. Similarly, if $v \in \tau$ then $d \leq |\omega| + |\tau| + 3$.

Suppose now that M has 2 layers $\mathcal{L}_1$ and $\mathcal{L}_2$ and let v be a vertex on $\omega = \omega(\mathcal{L}_1)$. Let μ be the common boundary of $\mathcal{L}_1$ and $\mathcal{L}_2$. Then, as above, the regions containing v contribute at least one to $|\omega|$ and at least $2(d - 3)$ to $|\mu|$. Let $\mu_i := \partial D_i \cap \mu$. Due to Proposition 2.2, $|\mu_i| \geq d(D) - 2 \geq 4 - 2$. (The last inequality is due to the $C(4)$ condition.) Hence, it follows from the $V(6)$ condition that $\mathcal{L}_2$ contains at least $\Sigma|\mu_i| - 2$ regions the boundary of which intersect $\bigcup_{i=2}^{r-1} \mu_i$ non-trivially. The contribution of these regions to $|\tau|$ is at least $2(d - 4)$. Therefore, $|\omega| + |\tau| \geq 2(d - 4) + 1$, hence $\frac{1}{2}(|\omega| + |\tau|) \geq d - \frac{7}{2}$, i.e., $d \leq \frac{1}{2}(|\omega| + (|\tau|) + \frac{7}{2} \leq |\omega| + |\tau| + 3$. The same argument shows that if $v \in \tau$ then $d(v) \leq |\omega| + |\tau| + 3$. Assume now that $v \in \mu$. If $\mathcal{L}_1$ has x regions which contain v on their boundary and $\mathcal{L}_2$ has y regions which contain v on their boundary then $x + y \geq d - 2$. Since $d(D) \geq 4$, for every region D of M, we get $|\omega| + |\tau| \geq 2(x + y) \geq 2d - 4$, hence $d \leq \frac{1}{2}(|\omega| + |\tau|) + 2 \leq |\omega| + |\tau| + 1$. Finally, assume M has at least three layers and follow the notation of the Main Theorem. Let $M_0 = M \backslash \mathcal{L}_k$ and let $M_1 = M \backslash \mathcal{L}_{-\ell}$. Then every inner vertex v of M is either an inner vertex of M_0 or an inner vertex of M_1 with $d_M(v) = d_{M_0}(v)$ or $d_M(v) = d_{M_1}(v)$, respectively and similarly for every boundary vertex v $d_M(v) = d_{M_0}(v)$ or $d_M(v) = d_{M_1}(v)$ respectively. Therefore, by the induction hypothesis either $d \leq |\omega(M_0)| + |\tau(M_0)| + 3 \leq |\omega| + |\tau| + 3$, where the last inequality follows from the Main Theorem, or $d \leq |\omega(M_1)| + |\tau(M_1)| + 3 \leq |\omega| + |\tau| + 3$.

The lemma is proved. $\qquad\square$

We are now ready to solve the conjugacy problem. Let u and w be cyclically reduced words in F. Suppose that u and w are conjugate in G. Then, there exists an annular $\mathcal{R}$-diagram M, which has u and w as boundary labels. Let U and W be freely reduced cyclically reduced words in F_0, which contain no subwords Z, with $\pi(Z) = 1$ in G, and let $\pi(U) = u$, $\pi(W) = w$. Since U and W contain no non-empty subwords which are one in G, we may assume that M has connected interior. By the Main Theorem M contains simple closed curves $\omega_1, \ldots, \omega_s$ and

$\nu_1, \ldots, \nu_t$ homotopic to the boundary of M with $u = \Phi(\omega_1)$ and $w = \Phi(\nu_1)$ such that for some words p_i and q_j which are labels of common edges of regions in M, the following hold

(i) $\|\omega_i\| \leq r\|u\|$, $i = 1, \ldots, s$ and $p_i \Phi(\omega_i) p_i^{-1} \Phi(\omega_{i-1}) \in N$;
(ii) $\|\nu_j\| \leq r\|w\|$, $j = 1, \ldots, t$ and $q_j \Phi(\nu_j) q_j^{-1} \Phi(\nu_{j-1}) \in N$;
(iii) there exists a word p in $A * B$, which is a piece in M such that
$p\Phi(\nu_t) p^{-1} \Phi(\omega_s)^{-1} \in N$.

Let $p_i = \pi(P_i)$ and let $p = \pi(P)$ where P_i and P are reduced words in F_0, which are products of words from $\mathrm{Supp}(\mathcal{R})$. Construct the diagram M_1 as above. By Lemma 3.5 $|U|_{F_0} \leq k_1(rL + 2f_0(\delta L))$, where $k_1 = \|u\|$, and by Lemma 3.6

$$\delta \leq \|u\| + \|w\| + 3. \tag{1}$$

Hence

$$|\omega_i|_{F_0} \leq k_1 r(rL + 2f_0(\delta L)), \quad i = 1, \ldots, s. \tag{2}$$

By the same argument,

$$|\nu_j|_{F_0} \leq k_2 r(rL + 2f_0(\delta L)), \quad j = 1, \ldots, t, \tag{3}$$

where $k_2 = \|w\|$.
Also,

$$|P| \leq rL + 2f_0(\delta L). \tag{4}$$

Therefore, following [20, p. 265], we proceed as follows: let $k = (k_1 + k_2)r(rL + 2f_0(\delta L))$; let $\mathcal{U} = \{Z \in F(X_1 \cup X_2) \mid |Z|_{F_0} \leq k\}$; let $\mathcal{Q} = \{Z \in F \mid |Z|_{F_0} \leq rL + 2f_0(\delta L)\}$. Clearly, by (2) and (3) $U, V \in \mathcal{U}$ and by (4) $P \in \mathcal{Q}$. Since G has solvable word problem, we may list all the words in $\mathcal{U}$ which are conjugate to U in G by a word from $\mathcal{Q}$. Denote this set by $\mathcal{U}_0$. Now list all the words in $\mathcal{U}$ the images of which are conjugate in G to the images of words in $\mathcal{U}_0$ by the images of a words from $\mathcal{Q}$. Denote this set by $\mathcal{U}_1$. Define similarly $\mathcal{Q}_i$ and $\mathcal{U}_i$ for $i \geq 2$. Since $\mathcal{U}$ and $\mathcal{Q}$ are finite, each $\mathcal{U}_i$ is finite, and $\widehat{\mathcal{U}} := \cup \mathcal{U}_i$ is finite and $\widehat{\mathcal{Q}} = \cup \mathcal{Q}_i$ is finite. If $\mathcal{U}_{i+1} = \mathcal{U}_i$ then the procedure stops, because then $\mathcal{U}_{i+2} = \mathcal{U}_i$, hence there are no more elements in $\mathcal{U}$ which are conjugate to U by a product of elements from $\mathcal{Q}$. If $\mathcal{U}_{i+1} \neq \mathcal{U}_i$ then $|\mathcal{U}_{i+1}| > |\mathcal{U}_i|$ and since $\mathcal{U}_j \subseteq \mathcal{U}$ for every $j \geq 0$ and $\mathcal{U}$ is finite, there is at most a finite number $(|\mathcal{U}|)$ of $\mathcal{U}_i$ with $|\mathcal{U}_{i+1}| > |\mathcal{U}_i|$. Therefore, after $|\mathcal{U}|$ steps at most, we must get $\mathcal{U}_{i+1} = \mathcal{U}_i$. Now u is conjugate to w if and only if $\widehat{\mathcal{U}} \cap \widehat{\mathcal{Q}} \neq \emptyset$. Since the word problem in G is solvable, this is decidable.

The theorem is proved. $\qquad\square$

3.3. The Proofs of Theorems 0.2 and 0.3

Since the proofs of Theorems 0.2 and 0.3 are much the same, we shall prove the word problem for HNN-extensions and the conjugacy problem for amalgamated free products.

3.3.1. The word problem for HNN-extensions. As in the proof of Theorem 0.1 we show that G has a finite free presentation $\mathcal{P} = \langle X_0|\mathcal{R}_1\rangle$ with a recursive isoperimetric function. We propose to construct such a free presentation. Write A_{-1} for A and write A_1 for B.

Let $\langle X_H|\mathcal{R}_H\rangle$ and $\langle t|-\rangle$ be free finite presentations of H and $\langle t\rangle$ respectively, and let $\psi : A_{-1} \to A_1$ be the isomorphism which identifies $t^{-1}A_{-1}t$ with A_1. Let $C = A_1 \cup A_{-1}$. Then F has a free presentation $\mathcal{P}_0 = \langle X_0|\mathcal{R}_0\rangle$ where $X_0 = X_H \dot\cup \{t\}$ and $\mathcal{R}_0 = \mathcal{R}_H \cup \{t^{-1}at\psi(a)^{-1}|a \in A_{-1}\}$. Let $\mathrm{Supp}(\mathcal{R}_0)$ and $\mathrm{Supp}_M(\mathcal{R}_0)$, where M is an $\mathcal{R}$-diagram, be defined as in the proof Theorem 0.1 with t added and for $R \in \mathcal{R}$ denote by $\bar{R}$ the preimage of R in $F(X_0)$, as a product of elements of $\mathrm{Supp}(\mathcal{R})$, as in the proof of Theorem 0.1. Then G has a finite free presentation $\mathcal{P} = \langle X_0|\mathcal{R}_1\rangle$ where $\mathcal{R}_1 = \mathcal{R}_0 \cup \bar{\mathcal{R}}$. Let $g_{\mathcal{R}_0}$ be an isoperimetric function for $\mathcal{P}_0$ and let $g_{\mathcal{R}}$ be an isoperimetric function for $\mathcal{P}$. Finally, let f be an isoperimetric function for the HNN-presentation of G by $\langle F | \mathcal{R}\rangle$. Since due to [14] a simply connected $V(6)$-diagram has quadratic isoperimetric function, hence

$$f \quad \text{is at most quadratic.} \qquad (*)$$

In what follows we construct the diagrams over the presentation $\mathcal{P}$ as composed of diagrams over $\mathcal{R}_0$ and diagrams over $\mathcal{R}$. If u is a cyclically reduced word in F with image 1 in G, then by Lemma 3.1 there is a connected, simply connected $\mathcal{R}$-diagram $M_{\mathcal{R}}$ over F with a boundary label u. Let Φ be the labeling function of $M_{\mathcal{R}}$. Starting with $M_{\mathcal{R}}$, we construct an $\mathcal{R}_1$-diagram $M_{\mathcal{R}_1}$ over $F(X_0)$, which is an $\mathcal{R}_1$-diagram for a preimage U_0 of u, in $\langle X_0|-\rangle$. Let $R_i \in \mathcal{R}$ and let $z_1^{(i)} \cdots z_\ell^{(i)}$ be the normal form for R_i in F relative to a fixed set of coset representatives. As explained in [20], the edges of $M_{\mathcal{R}}$ are labeled by reduced words in F, rather than normal forms. As a result, if μ is a common edge of two adjacent regions D_1 and D_2 and μ is labeled by W_1 as a subword of $\Phi(\partial D_1)$ and μ is labeled by W_2 as a subword of $\Phi(\partial D_2)$, then $W_1 \overset{=}{_F} W_2$, but W_1 and W_2 may be different reduced words. The relation between W_1 and W_2 is given in Lemma 3.3. Accordingly, in $M_{\mathcal{R}_0}$ we have to replace the path μ by the $\mathcal{R}_0$-diagram L'_μ which has a boundary cycle $\mu_1\mu_2^{-1}$, where $\Phi(\mu_1) = W_1$ and $\Phi(\mu_2) = W_2$ and L'_μ is either filled in with $\mathcal{R}_0$-diagrams M_i with boundary label $c_ig_ic'_ig'_i$, where $c_i, c'_i \in C$, $g_i, g'_i \in H$, or by single regions with boundary labels $t^{-1}c_itc'_i$, $c_i, c'_i \in C$. In order to estimate the number of regions in L'_μ we have to know $|W_1|_{F_0}$ and $|W_2|_{F_0}$. Again, since we are using reduced words and not a normal form we cannot know W_1 and W_2. To resolve this difficulty, we introduce an "imaginary region" D'_i, inside each region D_i, which has it boundary label in cyclically reduced normal form $z_1^{(i)} \cdots z_{\ell_i}^{(i)}$ and connect the endpoints of the edges of length one with those on ∂D.

If D is a region in $M_{\mathcal{R}}$, with $\Phi(\partial D)$ a cyclic conjugate of $R_i \in \mathcal{R}$, then due to Lemma 3.3, D gives rise to an annular $\mathcal{R}_0$-diagram $\mathcal{A}(D)$ with outer boundary ∂D and inner boundary $v_0\omega_D v_0$, where $\omega_D = \omega_1 v_1 \omega_2 v_2 \cdots \omega_{\ell-1}$, with $\Phi(\omega_D) = z_1^{(i)} \cdots z_{\ell_i}^{(i)}$ and for $i = 0,\ldots,\ell_i - 1$, the vertex v_i is connected to an appropriate vertex w_i on ∂D by a path γ_i with $\Phi(\gamma_i) \in C$ such that $w_0\sigma_1 w_1\sigma_2 \ldots \sigma_{\ell-1}w_0$ is a

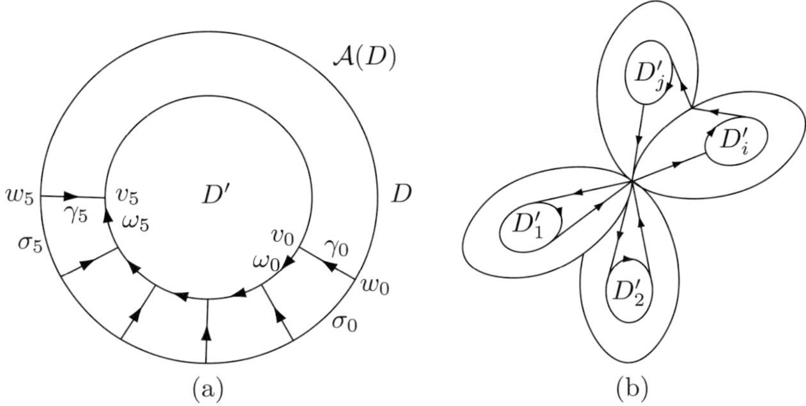

FIGURE 19

boundary cycle of D and $\Phi(\sigma_i) \in H \cup \{t, t^{-1}\}$ such that $\Phi(\omega_D)$ is in normal form (see Fig. 19(a)).

Suppose $M_{\mathcal{R}}$ has s regions: $D_1, \ldots, D_s$. Carrying out the above describe construction for all the regions of $M_{\mathcal{R}}$, we get an $\mathcal{R}_0$-diagram $M_{\mathcal{R}_0}$ with outer boundary $\partial M_{\mathcal{R}}$ and $s(:= |M_{\mathcal{R}}|)$ holes $D'_1, D'_2, \ldots, D'_s$, where each D'_i has a boundary label $(z_1^{(i)} \cdots z_2^{(i)} \ell)^{\pm 1}$ $i = 1, \ldots, s$. (See Fig. 19(b)).

We shall estimate the number of regions in $M_{\mathcal{R}_0}$ in two steps by estimating separately the contributions of the edges and vertices of $M_{\mathcal{R}}$ to the number of regions of $M_{\mathcal{R}_0}$. Let $R_0 \in \mathcal{R}$ with $|\bar{R}_0|_{F_0}$ maximal.

Call a path in $M_{\mathcal{R}}$ *integral*, if its endpoints are primary vertices. Every path μ in $M_{\mathcal{R}}$ has a decomposition into $\mu = v_0 \mu_0 q_1 \mu_1 q_2 \mu_2 v_1$, where q_1 and q_2 are primary vertices and either μ_0 (μ_2) is empty or v_0 (v_1) is a secondary vertex. Denote $[\mu] = \mu_1$. We call $[\mu]$ *the integral part of* μ. Clearly $[\mu]$ is well defined.

Let D_1 and D_2 be two regions in $M_{\mathcal{R}}$ with a common boundary path π. Then ∂D_1 has a subpath μ containing primary vertices w_i and w_j such that $\mu = \mu_0 w_i [\mu] w_j \mu_1$ and μ_0 and μ_1 are either empty or are half edges and ∂D_2 has a subpath ν containing primary vertices w_k and w_ℓ such that $\nu = \nu_0 w_k [\nu] w_\ell \nu_1$ and ν_0 and ν_1 are either empty or are half edges with $\Phi(\mu) \underset{F}{=} \Phi(\nu)$, such that $o(\mu) = o(\nu) = w_0$ and $t(\mu) = t(\nu) = w_1$, and such that $w_0 \mu w_1 \nu^{-1}$ is a boundary cycle of an $\mathcal{R}_0$-diagram L'_π (see Fig. 20). Also, $M_{\mathcal{R}_0}$ contains paths ζ_1 and ζ_2 with $\Phi(\zeta_1)$, $\Phi(\zeta_2) \in C$ $(C = A \cup B)$, which connect w_i to w_k and w_j to w_ℓ, respectively.

By the construction of the annular diagrams $\mathcal{A}(D_1)$ and $\mathcal{A}(D_2)$, $\partial D'_1$ contains vertices v_i and v_j and $\mathcal{A}(D_1)$ contains paths γ_i and γ_j connecting v_i to w_i and v_j to w_j respectively, with $\Phi(\gamma_i)$, $\Phi(\gamma_j) \in C$. Similarly, $\partial D'_2$ contains vertices u_k and u_ℓ and $\mathcal{A}(D_2)$ contains paths γ_k and γ_ℓ with $\Phi(\gamma_k)$, $\Phi(\gamma_\ell) \in C$ which connect u_k with w_k and u_ℓ with w_ℓ. Therefore, if μ' is the boundary path of D'_1 with $o(\mu') = v_i$ and $t(\mu') = v_j$ and similarly, if ν' is the boundary path

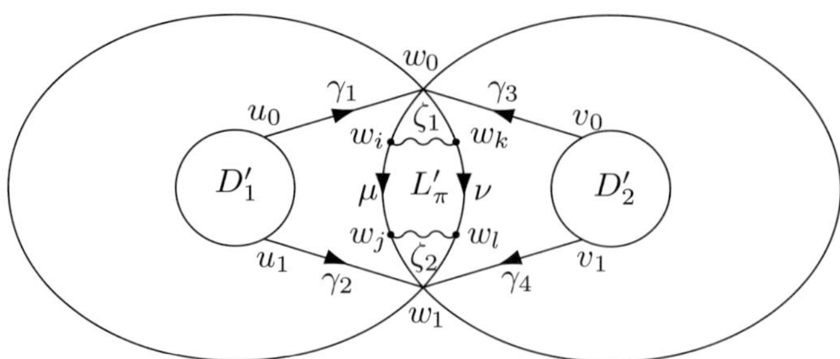

FIGURE 20

of D_2' with $o(\nu') = u_k$ and $t(\nu') = u_\ell$, then the regions in $M_{\mathcal{R}_0}$ added to $M_{\mathcal{R}}$ due to the integral part of the common piece π, are the regions of the diagram L_π with boundary cycle $\mu'\gamma_i\gamma_{i,k}\gamma_k^{-1}\nu'\gamma_\ell\zeta_{\ell,j}^{-1}\gamma_j^{-1}$, where $\gamma_{i,k}$ is the boundary path of D_1' with endpoints v_1 and v_k and similarly $\zeta_{k,j}$ is the boundary path of D_2' with endpoints u_j and u_ℓ. Consequently, the contribution $E(\pi)$ of $[\pi]$ to $|M_{\mathcal{R}_0}|$ is estimated by $E(\pi) := |L_\pi| \le g_{\mathcal{R}_0}\big(\Phi(\mu'\gamma_i\zeta_{i,k}\gamma_k^{-1}\nu'\gamma_\ell\zeta_{\ell,j}^{-1}\gamma_j^{-1})\big)$. For an integral path θ in M with $\Phi(\theta) = q_1 \cdots q_p \in F$ in reduced form define $|\theta|_{F_0} := |Q_1 \ldots Q_p|$, where $Q_i \in \mathrm{Supp}(R)$ with $\varphi(Q_i) = q_i$, $i = 1, \ldots, p$, φ the natural projection of $F(X_0)$ on F. Then, $|\mu'|, |\nu'| < |R_0|$, where R_0 is a longest relator hence if $\Phi(\gamma_\alpha) = c_\alpha$, $\alpha \in \{i, j, k, \ell\}$ and if c_π is the length of the longest label of a connecting path $\gamma_i\zeta_{i,k}\gamma_k^{-1}$ of the boundary of D_1' with the boundary of D_2' then $E(\pi) \le g_{\mathcal{R}_0}(2|\bar{R}_0| + c_\pi)$. Thus

$$E(\pi) \le g_{\mathcal{R}_0}(2|\bar{R}_0| + c_\pi) \ , \ \textit{for every piece} \ \ \pi \ \ \textit{of} \ M_{\mathcal{R}}. \tag{1}$$

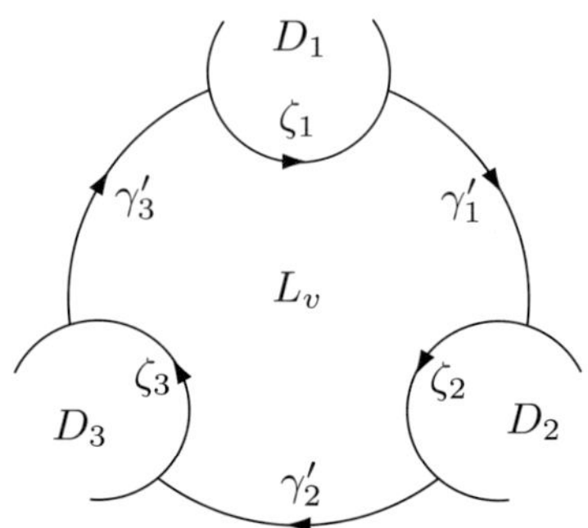

FIGURE 21

Now consider the contribution of the vertices. If v is a primary vertex then no new regions are added. So let v be a secondary vertex of $M_{\mathcal{R}}$ with valency d (see Fig. 21). Let v be on the common boundary of $D_1, \ldots, D_d$. Then, by the construction of $\mathcal{A}(D_i)$, v gives rise to an $\mathcal{R}_0$-diagram L_v with boundary cycle $\zeta_1\gamma_1'\gamma_2\zeta_2\gamma_2'\cdots\zeta_d\gamma_d'$, where $\phi(\zeta_i) \in \{z_j^{(i)}\} \cup \{r_j\}$, $i = 1, \ldots, d$, $u = r_1 \ldots r_q$ and

$\Phi(\gamma_i), \Phi(\gamma_i') \in C$. Hence, if $z_v = \max\limits_{j} \{|\bar{z}_j^{(i)}|, |\bar{r}_j|\}$ and $e_v = \max\limits_{i} \{|\Phi(\gamma_i)|, |\Phi(\gamma_i')|\}$, then the contribution $E(v)$ of v to the number of regions in $M_{\mathcal{R}_0}$ added to $M_{\mathcal{R}}$ is estimated by

$$E(v) \le g_{\mathcal{R}_0}\big(d(z_v + 2e_v)\big)\,,\ \textit{for every vertex}\ \ v\ \ \textit{with valency}\ \ d. \quad (2)$$

We are now ready to estimate $|M_{\mathcal{R}_0}|$. Let notation be as above. If $M_{\mathcal{R}}$ contains a edges (pieces + boundary edges) and b vertices such that the maximal valency of these vertices is d_0, then $|M_{\mathcal{R}_0}| \le a g_{\mathcal{R}_0}(2|\bar{R}_0| + e_\ell) + b g_{\mathcal{R}_0}(d_0(z + 2e_p))$, where $e_\ell = \max c_\pi$, π runs over the edges of $M_{\mathcal{R}}$, $e_p = \max e_v$ and $z = \max z_v$, where v runs over all the vertices of $M_{\mathcal{R}}$. Let $e = \max\{e_\ell, e_p\}$. Clearly $z \le |\bar{R}_0| + |U_0|$, hence if we define $\xi = \max\{2|\bar{R}_0|, |\bar{R}_0| + |U_0|\}$ then $|M_{\mathcal{R}_0}| \le a g_{\mathcal{R}_0}(\xi + e) + b g_{\mathcal{R}_0}(d_0(\xi + e))$. We estimate e, d_0, a and b in terms of $|\bar{R}_0|$ and $|U_0|$ via recursive functions. It follows easily from Lemma 3.6 that $d_0 \le \|u\|$, hence $d_0 \le \xi$. Also $a \le |\bar{R}_0| f(\|u\|)$ and $b \le |R_0| f(\|u\|)$. So it remains to estimate e.

Let $a_1, a_2 \in H$ and suppose that for some $c_1 \in A_i, c_2 \in A_j, i, j \in \{1, -1\}$ we have $c_1 a_1 c_2 = a_2$. Due to condition (iv) c_1 and c_2 are uniquely defined by a_1 and a_2: if $c_1' a_1 c_2' = a_2, c_1' \in A_i, c_2' \in A_j$, then $c_1' c_1^{-1} a_2 c_2^{-1} c_2' = a_2$, hence

$$a_2^{-1}(c_1' c_1^{-1})a_2 = c_2'^{-1} c_2 \quad (3)$$

Assume first the C is malnormal. It follows from Lemma 3.3 and the assumption that $\Phi(\omega_D)$ is in normal form that if $i = j = 1$ then $a_1, a_2 \notin A_1$ and if $i = j = -1$ then $a_1, a_2 \notin A_{-1}$. Hence, it follows from (3) by malnormality that $c_1 = c_1'$ and $c_2 = c_2'$. If C is weakly malnormal and $R \in \mathcal{R}$ is absolutely reduced then it follows easily from (3) that $c_1 = c_1'$ and $c_2 = c_2'$. Consequently, in both cases the map $(a_1, a_2) \mapsto |\bar{c}_1| + |\bar{c}_2|$ is a function which we denote by η. Thus $\eta : Q \to \mathbb{N}$, where $Q = \{(a_1, a_2) \in H \times H \mid A_i a_1 A_j = A_i a_2 A_j\}$, $i, j \in \{1, -1\}$. Here $\bar{c}_1$ is the shortest preimage of c_1. Moreover, η is a recursive function since the generalised word problem for A and A_{-1} in H is solvable: consider $a_2^{-1} x a_1$ and substitute for x elements of C. For each x check whether $a_2^{-1} x a_1$ is inside C. The first (and hence the only) value of x for which this happens is c_1. Now c_2 is given by $c_2 = a_1^{-1} c_1^{-1} a_2$. Let $E_0 = \{(x, y) \mid x, y \in \mathrm{Supp}(\bar{\mathcal{R}}) \cup \mathrm{Supp}(U_0), A_i x A_j = A_i y A_j\}$ and let $e_0 = \max\{\eta(x, y) \mid (x, y) \in E_0\}$. Then $e \le e_0$. Since η is recursive, e_0 is (recursively) computable. Thus

$$|M_{\mathcal{R}_0}| \le |\bar{R}| f(\|u\|) g_{\mathcal{R}_0}(\xi + e_0) + |\bar{R}| f(\|u\|) g_{\mathcal{R}_0}(\xi(\xi + e_0))$$

$$= |\bar{R}| f(\|u\|) \left[g_{\mathcal{R}_0}(\xi + e_0) + g_{\mathcal{R}_0}(\xi(\xi + e_0)) \right],$$

hence

$$|M_{\mathcal{R}_1}| = |M_{\mathcal{R}_0}| + |M_{\mathcal{R}}| \quad (4)$$

$$\le |\bar{R}| f(\|u\|) \left[g_{\mathcal{R}_0}(\xi + e_0) + g_{\mathcal{R}_0}(\xi(\xi + e_0)) \right] + f(\|u\|).$$

Now, by the above discussion, the solvability of the generalised word problem for A_1 and A_{-1} in H implies that e_0 is defined and computable. Hence if we define $g : \mathbb{N} \to \mathbb{N}$ by $g(x) = |\bar{R}_0| f(x) \left[g_{\mathcal{R}_0}(x + e_0) + g_{\mathcal{R}_0}(x^2 + x e_0) + 1 \right]$, then $g(x)$ is a recursive function since, by assumption, $g_{\mathcal{R}_0}$ is, and by (1), $f(x)$ is a linear or a quadratic function. Thus by (4), $|M_{\mathcal{R}_1}| \leq g(|U_0|)$. This implies that G has a solvable word problem.

3.3.2. The conjugacy problem for amalgamated free products. We turn now to the solution of the conjugacy problem. Let $F = A \underset{C}{*} B$ and assume that A and B have free presentations $\mathcal{P}_A = \langle X | \mathcal{R}_A \rangle$ and $\mathcal{P}_B = \langle Y | \mathcal{R}_B \rangle$, respectively. Let F_0 be the free group freely generated by $(X \cup Y)$. Then F has free presentation $\mathcal{P}_0 = \langle X \cup Y | \mathcal{R}_0 \rangle$, where $\mathcal{R}_0 = \mathcal{R}_A \cup \mathcal{R}_B \cup \mathcal{R}_C$, $\mathcal{R}_C = \{\alpha(c)\beta(c)^{-1} | c \in C\}$. Consequently, G has free presentation $\mathcal{P} = \langle X \cup Y | \mathcal{R}_0 \cup \bar{\mathcal{R}} \rangle$, where $\bar{\mathcal{R}}$ is the preimage of $\mathcal{R}$ in F_0 defined as in the proof of Theorem 0.1. Let $u = a_1 b_1 \ldots a_n b_n$ and $v = a_1' b_1' \ldots a_m' b_m'$ be cyclically reduced words in F, which are $\mathcal{R}_0 \cup \bar{\mathcal{R}}$ - reduced in the sense that they do not contain more than a half of a non-empty null homotopic subword. We can make this assumption because the word problem is solvable for $\mathcal{P}$. Suppose u is conjugate to v in G. Then by [20, Lemma 5.2, p. 254] and remarks on page 279, line (-11) on the geometry of conjugacy diagram, there exists an annular R-diagram M with disjoint boundary cycles ω and τ such that $\Phi(\omega) = u$ and $\Phi(\tau) = v$. By the Main Theorem M has layer decomposition $\Lambda = (\mathcal{L}_i)$ such that $|\omega(\mathcal{L}_i)| \leq |\omega| + |\tau|$, for every $\mathcal{L}_i$ in Λ. We imitate the proof of Theorem 0.1(b). To this end we construct for every region D of M the annular diagram $\mathcal{A}(D)$ over F_0, as in part 3.3.1 and obtain the diagram M', as in part 3.3.1. Thus, with the notation of part 3.3.1, M' is an annular $\mathcal{R}_0$-diagram over F_0 with disjoint boundary cycles ω' and τ' labeled $U_1 V_1 \ldots U_n V_n$ and $U_1' V_1' \ldots U_m' V_m'$, respectively, where $U_i, U_i', V_i, V_i' \in \mathrm{Supp}(\mathcal{R}_0)$ and $\pi(U_i) = a_i$, $\pi(U_i') = a_i'$, $\pi(V_i) = b_i$ and $\pi(V_i') = b_i'$. We need to estimate $|\omega(\mathcal{L}_i)|_{F_0}$ and $|\partial D_1 \cap \partial D_2|_{F_0}$ for arbitrary regions D_1 and D_2 in M'. This is given in the lemma below, where we use the notation of part 3.3.1, which is the extension of Lemma 3.5 to amalgamated free products. Let $e_1 = \max\{\eta(x, y) \mid (x, y) \in E_1\}$, where $E_1 = \{(x, y) \mid x, y \in \mathrm{Supp}(\bar{R}) \cup \mathrm{Supp}(U_0) \cup \mathrm{Supp}(V), \ CxC = CyC\}$.

Lemma 3.7. *Let $\Lambda = (\mathcal{L}_i)$ be a layer decomposition of an annular $V(6)$ $\mathcal{R}$-diagram over F with connected interior and let μ be a boundary cycle or a simple boundary path of one of the layers $\mathcal{L}_i$ of Λ or a common edge of neighboring regions. Follow the notation of part 3.3.1 and let g be an isodiametric function for $\mathcal{P}$. Then*

$$|\mu|_{F_0} < |\mu|_F \, g\big(3(|\omega| + |\tau| + 3)(2e_1 + |\bar{R}_0|)\big)$$

Proof. Assume first that $v_1 \mu v_2$ is a common boundary path of adjacent regions D_1 and D_2. Replace $v_1 \mu v_2$ with the diagram $L := L_{v_1} \cup L_\mu \cup L_{v_2}$, defined in part 3.3.1. L has boundary cycle $\theta_1 \gamma_1 \zeta_1 \gamma_3^{-1} \theta_2 \gamma_4 \zeta_2 \gamma_1^{-1}$, where θ_i is a boundary path of D_i', $i = 1, 2$, γ_1, γ_2 and γ_3, γ_4 are connecting paths inside $\mathcal{A}(D_1)$ and $\mathcal{A}(D_2)$, respectively, with $\Phi(\gamma_i) \in C$ and ζ_i are connecting paths between $\mathcal{A}(D_i)$ and $\mathcal{A}(D_j)$ where D_i and D_j are adjacent regions containing v_1 or v_2 on their common boundary, such that $\Phi(\zeta_i)$ are alternating products of elements of length at most l, from $\mathrm{Supp}(\mathcal{R})$, where $l \leq d(v_1), d(v_2)$ and labels in C.

Let v_1' and v_2' be vertices in L_{v_1} and L_{v_2}, respectively. Since L is connected, there is a shortest (not necessary unique, simple) path μ' in L connecting v_1' and v_2'.

Now L is an $\mathcal{R}_0$-diagram hence every simple path inside $\mathcal{R}_0$ has length (in F_0) bounded above by $f(|\partial L|_{F_0})$. Let us estimate $|\partial L|_{F_0}$. Like in part 3.3.1, $|\gamma_i| \le e_1$, hence if r is the length of the longest relator in $\mathcal{R}$ and $d(v_1), d(v_2) \le \delta$ then

$$|\partial L|_{F_0} < 2\big(r + \delta(r + e_1)\big) + 4e_1 < 3\delta(r + 2e_1).$$

By Lemma 3.6 $\delta \le |\omega|_F + |\tau|_F + 3$, where ω and τ are the outer and inner boundaries of $M_{\mathcal{R}}$, respectively. Consequently, $|\partial L|_{F_0} < 3(|\omega|_F + |\tau|_F + 3)(r + 2e_1)$. Now, let μ be any path in $M_{\mathcal{R}}$ with $\mu = \mu_1 \mu_2 \cdots \mu_k$, such that each μ_i is a common boundary of two adjacent regions. Let $\mu' = \mu'_1 \cdots \mu'_k$, where μ'_i is obtained from μ as above. Then $k \le |\mu|_F$ and hence $|\mu'|_{F_0} < |\mu|_F g\big(3(|\omega|_F + |\tau|_F + 3)(r + 2e_1)\big)$, as stated.

The lemma is proved. $\qquad\qquad\qquad\qquad\qquad\qquad\qquad\qquad\qquad\qquad\quad\square$

We are now in a position to imitate the solution of the conjugacy problem in Theorem 0.4. Define sets $\mathcal{Q}$ and $\mathcal{U}$ as in the proof of Theorem 0.1, this time the bounds on the length of the words in $\mathcal{Q}$ and $\mathcal{U}$ are provided by the last lemma.

The theorem is proved. $\qquad\qquad\qquad\qquad\qquad\qquad\qquad\qquad\qquad\qquad\qquad\quad\square$

3.4. The Proof of Theorem 0.4

(a) Suppose that ν_K is not an injection. Then there exists a connected, simply connected reduced $\mathcal{R}$-diagram M, which we may assume to have connected interior and more than one region, with reduced boundary label $k \in K$, $k \ne 1$. It follows from [14] by an easy induction argument on $|M|$, that $|\partial M| \ge |\partial D|$ for every region D of M. Since M satisfies the condition $C(4)$, hence $|\partial D| \ge 4$. But then $\|\Phi(\partial D)\| \ge 4$ since M has reduced boundary cycle, violating $\|k\| = 1$.

(b) Since $\|c\| = 1$ and $\|d\| = 1$ it follows easily that if A is an annular conjugacy diagram with $\Phi\big(\omega(A)\big) = c$ and $\Phi\big(\tau(A)\big) = d$ then $|\omega(A) \cap \partial D| = 1$ and $|\tau(A) \cap \partial D| = 1$. Therefore, by part (iv) of the Main Theorem $d(D) = 4$ for every region D of A and $d(v) = 4$ for every inner vertex v of A and $d(v) = 3$ for every boundary vertex of A. Also, every layer of A has annular interior. It follows that every layer contains the same number n of regions, hence if A contains m layers and denote the ith region in the jth layer by $D_{i,j}$ then D_{ij} has a boundary cycle labeled $R_{i,j}$ as given in the theorem.

(c) If $\mathcal{R}$ satisfies the condition $C(5)$ then it is impossible for $R \in \mathcal{R}$ to be a product of four pieces. Consequently (2) implies (1). But if condition (1) holds then in particular we cannot have the situation described in part (b). Therefore, the result follows from part (b).

(d) Suppose $F = A \underset{C}{*} B$ and $L \subseteq A$ is malnormal in A. Assume $\nu(L)$ is not malnormal in G. Then there are elements u, v in L, $g \in F \backslash L$ such that $\nu(g)^{-1} \nu(u) \nu(g) = \nu(v)$. Consequently $\nu(u)$ and $\nu(v)$ are conjugate in G while u and v are not conjugate in F, contradicting part (c). For the cases $F = A * B$ and $F = \langle G, t | t^{-1} A_{-1} t = A_1 \rangle$ similar argument applies.

The theorem is proved. $\qquad\qquad\qquad\qquad\qquad\qquad\qquad\qquad\qquad\qquad\qquad\quad\square$

3.5. The Proof of the statements in Example 1

There are no general necessary and sufficient conditions for a diagram to satisfy a small cancellation condition. There are, however, obvious sufficient conditions, like requiring that the length of the pieces is uniformly bounded and the relations are long enough. Also, if $F = A * B$ then the requirement that there are no relations of length three in the factors A and B guarantees that every inner vertex has valency at least four. A different type of sufficient conditions arise by applying word combinatorics. See [15]. See also [4]. In Example 1 we shall use the simplest criteria for the $C(p)$ condition, namely, we shall require that every piece has length 1. More precisely, we observe that the endpoints of every piece in a diagram over F are secondary vertices. Therefore, every piece contains at least two half edges which contain the endpoints of the piece. Hence, if $R = a_1 b_1 \ldots a_n b_n$ in normal form in F ($b_i \in \langle t \rangle$ if $F = \langle H, t | t^{-1} U t = V \rangle$) such that $a_i \neq a_j^{\pm 1}$ and $b_i \neq b_j^{\pm 1}$ for $i \neq j$ then every piece has length 1. Due to the efficient solution of the word problem in the appropriate quotients of F we can produce this relatively easily. Also, for the condition $T(4)$ we can show that $xyz \neq 1$ in F for components $x^{\pm 1}, y^{\pm 1}, z^{\pm 1}$ of the defining relators.

$$\mathcal{R}_1 \quad and \quad \mathcal{R}_2 \quad satisfy \ the \ condition \quad C(6) \tag{1}$$

Since all the components of R_1 in A_0 and in A_1 are different, it follows that no component of R_1 is a piece (but is the product of two half-edges, as explained above). For the same reason, $\mathcal{R}_2$ satisfies the condition $C(6)$.

$$\mathcal{R}_3 \quad satisfies \ the \ condition \quad C(4)\&T(4) \tag{2}$$

First, R_3 is in reduced form, hence its components are $\bar{x}_1 \bar{b}, \bar{y}_2 \bar{c}^2 d, \bar{x}_2^{-1} \bar{a}^{-2}$, $\bar{y}_1 \bar{d}^{-3}$, each belonging to a factor of F. Now, each component occurs exactly once in R_3, hence no component is a piece. To see this, assume by way of contradiction that z_1 and z_2 are distinct components of R_3 in the same factor such that $z_1 = c_1 z_2^{\pm 1} c_2$ for some $c_1, c_2 \in C$. Suppose both z_1 and z_2 belong to A. Since A contains no elements of order two we have $c_1 \bar{x}_1 \bar{b} c_2 = (\bar{x}_2^{-1} \bar{a}^{-2})^{\pm 1}$ for suitable $c_1, c_2 \in A_1$. Hence, either $c_1 \bar{x}_1 \bar{b} c_2 \bar{a}^2 \bar{x}_2 = 1$ in A or $c_1 \bar{x}_2 \bar{b} c_2 \bar{x}_2^{-1} \bar{a}^{-2} = 1$ in A. But this contradicts (1) since $x_1^{\pm 1}$ and $x_2^{\pm 1}$ occur in R_1 at least in the third power, while in the last equation they appear in the first order. Consequently, each component occurs exactly once in R_3, hence no component is a piece and as a result every piece is the product of two half segments, as above. A similar argument applies for the components of R_3 which belong to B. Therefore, the condition $C(4)$ is satisfied by $\mathcal{R}_3$. Now we claim that every inner vertex of an $\mathcal{R}_3$-diagram has valency at least four. Suppose by way of contradiction that v is an inner vertex of a reduced $\mathcal{R}_3$-diagram with valency three. Then there are components z_1, z_2 and z_3 of R_3 in the same factor and there are c_1, c_2, c_3 in C such that $c_1 z_1^{\pm 1} c_2 z_2^{\pm 1} c_3 z_3^{\pm 1} = 1$ in A, say. But the components of R_3 from $\bar{A}_0$ are all in the first power while the components of R_1 from A_0 are all in the fourth power at least, violating (1). A similar argument applies to B. Consequently, $\mathcal{R}_3$ satisfies $C(4)\&T(4)$.

$$\mathcal{R}_4 \text{ satisfies the condition } C(6) \tag{3}$$

Let $G_1 = \langle A \underset{C}{*} B \mid R_3 \rangle$ and denote by $\tilde{A}$ and $\tilde{B}$ the images of A and B in G_1, respectively.

Let $W_1 = x_2^3 y_2^4$, $W_2 = x_1^2 x_2^{-2}$ and $W_3 = y_1^3$ be elements of $A_0 * B_0$ and for $i = 1, 2, 3$ let g_i be the image of W_i in G_1. We claim that

(i) $g_1 \notin \tilde{B}_0 \tilde{A}_0$, $g_2 \notin \tilde{B}_0$ and $g_3 \notin \tilde{A}_0$

(ii) $g_i \neq g_j$ for $1 \leq i \neq j \leq 3$

For suppose $g_1 \in \tilde{B}_0 \tilde{A}_0$. Then there are elements $b \in B_0$ and $a \in A_0$ such that $g_1 = \tilde{b}\tilde{a}$. Consequently, $\bar{x}_2^3 \bar{y}_2^4 \bar{a}^{-1} \bar{b}^{-1} \in N_3$, where N_3 is the normal closure of R_3 in $A \underset{C}{*} B$. Hence, by Lemma 3.1 there exists a connected simply connected $\mathcal{R}_3$-diagram M over $A \underset{C}{*} B$ with boundary label $\bar{x}_2^3 \bar{y}_2^4 \bar{a}^{-1} \bar{b}^{-1}$. But $\mathcal{R}_3$ satisfies the condition $C(4)\&T(4)$, hence it follows easily from the structure theorem of simply connected $C(4)\&T(4)$ diagrams in [14] that $\|\partial M\| \geq 4$ and $\|\partial M\| = 4$ if and only if M consists of a single region. But $\bar{x}_2^3 \bar{y}_2^4$ is not a subword of a cyclic conjugate of $R_3^{\pm 1}$ in $A \underset{C}{*} B$ since $\bar{c}_1 \bar{x}_2^3 \bar{c}_2 \neq z$, for every $z^{\pm 1} \in \{\bar{x}_1 \bar{b},\ \bar{y}_2 \bar{c}^2 \bar{d},\ \bar{x}_2^{-1} \bar{a}^{-2},\ \bar{y}_1 \bar{d}^3\}$ and every $c_i \in A_1$, $i = 1, 2$, because x_2 occurs in R_3 in the first power while in $\bar{c}_1 \bar{x}_2^3 \bar{c}_2$ in the third power. Consequently, $g_1 \notin \tilde{B}_0 \tilde{A}_0$ and the same argument proves the rest of (i) and (ii). Now let $R_4 = g_1 t^2 g_2 t^{-3} g_3 t$. Then R_4 is in (absolutely) reduced form due to (i) and $g_i \neq g_j$ for $1 \leq i \neq j \leq 3$, due to (ii). Therefore, as explained in the previous case for $\mathcal{R}_3$, $\mathcal{R}_4$ satisfies the condition $C(6)$, as required.

The statements are proved. $\qquad\square$

Acknowledgments

I am very grateful to the referee for his useful remarks, which improved the presentation of the work.

References

[1] M. Bestvina, M. Feighn: A combination Theorem for neatively curved groups. J. Differential Geom. 35 (1992) pp. 85–101.

[2] S. Brick and J.M. Corson: Annular Dehn functions of groups. Bull. Austr. Math. Soc. 58 (1998), 453–464.

[3] M. Bridson: Geodesics and Curvature in metric simplicial complexes pp. 373-463 in "Group Theory from a Geometrical Viewpoint", edited by E. Ghys, A. Haefliger and A. Verjovsky, Word Scientific (1991).

[4] A.J. Duncan and J. Howie: One relator products and high powered relators p. 48–74 in "Geometric Group Theory" Vol. 1, edited by G.A. Niblo and M.A. Roller, London Mathematical Society Lecture Notes Series, No. 181 (1993).

[5] A.J. Duncan and J. Howie: Weinbaum's Conjecture on unique subwords of non-periodic words, Proc. Amer. Math. Soc. 115, pp. 947–958 (1992).

[6] D.B.A. Epstein et al.: Word processing in groups; Jones and Bartlett, London, 1992.

[7] P.A. Firby and C.F. Gardiner: Surface Topology. New York, NY: Ellis Horwood, 1982.

[8] S. Gersten: Isoperimetric and isodiametric functions of finite presentations pp. 79–96 in Geometric Group Theory Vol. 1, edited by G.A. Niblo and M.A. Roller, London Mathematical Society Lecture Notes Series, No. 181 (1993).

[9] R. Gitik: On the combination theorem for negatively curved groups; International Journal of Algebra and Computation, Vol. 6, 1996, pp. 751–760.

[10] M. Gromov: Hyperbolic Groups, In "Essays in Group Theory", S.M. Gersten Editor, Math. Sci. Res. Inst. Publ. vol. 8, Springer Verlag, Berlin, 1987, 76–263.

[11] F. Grunewald: Solution of the Conjugacy Problem in certain arithmetic groups. Word Problems II, pp. 101–139, Studies in Logic and Foundation of Math, 95, North Holland, Amsterdam, 1980.

[12] G. Huck and S. Rosebrock: Cancellation diagrams with non-positive curvature in "Computational and Geometric Aspects of Modern Algebra", Edited by M. Atkinson et al. LMSLNS No, 275, pp. 128–149, 2000.

[13] D. Hurwitz: The conjugacy problem in groups – a survey, Contemporary Mathematics, 33 (1987), pp. 278–298.

[14] A. Juhász: Small Cancellation Theory with a unified small cancellation condition I, J. London Math. Soc. (2) 40 (1989) pp. 57–80.

[15] A. Juhász: On powered one-relator amalgamated free products, To appear in International Journal of Algebra and Computation.

[16] A. Juhász. On powered one-relator HNN-extensions of groups, To appear in Journal of Group Theory.

[17] I. Kapovich: A non-quasiconvexity embedding theorem for hyperbolic groups. Math. Proc. of the Cambridge Phil. Soc. Vol. 127, pp. 461–480.

[18] O. Kharlampovich and A. Myasnikov: Hyperbolic groups and free constructions. Trans. Amer. Math. Soc. 350, pp. 571–613, 1998.

[19] R.C. Lyndon: On Dehn's Algorithm, Math. Ann. 166, 208–228, 1966.

[20] R.C. Lyndon P.E. Schupp: Combinatorial Group Theory, Springer 1977.

[21] S.J. Pride: Complexes and the dependence problems for hyperbolic groups, Glasgow Math. J. 30 (1988) 155–170.

[22] P.E. Schupp: On Dehn's algorithm and the conjugacy problem, Math. Ann. 178, 119–130 (1968).

[23] P.E. Schupp: Small cancellation theory over free products with amalgamation, Math. Ann. 193, 255–264 (1971)

Arye Juhász
Department of Mathematics
Technion – Israel Institute of Technology
Haifa 32000, Israel

Solution of the Membership Problem for Magnus Subgroups in Certain One-Relator Free Products

Arye Juhász

Introduction

Let $\mathcal{P} = \langle X | \mathcal{R} \rangle$ be a presentation of a group G. The most fundamental decision problem for $\mathcal{P}$ is the Word Problem which asks for an algorithm to decide whether a given word on X represents the neutral element 1 of the group G. This problem is known to be unsolvable in general, however, for important classes of groups, it has been solved (see [L-S]). The membership problem (or generalized word problem) for $\mathcal{P}$ and a subgroup H of G asks for an algorithm to decide whether a given word on X represents an element of H. Thus, the word problem is the membership problem for the trivial subgroup, $\{1\}$, of G. One of the classes of groups in which the word problem has been solved is the class of groups presented by a single defining relator R (one-relator groups). The word problem for one-relator groups was solved by W. Magnus [M]. An important ingredient of his proof, which is the basis for many classical and more recent results, is focusing on subgroups of G which are generated by proper subsets of X, which miss at least one letter from X that occurs in $R^{\pm 1}$. These subgroups are called *Magnus Subgroups*. In the course of the solution of the word problem, Magnus solved the Membership Problem for Magnus Subgroups. For more on the Membership Problem see [K-M-W] and references therein. See also [L-S].

One-relator free products are natural generalizations of one-relator groups; we replace the infinite cyclic factors $\langle x_i \rangle$, $i = 1, \ldots, n$ of the free group $\langle X | - \rangle :=$ $\langle x_1 \rangle * \langle x_2 \rangle * \cdots * \langle x_n \rangle$ by arbitrary groups. These groups were introduced and their study was initiated by J. Howie in [H]. Relatively little is known on these groups. It is not even known whether the word problem is solvable in this class of groups, (provided it is solvable in the free product). Yet, under suitable assumptions, the word problem and several other problems are solvable (see [D-H] and references

---

This article originates from the Barcelona conference.

there). One such assumption is that the group has a presentation which satisfies a small cancellation condition (see [L-S, Ch. V]). While these conditions directly imply solution of the word problem, they do not directly imply solution of the Membership Problem for any subgroup of G, other than $\{1\}$.

It is natural to generalise the notion of Magnus Subgroups to one-relator products. Let $n \geq 2$ be a natural number and for i, $1 \leq i \leq n$, let G_i be non-trivial groups. Let $F = G_1 * \cdots * G_n$ be their free product. Let R be a cyclically reduced word in F. Denote by $\ll R \gg$ the normal closure of R in F. Let $G = F / \ll R \gg$ and let $\pi : F \to G$ be the natural projection. Without loss of generality, assume that R contains a letter from each component G_i. Then for every proper subset S of $\{1, \ldots, n\}$ the Magnus Subgroup M_S corresponding to S is $\pi_{i \in S} (* G_i)$. In [J2] we proved a strengthened Feiheitssatz for one-relator free products with small cancellation. In the present work we use some of the ideas from [J2] together with additional ideas, to solve the Membership Problem for Magnus Subgroups in one-relator products with a different small cancellation condition. Our main result is the following. (For unexplained terms see [L-S, Ch. V]. See also Subsection 1.1 below.)

Main Theorem. *Let $F = G_1 * \cdots * G_n$, $n \geq 2$, $G_i \neq \{1\}$, $i = 1, \ldots, n$ groups with solvable word problems and let $R = a_{i_1} \ldots a_{i_m}$ in normal form, $m \geq 2$, $a_{i_j} \in G_{i_j}$, $i_j \in \{1, \ldots, n\}$. Assume that R is cyclically reduced and that for each a_{i_j} the following holds:*

If a_{ij} has finite order and $a_{ij} \in G_k$, then R contains at least two elements of G_k and a_{i_j} has order different from two.

Suppose moreover that R has no cyclic conjugate of the form $UaU^{-1}W$, where $a \in G_k$ for some k, W cannot be expressed as a product of less than 3 pieces, U is a piece and F has a component G_ℓ such that U contains an element of G_ℓ while W contains none.

If the (relative) presentation $\mathcal{P} = \langle F|R \rangle$ satisfies the small cancellation condition $C'(^1/_4) \& T(4)$ then the Membership Problem is solvable for the Magnus Subgroups of the group G presented by $\mathcal{P}$.

Remark 1. With a deeper analysis on diagrams it can be shown that the element $a \in G_k$ in the "moreover" part of the assumption in fact has finite order. Hence, if F is torsion free then the only assumption is that $\mathcal{R}$ satisfies $C'(1/4) \& T(4)$.

Remark 2. If the last part of the "moreover" condition is violated then the statement of the theorem is still valid for those Magnus Subgroups which omit the components like G_ℓ.

Remark 3. The theorem holds true also for other types of small cancellation conditions.

The two main ingredients of the proof of the Main Theorem are analysis of van Kampen diagrams and word combinatorics. The principal achievement in the paper is Proposition 2.2 which allows to deduce that the so-called "Greendlinger-regions" not only share a large portion of their boundaries with the boundary of the given diagram, as usual in small cancellation theory, but *some* of them have a "small neighborhood" the common boundary of which with the boundary of the diagram contains a letter from each component G_i, a property that usually is not shared by a single Greendlinger-region. (For the precise statement see Proposition 2.2.) This is proved by using massive word combinatorics, in a systematic way, analysing the possible piece-structures of the relator. Then it uses a detailed description of the structure of small cancellation diagrams, as developed in [J1].

The work is organized as follows. In Section 1 we introduce notation and recall the necessary results from [J1] on small cancellation theory. In Section 2 we develop the necessary word combinatorics. This is the main technical body of the work. In Section 3, relying on the results of Section 1 and Section 2, we prove the Main Theorem.

1. Preliminary results on words and diagrams

1.1. Words

Let $F = G_1 * \cdots * G_n$, $n \geq 2$, the free product of non-trivial groups G_i, $i = 1, \ldots, n$. We call the G_is the *components of F*. Recall from [L-S, pp. 174-176] that if G_i for $i = 1, \ldots, n$ has free presentation $\langle Y_i \mid S_i \rangle$ such that the Y_is are disjoint, then F has free presentation $\langle Y \mid S \rangle$, where $Y = \bigcup_{i=1}^{n} Y_i$ and $S = \bigcup_{i=1}^{n} S_i$. The elements of $F \setminus \{1\}$ can be uniquely presented by finite sequences of non-trivial elements of the components, such that adjacent elements of the sequence come from different components. More precisely, we call the elements of G_i, $i = 1, \ldots, n$, *letters* and the sequences of elements, *words*. For a letter $a \in G_i$ denote $\alpha(a) = i$. Thus, if $1 \neq W \in F$ then W can be uniquely expressed as a word: $W = b_{i_1} \cdots b_{i_k}$, $k \geq 1$, $1 \neq b_{i_j} \in G_{i_j}$ and $\alpha(b_{i_j}) \neq \alpha(b_{i_{j+1}})$ for $j = 1, \ldots, k - 1$. We call this presentation of W its *normal form*, call k its *length* and denote it by $|W|$.

Let $R = a_{i_1} \cdots a_{i_m} \in F$, $a_{i_j} \in G_{i_j}$ in normal form. R is *cyclically reduced* if either $m = 1$ or $m \geq 2$ and $i_1 \neq i_m$. R is *weakly cyclically reduced* if either it is cyclically reduced or $i_1 = i_m$, but $a_{i_m} a_{i_1} \neq 1$ in G_{i_1}.

The *symmetric closure of R in F* is the collection of all the weakly cyclically reduced conjugates of R and R^{-1} in F. These are obtained by forming the m cyclically reduced cyclic conjugates R_j of R, where $R_j = a_{i_j} \cdots a_{i_m} a_{i_1} \cdots a_{i_{j-1}}$ and the cyclically reduced cyclic conjugates R_j^{-1} of R^{-1} and then conjugating each of them by arbitrary letters from G_{i_j}. We denote this set by $\mathcal{R}$. Thus

$$\mathcal{R} = \left\{ c^{-1} R_j^\varepsilon c \mid c \in G_{i_j}, \ j = 1, \ldots, m, \ \varepsilon \in \{1, -1\} \right\}.$$

In particular, if one of the components has infinite order, then $\mathcal{R}$ contains infinite number of elements. This is in contrast with the symmetric closure of an element $W \neq 1$ in free groups, where it contains always $2m$ elements, where $m = |W|$. This is one of the sources of difficulties when trying to solve decision problems in quotients of free products. Another one is given in Remark 4 below.

In this subsection $F = G_1 * \cdots * G_n$, $n \geq 2$, $G_i \neq \{1\}$ and all words are in F.

Let U and V be reduced words in F. We say that the product UV is *reduced as written* if either the last letter of U and the first letter of V are in different components, or there is no cancellation between U and V, however the last letter of U and the first letter of V may come from the same component (consolidation).

Remark 4. If F is a free group and U and V are reduced words such that UV is reduced as written, then $|UV| = |U| + |V|$. In contrast, if F is a free product such that the last letter of U and the first letter of V belong to the same component then $|UV| = |U| + |V| - 1$, while $|UV| = |U| + |V|$, if they are in different components. this makes it difficult to control the length of a word U which is the product of given subwords $U_1, \ldots, U_k$. Also, while over a free group a non-empty word and its inverse cannot overlap, this may happen over a free product, even if no component contains elements of order two. (See Lemma 1.2(b).)

Another problem with a decomposition of a reduced word W to $W = UV$ reduced as written, is that in contrast with decompositions over free groups if we are given the length of U, the length of V and that $|W| = |U| + |V| - 1$, this data doesn't define U and V uniquely, because we may choose the last letter of U as an arbitrary element (not equal to 1) of a component G_i of F and then adjust the first letter of V, which is in the same component, accordingly. For these reasons we introduce "integral subwords" below.

Denote by $\mathcal{H}(W)$ the set of initial subwords of W and by $\mathcal{T}(W)$ the set of terminal subwords of W. Also, for a reduced non-empty word W we denote by $h(W)$ the first letter of W and by $t(W)$ the last letter of W.

Definition 1.1.

(a) Let W be a reduced word and let U be a proper subword of W: $W = W_1 U W_2$, reduced as written. Say that U is *integral in* W if one of the following holds:
   (i) $W_1 = 1$ and $\alpha\big(t(U)\big) \neq \alpha\big(h(W_2)\big)$ (i.e., the last letter of U and the first letter of W_2 are in different components G_i);
   (ii) $W_2 = 1$ and $\alpha\big(h(U)\big) \neq \alpha\big(t(W_1)\big)$;
   (iii) $W_1 \neq 1$ and $W_2 \neq 1$ and $\alpha\big(t(U)\big) \neq \alpha\big(h(W_2)\big)$ and $\alpha\big(h(U)\big) \neq \alpha\big(t(W_1)\big)$.
   Clearly, every subword U of W contains a unique maximal integral subword of length greater than or equal to $|U| - 2$. Denote it by $[U]$.
(b) Let W be a reduced word and suppose that W has a decomposition $\tau(W)$: $W_1 \cdots W_k$, $k \geq 2$, $W_i \neq 1$, reduced as written. The *integral skeleton* of $\tau(W)$ is the sequence $\big([W_1], \ldots, [W_k]\big)$.
(c) Let $\tau_1(W)$ and $\tau_2(W)$ be two decompositions of W into subwords. Say that $\tau_1(W)$ *is equivalent to* $\tau_2(W)$ if they have the same integral skeleton. Clearly, this is an equivalence relation.

Remark 5. Observe that although W may have infinite number of decompositions, it has only a finite number of inequivalent decompositions, representatives of which can be effectively written down via the possible integral skeletons.

(d) Let W be a reduced word. W is an $\mathcal{R}$-*word* if W has a decomposition to a product of subwords W_i of elements of $\mathcal{R}$, such that if $|W_i| = 1$ then W_i is integral in W or $|W_i| \geq 2$.

We have the following well-known results.

Lemma 1.2.

(a) *Let A, B, C be reduced words such that AB and BC are reduced as written. If $|AB| \geq 2$ and $AB = BC$ then $A = KL$, $C = LK$ and $B = (KL)^\beta K$, $\beta \geq 0$.*
(b) *Let Z be a reduced word which contains no letters of order 2.*
    (i) *If for some reduced words V and U, such that ZU and $Z^{-1}V$ are reduced as written, and we have $ZU = Z^{-1}V$, then $|Z| = 1$ and $V = Z^2U$. Moreover, Z and the first letter of V and U respectively, are in the same component G_i.*
    (ii) *If for some reduced words U and V such that UZ and $Z^{-1}V$ are reduced as written, and we have $UZ = Z^{-1}V$, then $U = Z^{-1}a$ and $V = aZ$, where a is a letter and the first letter of Z and a are in the same component G_i.*

We introduce below the key notion of the work.

Definitions and notations.

(a) Let $W \in F$, $W = a_{i_1} \ldots a_{i_k}$, $a_{i_j} \in G_{i_j}$ reduced as written. Define
$$\mathrm{Supp}(W) = \{i_1, \ldots, i_k\} \subseteq \{1, \ldots, n\}.$$

(b) Let W_1 and W_2 be reduce words in F. W_2 *majorises* W_1 if $\mathrm{Supp}(W_2) \supseteq \mathrm{Supp}(W_1)$. In this case write $W_2 \succ W_1$. If $W_1 \succ W_2$ and $W_1 \succ W_3$ we shall write $W_1 \succ W_2, W_3$. Also, if $\mathrm{Supp}(W_1) \cup \mathrm{Supp}(W_2) \supseteq \mathrm{Supp}(W_3)$ we shall write $W_1, W_2 \succ W_3$.

(c) For W_1 and W_2 in part (b) define $W_1 \sim W_2$ if $W_1 \prec W_2$ and $W_2 \prec W_1$. Thus $W_1 \sim W_2$ if and only if $\mathrm{Supp}(W_1) = \mathrm{Supp}(W_2)$.
Clearly "$\sim$" is an equivalence relation, which contains the equality of elements in F.

The following lemma is immediate from the definition, hence its proof is omitted.

Lemma 1.3.

(a) *If A is a subword of B then $A \prec B$.*
(b) *If $A \prec B$ then $A^{\pm 1} \prec B^{\pm 1}$.*
(c) *If $A \sim B$ and $A \prec C$ then $B \prec C$.*
(d) *If $A = P_1 \ldots P_m$, reduced as written and $P_i \sim Q$ for $i = 1, \ldots, m$ then $A \sim Q$.*
(e) *If $A \succ P_1, \ldots, P_m$ then $A \succ W(P_1, \ldots, P_m)$, for every word W on $P_1, \ldots, P_m$.*

The following lemma is an immediate corollary of Lemma 1.2 and Lemma 1.3, hence we omit its proof.

Lemma 1.4. *Let A, B and C be as in Lemma 1.2 (a). Then $B \prec A \sim C \sim AB \sim BC$. If $\beta \geq 1$ then $B \sim A$.*

Finally, we need the following basic notions.

Definition 1.5.

(a) Let R be a weakly cyclically reduced word in F and let P be a subword of a cyclic conjugate of R. P is a *piece* in R (or a piece relative to the symmetric closure $\mathcal{R}$ of R) if R has distinct cyclic conjugates R_1 and R_2 such that $R_1 = PR_1'$, $R_2^\varepsilon = PR_2'$, reduced as written, for some $\varepsilon \in \{1, -1\}$. We call the two occurrences of P in R_1 and R_2^ε, respectively, a *piece-pair* and denote it by (P, P'), where $P' = P^\varepsilon$ is the occurrence of P^ε in R_2. We shall deliberately use both notations $\varepsilon(P_i')$ and ε_i for ε, if there are several pieces P_i, as convenient.

(b) A piece pair (P, P') as in part (a) of the definition is *right normalized* if $R_1^{-1} R_2$ is reduced as written.

1.2. Diagrams

For basic results on diagrams see [L-S, Ch. V]. We recall here some of the basic definitions from [L-S, p. 236 and pp. 274–276] for convenience. Let $\mathbb{E}^2$ denote the Euclidean plane. If $S \subseteq \mathbb{E}^2$ then ∂S will denote the boundary of S, the topological closure of S will be denoted by $\bar{S}$. A *vertex* is a point of $\mathbb{E}^2$. An *edge* is a bounded subset of $\mathbb{E}^2$ homeomorphic to the open unit interval. A *region* is a bounded set homeomorphic to the open unit disc. A *map* is a finite collection of vertices, edges and regions which are pairwise disjoint and satisfy:

(i) If e is an edge of M, there are vertices u and v (not necessarily distinct) in M such that $\bar{e} = e \cup \{u\} \cup \{v\}$.

(ii) The boundary, ∂D, of each region D of M is connected and there is a set of edges. $e_1, \ldots, e_n$ such that $\partial D = \bar{e}_1 \cup \cdots \cup \bar{e}_n$.

A *diagram over a group* F is an oriented map M and a function Φ assigning to each oriented edge e of M as a *label* an element $\Phi(e)$ of F such that if e is an oriented edge of M and e^{-1} the oppositely oriented edge, then $\Phi(e^{-1}) = \Phi(e)^{-1}$, and if $\mu = e_1 v_1 e_2 v_2 \cdots e_k$ is a path in M then $\Phi(\mu) = \Phi(e_1)\Phi(e_2)\cdots\Phi(e_k)$. We denote by Φ_M the labelling function of M over F. If M is fixed we shall write Φ for Φ_M.

In the case of diagrams M over free products the vertices are divided into two types, *primary* and *secondary*. The label on every edge of M will belong to a factor G_i of F with the labels on successive edges meeting at primary vertices belonging to different factors G_j, while the labels on the edges at a secondary vertex all belong to the same factor of F.

Definitions 1.6. Let M be a diagram over F.

(a) Two regions D_1 and D_2 in M are *neighbors* if $\partial D_1 \cap \partial D_2 \neq \emptyset$. They are *proper neighbors* if $\partial D_1 \cap \partial D_2$ contains a non-empty edge.

(b) A region D is a *boundary region* if $\partial D \cap \partial M \neq \emptyset$. A region D is a *proper boundary region* if $\partial D \cap \partial M$ contains a non-empty edge. A region of M which is not a boundary region is an *inner region*.

(c) Let M be a connected, simply connected map. M is a *simple one-layer* map, if the dual map M^*, obtained from M by putting in each region D a vertex D^* and connecting two vertices D_1^* and D_2^* by an edge if D_1 and D_2 are proper neighbors, is a straight line. (See Fig. 4(b).) In particular, M has connected interior, every region is a boundary region, each region has at most two proper neighbors and if M contains more than one region then M contains exactly two regions (D_1 and D_r on Fig. 4(b).) which have exactly one neighbor each. M is a *one-layer map* if it is composed from simple one-layer maps and paths in the way shown on Fig. 4(a).

We recall the main structure theorem from [J1], where it is proved in a more general setting. Observe that the condition $C'(^1/_4)\,\&\,T(4)$ implies the condition $W(6)$ in [J1].

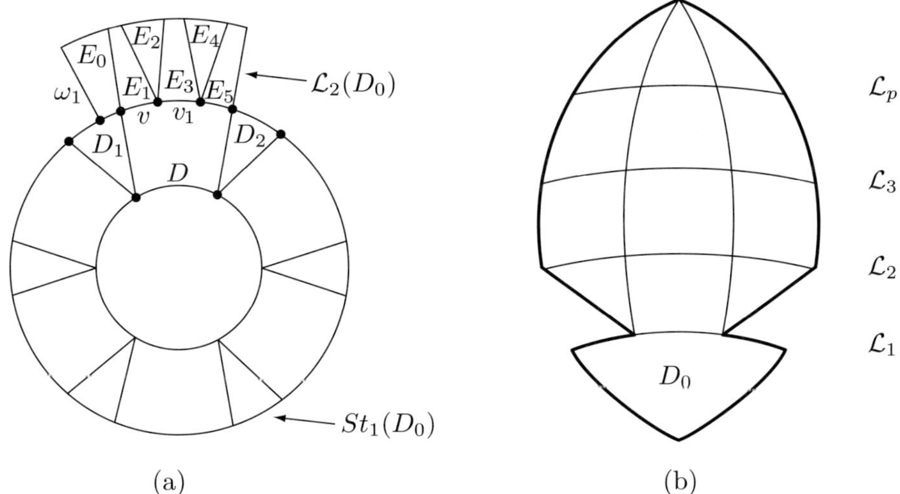

FIGURE 1

Theorem 1.7 (Layer Decomposition, [J1]). *See Fig. 1. Let M be a simply connected map (diagram) with connected interior and let D_0 be a region of M. Assume that M satisfies the condition $C'(^1/_4)\,\&\,T(4)$.*
Define $St_0(D_0) = D_0$ and for $i \geq 1$ let $St_i(D_0) = St_{i-1}(D) \cup \mathcal{L}_i(D_0)$ where $\mathcal{L}_0 = \{D_0\}$, $\mathcal{L}_i(D_0) = \langle D \text{ in } M \backslash St_{i-1}(D_0) | \partial D \cap \partial St_{i-1}(D_0) \neq \emptyset \rangle$. Let p be the smallest number such that $St_p(D_0) = M$ and assume that $p > 0$ (i.e., M contains more than one region). Then each of the following holds:

(a) *Every regular submap of $St_{i+1}(D_0)$ containing $St_i(D_0)$ is simply connected, $0 \leq i \leq p$. (A submap is regular if every edge is on the boundary of a region.)*

(b) *Every connected and simply connected submap of $\mathcal{L}_i(D_0)$ is a one-layer map.*

176   A. Juhász

When D_0 is fixed, we shall abbreviate $\mathcal{L}_i(D_0)$ by $\mathcal{L}_i$ and call $\Lambda(D_0) = (\mathcal{L}_0, \ldots, \mathcal{L}_p)$ a *layer decomposition of M*. We call D_0 the *center* of the layer decomposition.

(c) *For a region $D \in \mathcal{L}_i$, $i \geq 1$ denote by $\mathcal{A}(D)$ the set of regions E in $\mathcal{L}_{i-1}$, which have a non-trivial common edge with D, denote by $\mathcal{B}(D)$ the set of regions S in $\mathcal{L}_i$ with $\partial S \cap \partial D \neq \emptyset$. Also, let $a(D) = |\mathcal{A}(D)|$, $b(D) = |\mathcal{B}(D)|$. Then $a(D) \leq 1$ and $b(D) \leq 2$. In other words, D has at most two neighbors in $\mathcal{L}_i$ and at most one neighbor in $\mathcal{L}_{i-1}$.*

(d) *If $v \in \partial St_i(D_0)$ then v has valency at most three in $St_i(D_0)$.*

(e) *For regions D, E in M with $\partial D \cap \partial E \neq \emptyset$ we have that $\partial D \cap \partial E$ is connected.*

Remark 6. Let M be a connected simply connected map (diagram) with connected interior and let D_0 be a region in M. Let $\Lambda(D_0)$ be a layer decomposition of M with center D_0. Suppose that D_0 is a boundary region of M with a non-empty edge on ∂M. (See Fig. 1(b).) Then it follows from the above theorem that $\mathcal{L}_1(D_0)$ is not annular, hence simply connected, though not necessarily with connected interior. (See Fig. 2(a), where the interior of $\mathcal{L}_1$ is simply connected and connected and see 2(b), where the interior of $\mathcal{L}_1$ is not connected.) But then due to the simply connectedness of M, $\mathcal{L}_i$ is simply connected for every i.

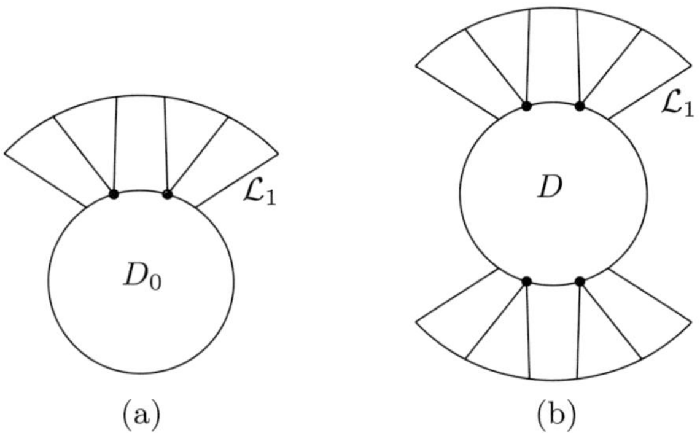

(a)   (b)

FIGURE 2

Remark 7. We could start the construction of the layers not with a region D_0, but with a vertex v and defining $St_0(v) = \{v\}$, $St_1(v)$ is the submap consisting of $\{D \mid v \in \partial D\}$ and for $k \geq 2$ define $St_k(v)$ like in Theorem 1.7. As we shall see below in certain situations we could start the layer structure with a path. (This is definitely not true for every simple path, but only for those which in some sense are "convex".) These are the transversals.

Remark 8. The notion of one-layer map is independent of any given layer structure. For example the map on Fig. 4(b) is a one-layer map, by definition. However, it also has a layer structure Λ with center D_1 for which $\mathcal{L}_1$ consists of E and D_3.

Another typical example is M_k on Fig. 4(a), which is a one-layer map. In its layer structure with center D_0 every layer consists of a single region, hence certainly each $\mathcal{L}_i$ is a one-layer map, however these layers $\mathcal{L}_i$ are "perpendicular" to the natural one-layer structure of M_k.

In the next definition we introduce special subdiagrams and regions, the boundaries of which share a large portion with the boundary of M.

Definition 1.8.

(a) Let $\Lambda(D_0)$ be a layer decomposition of M with D_0 a boundary region of M with a non-empty edge on ∂M. If $\Lambda(D_0) = (\mathcal{L}_0, \mathcal{L}_1, \ldots, \mathcal{L}_p)$ then the closure of every connected component P of the interior of $\mathcal{L}_p$ is a *peak*. We say that P *is a peak relative to* D_0. (See Fig. 1(a), where $p = 2$ and $\mathcal{L}_2$ is a peak and see Fig. 1(b), where $\mathcal{L}_p$ is a peak.) Related to peaks is the following notion.

(b) A boundary region D of M is a *k-corner region* for $k = 1, 2$ if each of the following holds:
    1) $\partial D \cap \partial M$ is connected and
    2) D has k neighbors.

Example 1.9. Let M be a diagram of a $C'(^1/_4)\&T(4)$ presentation. Let P be a peak, depicted on Fig. 3(a). Then its extremal regions D_1 and D_k are 2-corner regions because $a(D_1) \leq 1$ and $b(D_1) \leq 1$, due to being extremal. If P is a peak consisting of a single region E, then E is a 1-corner region due to Theorem 1.7. Also, if $a(D_{k-1}) = 0$ then D_{k-1} in Fig. 3(a) is a 2-corner region.

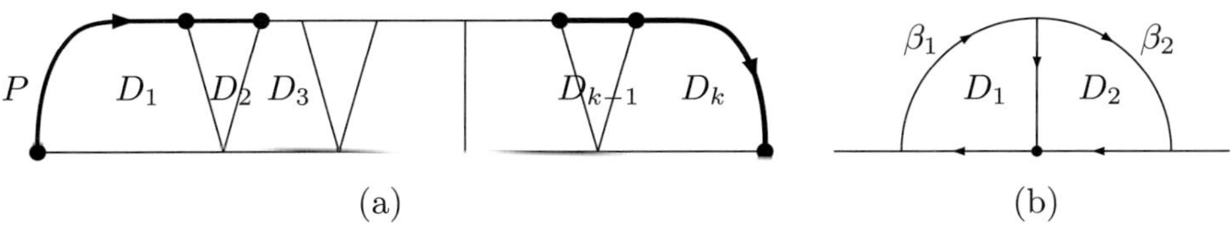

$$\text{(a)} \qquad\qquad\qquad\qquad\qquad \text{(b)}$$

FIGURE 3

D is a *weak corner region* if 2) above holds with $k = 2$ and 1) is replaced by 1') $\partial D \cap \partial M = \mu \cup \{z\}$, where μ is a simple path and z is a vertex, disjoint from $\bar{\mu}$. Thus, if D is a weak corner region as above and ν is the complement of μ on ∂D such that $u\mu v\nu^{-1}$ is a boundary cycle of D (where u and v are vertices) then $|\Phi(\nu)| < \frac{1}{2}|R|$. Similarly, if D is a 2-corner region with $\partial D \cap \partial M = \mu$ and ν is again the complement of μ on ∂D, then $|\Phi(\nu)| < \frac{1}{2}|R|$. This allows us to carry out Dehn reductions as follows.

Let ζ be a boundary path of M which contains μ. Suppose first that u and v are primary vertices. Then $\Phi(\mu)$ is an integral subword of $\Phi(\zeta)$ and also of the cyclic word $R^* (= \Phi(\partial D))$. Just as with free presentation we may carry out effectively a

Dehn reduction by replacing $\Phi(\mu)$ with $\Phi(\nu)$, since we know the words $\Phi(\mu)$ and $\Phi(\nu)$. Notice that this would not be the case if one of the vertices u and v would be a secondary vertex, say $\mu = u\mu_0 v_1 \mu_1 v$, $|\Phi(\mu_0)| = 1$, v_1 a primary vertex and u a secondary vertex, because then we could not know what $\Phi(\mu_0)$ is.

Suppose now that u and v are secondary vertices. Then $u\mu v = u\mu_1 u_1 [\mu] v_1 \mu_2 v$, where u_1 and v_1 are primary vertices, $|\mu_1| = |\mu_2| = 1$ and $\Phi([\mu])$ is the integral part of $\Phi(\mu)$ in $\Phi(\zeta)$ and R^*. Let $\hat{\nu} = u_1 \mu_1^{-1} uvv \mu_2^{-1} v_1$. Since u and v are secondary vertices, $|\hat{\nu}| = |\nu|$. But then due to the $C'(^1/_4)$ condition $|\hat{\nu}| = |\nu| < \frac{1}{2}|R|$ and since $\Phi(\hat{\nu})$ and $\Phi([\mu])$ are integral subwords of R^* we have $|[\mu]| + |\hat{\nu}| = |R|$. Hence $|\hat{\nu}| = |\nu| < \frac{1}{2}|R| < |R| - |\hat{\nu}| = |[\mu]|$. Therefore, we can replace the integral subword $\Phi([\mu])$ of R^* and of $\Phi(\zeta)$ with $\Phi(\hat{\nu})$ and shorten $\Phi(\zeta)$. If one of u and v is primary and the other secondary then a similar argument applies. We summarise this in the following lemma.

Lemma 1.10. *Let M be an $\mathcal{R}$-diagram and let D be a k-corner region with $k \leq 2$. Let $\mu = \partial D \cap \partial M$ and let $u\mu_1 u_1 [\mu] u_2 \mu_2 vv$ be a boundary cycle of D, where u, u_1, u_2 and v are vertices, $|\mu_1|, |\mu_2| \leq 1$, u_1 and u_2 primary vertices and if $\mu_1 \neq \emptyset$ then u a secondary vertex and similarly, if $\mu_2 \neq \emptyset$ then v a secondary vertex. Let ζ be a boundary path of M which contains μ, $\zeta = \zeta_1 w_1 \beta_1 u\mu v \beta_2 w_2 \zeta_2$, w_1, w_2 primary vertices, $\Phi(\beta_1 u\mu_1)$ and $\Phi(\mu_2 v\beta_2)$ integral letters in $\Phi(\zeta)$. Let $\zeta' = \zeta_1 w_1 \beta_1 uvv \beta_2 w_2 \zeta_2$. Then*

(i) $\Phi(\zeta') \stackrel{=}{_G} \Phi(\zeta)$

(ii) $|\zeta'| < |\zeta|$

(iii) $\Phi(\zeta') = \Phi(\zeta_1)\Phi(\beta_1 \mu_1)\Phi(\mu_1^{-1} \nu_1)\Phi([\nu])\Phi(\nu_2 \mu_2^{-1})\Phi(\mu_2 \beta_2)\Phi(\zeta_1),$

where $\nu = \nu_1 [\nu] \nu_2$ and all the components are integral in $\Phi(\zeta')$, $\Phi(\beta_1 \mu_1)$ and $\Phi(\mu_2 \beta_2)$ are integral letters in $\Phi(\zeta)$ or empty and $\Phi(\mu_1^{-1} \nu_1)$ and $\Phi(\nu_2 \mu_2^{-1})$ are integral letters of R^*, or empty. In particular, $\Phi(\zeta')$ can be obtained effectively from $\Phi(\zeta)$ and R^*.

Suppose now that no further Dehn reductions can be carried out. The next lemma describes the diagram we get.

Lemma 1.11. *Let V be a reduced word representing the element $\bar{V} \in G$, $\bar{V} \neq 1$. If V cannot be shortened by the Dehn algorithm then there exists a shortest representative U of $\bar{V}$ and a one-layer diagram M, such that M has a boundary cycle $v\mu wv^{-1} v$, v, w vertices with $\Phi(\mu) = V$ and $\Phi(\nu) = U$. Moreover, if M_1 and M_2 are connected components of the interior of M such that $\partial M_1 \cap \partial M_2 \neq \emptyset$ then $\partial M_1 \cap \partial M_2$ is a vertex with valency four. All other vertices of ∂M_1 and ∂M_2 which are not common with ∂M_3, for some connected component M_3 of the interior, $M_3 \neq M_1$ and $M_3 \neq M_2$, have valency three. See Fig. 4(a).*

Proof. Let U be a shortest representative. Then $VU^{-1} \stackrel{=}{_G} 1$, hence there exists a connected, simply connected van Kampen diagram M with boundary cycle $v\mu wv^{-1} v$ such that $\Phi(\mu) = V$ and $\Phi(\nu) = U$. Suppose that M contains at least 3-corner regions. Then at least for one, say D, we have $\partial D \cap \partial M \subseteq \mu$, since ν is

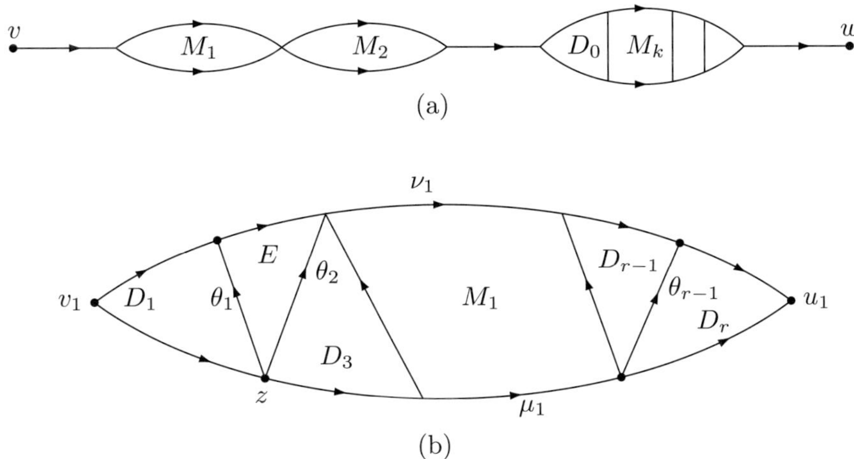

FIGURE 4

shortest and if such a region D would satisfy $\partial D \cap \partial M \subseteq \nu$ then we could Dehn-reduce $\Phi(\nu)$, via D, violating the minimality of $|\Phi(\nu)|$. However, by assumption V cannot be shortened by a Dehn reduction, therefore M cannot contain such a corner region D. Consequently, either M consists of a single region, in which case the result holds or M contains 2-corner regions which contain v and w on their boundary, respectively. But such a diagram is a one-layer diagram. (See [J1].) Finally, let M_1 be the closure of a connected component of the interior of M and let $v_1\mu_1 u_1\nu_1^{-1}$ be a boundary cycle, where $\mu_1 = \partial M_1 \cap \mu$ and $\nu_1 = \partial M_1 \cap \nu$, u_1, v_1 vertices. See Fig. 4(b). If μ_1 contains a vertex z, $z \neq v_1$, $z \neq u_1$ with valency greater than three then there exists a region E in M, which contains z on its boundary and has at most two neighbors in M such that $\partial E \cap \nu_1$ is the complement of two pieces (which are the common boundaries of E and its neighbors in M_1, θ_1 and θ_2). Therefore ν can be Dehn reduced via E. But $|\Phi(\nu)|$ is shortest in its class mod N, where $N = \ll R \gg$, hence it cannot be shortened, a contradiction. By the same argument, if ν contains a vertex q, $q \neq v_1$, $q \neq u_1$ with valency at least four then M_1 contains a region D via which we can Dehn-reduce μ, contrary to assumption. Consequently neither μ_1 nor ν_1 may contain vertices different from u_1 and v_1 with valency greater than three in M_1.

The lemma is proved. □

Remark 9. Let M be a connected, simply connected diagram with no vertices of valency 1. Let μ be a boundary path of M and let $v \in \mu$ be a vertex. Say that v is a *double point* of μ if v is met more than once when traversing along μ. (Actually it may be a multiple point, but from our point of view we only need it is not a regular point.) It is well known (see [L-S, Ch. V]) that if μ has a double point then μ encloses one or more connected components of the interior of M. Hence, if μ

contains a double point v then μ contains a peak relative to v. Consequently, due to Lemmas 1.10 and 1.11 we have

$$\textit{if} \quad \Phi(\mu) \textit{ is Dehn reduced then } \mu \textit{ contains no double points.} \qquad (*)$$

Let $\Lambda(D)$ be a layer decomposition of M and let $\omega_i = \partial \mathcal{L}_i \cap \partial \mathcal{L}_{i+1}$, $i = 1, \ldots, p-1$, see Fig. 1(a), showing ω_1. Then due to Theorem 1.7 ω_i has the property that $\partial E \cap \omega_i$ is either a vertex or an edge, for every region E of $\mathcal{L}_{i+1}$. (Observe that this is not true for regions in $\mathcal{L}_i$. If K is a region of $\mathcal{L}_i$ then $\partial K \cap \omega_i$ may contain more than one edge.) It turns out that this property of ω_i is responsible for the existence of peaks in M. We develop this idea below in order to produce peaks on specific places along ∂M.

Let $\Lambda(D)$ be a layer structure of M where D may be a single vertex. Let z be a vertex in M such that z is not in the last layer. Then z is on the common boundary ω_i of $\mathcal{L}_i(D)$ and $\mathcal{L}_{i+1}(D)$ for some i. Suppose that $\mathcal{L}_{i+1}(D)$ contains a vertex $w_1 \in \omega_i$ which has valency at least four in $\mathcal{L}_{i+1}(D)$. See Fig. 5(a). For example, if $\mathcal{L}_{i+1}(D)$ contains at least three regions, then, due to the $C'(^1/_4)\,\&\,T(4)$ condition, it follows from Theorem 1.7(c) that $\mathcal{L}_{i+1}(D)$ contains such a vertex w_1. Let u be the initial vertex of ω_i and let $u\theta_1 w_1$ be the subpath of ω_i which starts at u and terminates at w_1. Then, as observed above, due to Theorem 1.7(c) θ_1 satisfies condition $(**)$ below, with $\mu = \theta_1$

$$\textit{when traversing along } \mu, \textit{ for every region } E \textit{ to the left of } \mu \textit{ with } \partial E \cap \mu \textit{ nontrivial, (i.e., contains an edge) we have that } \partial E \cap \mu \textit{ is connected and } \partial E \cap \mu \textit{ is an edge (i.e., does not contain a vertex with valency at least three in } M). \qquad (**)$$

Since w_1 has valency at least four in M, hence $\mathcal{L}_{i+1}(D)$ contains a region E with $\partial E \cap \omega_i = \{w_1\}$. Let $\theta_2 = \partial E \cap \partial F$, where F is the right-hand side neighbor of E in $\mathcal{L}_{i+1}$. Then $u\theta_1 w_1 \theta_2 w_2$ satisfies condition $(**)$ above, where w_2 is the endpoint of θ_2, different from w_1. Assume that w_2 is not a boundary vertex of M. Then $w_2 \in \omega_{i+1}$ and there are left most adjacent regions E_1 and F_1 in $\mathcal{L}_{i+2}(D)$ which contain w_2 on their boundary. Define $\theta_3 = \partial E_1 \cap \partial F_1$ and keep defining θ_i and w_i for $i \geq 3$ until reaching ∂M. Let $\theta = u\theta_1 w_1 \theta_2 w_2 \ldots \theta_e v$ where v is a boundary vertex of M and for $1 \leq i \leq e-1$, w_i are inner vertices of M. Then each initial segment of θ satisfies condition $(**)$ and $v \in \partial M$. This leads us to the following definition.

Definition 1.12. Let notation be as above. Let θ be a simple path in M with initial and terminal vertices u and v, respectively. Suppose that $u, v \in \partial M$ and $\theta \cap \partial M = \phi$ (i.e., the open path θ does not intersect ∂M). Call θ a *left transversal* if it satisfies condition $(**)$ above, with μ replaced by θ. If μ is a left transversal, then μ defines a submap M_μ of M by its boundary $u\mu v\zeta^{-1}$ where ζ is the boundary path of M which starts at u, terminates at v and is to the left of μ. See Fig. 5(a).

The proof of Theorem 1.7 (see [J1]) can be easily adapted to prove the next proposition. We omit its proof here.

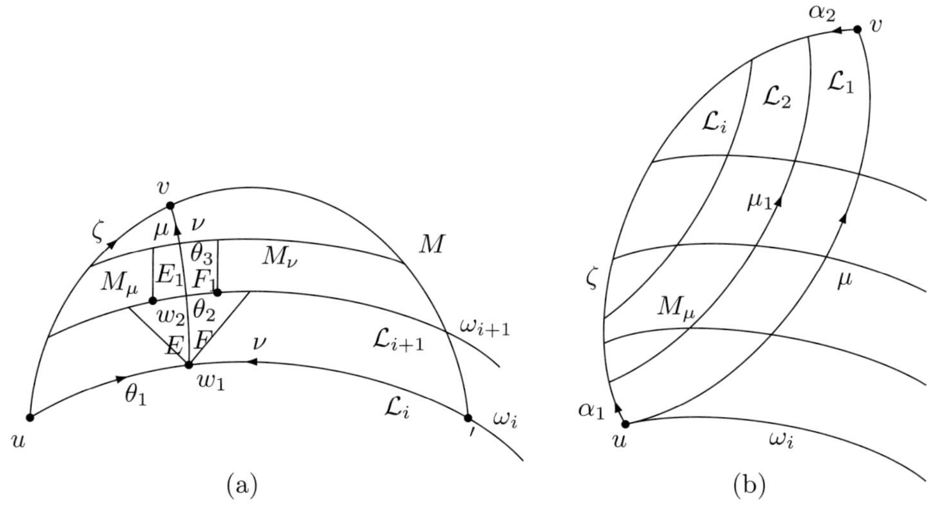

FIGURE 5

Proposition 1.13. *Let μ be a left transversal in M. Then M_μ has a layer structure relative to μ. In particular, M_μ has a peak relative to μ. More precisely, define $\mathcal{L}_0(\mu) = \mu$. Let $\mathcal{L}_1(\mu)$ be the submap of M_μ consisting of those regions of M_μ, the boundary of which intersects μ. Then $\mathcal{L}_1(\mu)$ is connected and simply connected, and μ is a boundary path of $\mathcal{L}_1(\mu)$. Let μ_1' be its complement on $\partial\mathcal{L}_1(\mu)$. Then $\mu_1' = \alpha_1\mu_1\alpha_2^{-1}$, where α_1 and α_2 are boundary paths of M and μ_1 is an inner path of M. Define $\mathcal{L}_2(\mu) = \mathcal{L}_1(\mu_1)$ and repeat this process until the last layer $\mathcal{L}_k(\mu)$ is reached. See Fig. 5(b). Then $\mathcal{L}_0(\mu), \ldots, \mathcal{L}_k(\mu)$ is a layer structure of M_μ relative to μ.*

Clearly, by the same method we may construct right transversals. Hence, if $\mathcal{L}_{i+1}(D)$ contains at least two vertices with valency at least four in $\mathcal{L}_{i+1}(D)$ or one vertex with valency at least five in $\mathcal{L}_i(D)$ and u, u' are endpoints of ω_i, say u to the left of u', then we may construct a left transversal μ with initial vertex u and a right transversal ν with initial vertex u'. See Fig. 5(a). It follows by an easy induction on the number of layers that $M_\mu \cap M_\nu = \emptyset$. Now, due to the $C'(^1/_4)\,\&\,T(4)$condition and Theorem 1.7(c), if $\mathcal{L}_{i+1}(D)$ contains at least four regions which have a non-empty common edge with ω_i, then $\mathcal{L}_{i+1}(D)$ contains at least two vertices with valency at least four. We summarize this in the following proposition.

Proposition 1.14. *Let M be a diagram, let D be either a boundary region of M such that $\partial D \cap \partial M$ contains a non-empty edge or a boundary vertex, and let $\Lambda(D)$ be a layer decomposition of M, with center D. Suppose that for some $i, i \geq 1$, $\mathcal{L}_{i+1}(D)$ contains at least four regions which have a non-empty edge on $\omega_i :=$ $\mathcal{L}_i(D) \cap \mathcal{L}_{i+1}(D)$ or a vertex with valency at least five in $\mathcal{L}_{i+1}(D)$. Then M contains a right transversal ν and a left transversal μ with $M_\mu \cap M_\nu = \emptyset$ such that ν starts at $t(\omega_i)$ and μ starts at $h(\omega_i)$. See Fig. 5(a).*

2. Piece configurations of 1-corner regions and 2-corner regions

In this section we assume that the conditions of the Main Theorem are satisfied.

2.1. 1-corner regions

Let D be a 1-corner region in M with neighbor E. Denote $\alpha = \partial D \cap \partial E$, let $P = \Phi(\alpha)$ and let $Q = \Phi(\partial D \cap \partial M)$. See Fig. 6(a). Then P is a piece and $vPuQ$ is a boundary label of D, where u and v are vertices.

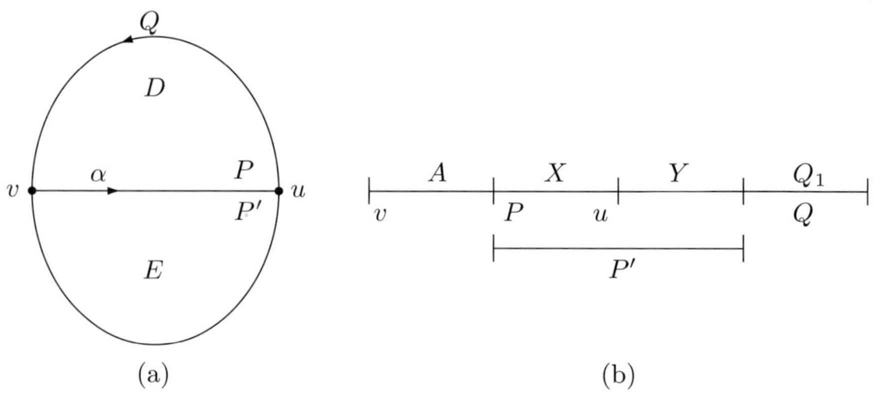

(a)   (b)

Figure 6

Lemma 2.1. *Let notation be as above. Then* $Q \succ P$.

Proof. Let (P, P') be the corresponding piece pair. Then one of the following holds:

1) P' is a subword of Q;
2) P' contains u as an inner vertex and
3) P' contains v as an inner vertex.

In case 1) $Q \succ P$, by Lemma 1.3 (a). Also, in cases 2) and 3), if $|P| = 1$ then $Q \succ P$. Hence assume that $|P| \geq 2$. In case 2) we have $P = AX$, $P' = XY$, $Q = YQ_1$, reduced as written, $Q_1 \in \mathcal{H}(Q)$. See Fig. 6(b). Applying Lemma 1.4 to the first two of these equations and remembering that P^{-1} cannot overlap P in more than one letter (see Lemma 1.2(b)), we get $A \sim Y \succ X$ and hence, by Lemma 1.3, $P \sim Y$. Applying Lemma 1.3 to the last equation implies $Q \succ P$. Finally, Case 3 is dual to Case 2. Hence in all the cases $Q \succ P$.

The lemma is proved. $\qquad\qquad\square$

2.2. 2-corner regions

Let D be a 2-corner region in M with neighbors E_r and E_ℓ. See Fig. 7. Denote $\alpha_1 = \partial D \cap \partial E_r$ and denote $\alpha_2 = \partial D \cap \partial E_\ell$. Let $v_0 = \alpha_1 \cap \partial M$, let $v_2 = \alpha_2 \cap \partial M$ and let $v_1 = \alpha_1 \cap \alpha_2$. Denote $P_1 = \Phi(\alpha_1)$, $P_2 = \Phi(\alpha_2)$ and $Q = \Phi(\partial D \cap \partial M)$. It is convenient and harmless to identify P_i with α_i and, similarly, P'_i with α'_i, $i = 1, 2$. Then $v_2 Q v_0 P_1 v_1 P_2 v_2$ is a boundary label of D, which we may assume to coincide with R, without loss of generality, hence P_1 and P_2 are pieces.

Let (P_1, P'_1) and (P_2, P'_2) be the corresponding piece pairs. Then P'_1 and P'_2 are subwords of the cyclic word R or R^{-1}, hence one of the following holds for each of P'_1 and P'_2:

<u>Case 1</u>	v_0 is an inner vertex of P'_1;	<u>Case 1'</u>	v_2 is an inner vertex of P'_2
<u>Case 2</u>	v_1 is an inner vertex of P'_1;	<u>Case 2'</u>	v_1 is an inner vertex of P'_2
<u>Case 3</u>	v_2 is an inner vertex of P'_1;	<u>Case 3'</u>	v_0 is an inner vertex of P'_2
<u>Case 4</u>	P'_1 is a subword of P_2;	<u>Case 4'</u>	P'_2 is a subword of P_1
<u>Case 5</u>	P'_1 is a subword of Q;	<u>Case 5'</u>	P'_2 is a subword of Q

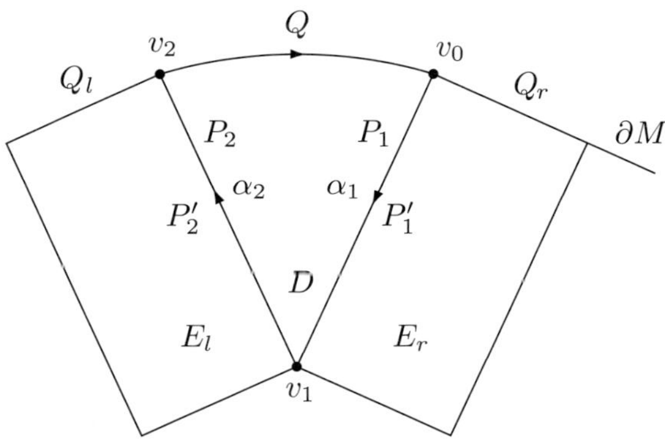

FIGURE 7

We propose to show that in most of the cases $Q \succ P_1 P_2$. (See precise statement below.) We see that for $i = 1, 2, 3, 4$ and 5, Case i' is the dual of Case i obtained by exchanging P_1 with P_2 and v_0 with v_2. Hence out of the 25 cases $(i, j'), 1 \le j', i \le 5$, only those with $j' \ge i$ have to be checked, because the rest is obtained by duality.

The following is the main result of this section.

Proposition 2.2. *Let notation be as above and assume that R satisfies the assumptions of the Main Theorem. Assume that the piece pairs (P_1, P'_1) and (P_2, P'_2) are* *right normalized (see Def. 1.5(b)). Let $Q_r = \partial E_r \cap \partial M$ and let $Q_\ell = \partial E_\ell \cap \partial M$. Then the following hold*

(a) *If both v_0 and v_2 have valency three and both Q_r and Q_ℓ are not pieces (i.e., the products of at least two pieces) then either $Q_\lambda Q \succ R$ or $QQ_\rho \succ R$. In particular, $Q_\lambda QQ_\rho \succ R$.*

(b) *Moreover,*

$\quad Q_\rho \succ P_1$ *in cases* $(1, j')$, $j' = 1, 2, 3, 4, 5$ *and cases* $(2, j')$, $j' = 2, 3, 4, 5$
$\quad Q \succ P_1, P_2$ *in cases* $(3, j)$, $j = 3, 4, 5$ *and cases* $(4, 5)$ *and* $(5, 5)$,
$\quad Q_\rho \succ P_1, P_2$ *and* $Q_\lambda \succ P_1, P_2$ *in case* $(4, 4)$.
Dually,
$\quad Q_\lambda \succ P_2$ *in cases* $(i, 1)$, $i = 1, 2, 3, 4, 5$ *and cases* $(i, 2)$, $i = 2, 3, 4, 5$
$\quad Q \succ P_1, P_2$ *in cases* $(i, 3)$, $i = 3, 4, 5$ *and case* $(5, 4)$.

Proof. As explained above it is enough to check the 15 cases $(i, j), 1 \leq i \leq j' \leq 5$.

Cases (1, j'). $j' = 1, 2, 3, 4, 5$. Since P_1 and P_1' overlap and R contains no elements of order two hence either $\varepsilon(P_1') = -1$ in which case $P_1 \prec Q$ follows from Lemma 1.2 (b), or $\varepsilon(P_1') = 1$. Then we have $P_1' = Q_1 X$, $Q_1 \in \mathcal{T}(Q)$ and $P_1 = XY$. Hence $P_1 = Q_1 X = XY$ and by Lemma 1.4 $Q_1 \sim X \sim Y$. Therefore by Lemma 1.3 $Q \succ Q_1 \succ P_1$ and $P_1 \prec Q$. Hence in all cases

$$P_1 \prec Q. \tag{$*$}$$

It is enough to show that

$$P_2 \prec Q. \tag{$**$}$$

Consider Cases $1'$–$5'$ in turn (i.e., $j' = 1, \dots, 5$).

Case 1$'$. This follows from the above argument with P_1 and P_2 interchanged hence $(**)$ follows from $(*)$.

Case 2$'$. Consider two subcases according as P_2' is a subword of $P_1 P_2$ or is not a subword of $P_1 P_2$. In both cases P_2 and P_2' overlap, hence if $\varepsilon(P_2') = -1$ then due to Lemma 1.2(b)(ii) $P_1 \sim P_2$. Hence we may assume that $\varepsilon(P_2') = 1$.

<u>Subcase 1.</u> P_2' is a subword of $P_1 P_2$. We have $P_1 = UX$, $P_2 = XY = YV$. Consequently, $P_1 \succ X$, $X \sim V \sim P_2$ by Lemma 1.4 and hence $P_2 \prec P_1$ by Lemma 1.3. Hence $(*)$ implies $Q \succ P_2$.

<u>Subcase 2.</u> P_2' is not a subword of $P_1 P_2$. We have $P_2' = Q_1 P_1 X$, $Q_1 \in \mathcal{T}(Q)$, $P_2 = XY$. Hence $(Q_1 P_1)X = XY$ and $Q_1 P_1 \sim Y \sim P_2$. But $Q_1 \prec Q$, by Lemma 1.3 and $P_1 \prec Q$ by $(*)$, hence $P_2 \sim Q_1 P_1$ implies that $P_2 \prec Q$.

Case 3$'$. We may assume that $v_1 \notin P_2'$, by Subcase 2 above. Then $P_2' = Q_1 X$, $Q_1 \in \mathcal{T}(Q)$ and $P_1 = XY$. Therefore $Q_1 \prec Q$ and due to $(*)$ $X \prec Q$. Consequently, $P_2' = Q_1 X \prec Q$, by Lemma 1.3 and $(**)$ follows.

Case 4$'$. If $\varepsilon(P_2') = 1$ then $P_2 \prec P_1$, hence $P_2 \prec Q$, due to $(*)$ and Lemma 1.3. If $\varepsilon(P_2') = -1$ then $P_2^{-1} \prec P_1$, hence we have $P_1 \prec Q$ and $P_2^{-1} \prec Q$.

Case 5$'$. Similar to Case 4$'$.

Case (2, 2′). Consider 4 subcases according to whether v_2 belongs or does not belong to P_1' and dually, v_0 belongs or does not belong to P_2'.

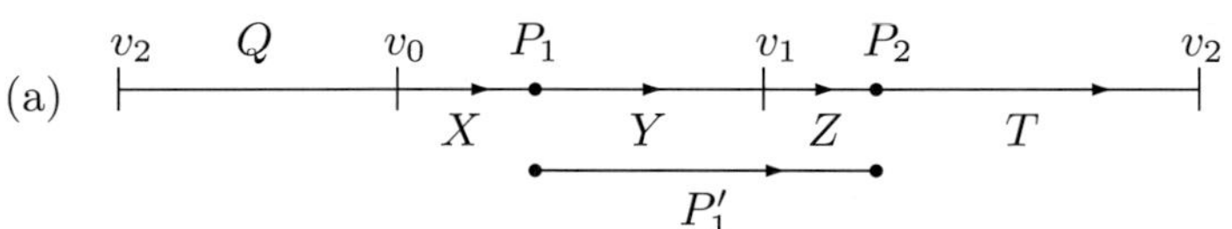

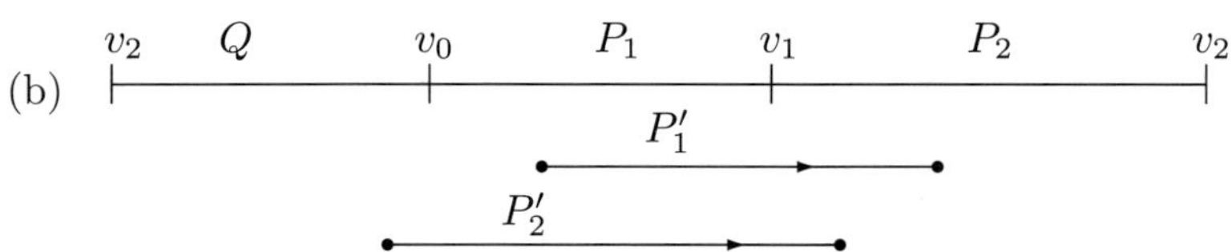

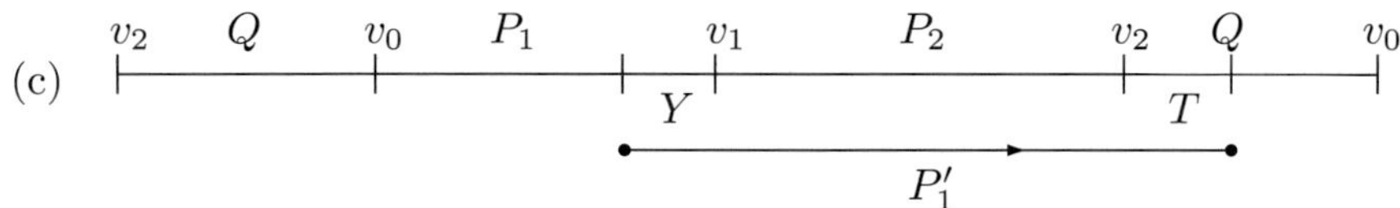

FIGURE 8

Subcase 1. $v_2 \notin P_1'$ and $v_0 \notin P_2'$. See Fig. 8(a). Then $P_1 = XY$, $P_1' = YZ$, $P_2 = ZT$. Suppose $d_M(v_0) = 3$. Fig. 8(a) shows the relative positions of P_1' and P_1 on ∂D, due to the assumption that $v_1 \in P_1'$. (P_1' has an occurrence on ∂D because E_r and D have the same boundary labels.) We see that in order to get P_1 from P_1' we have to turn P_1' along ∂D anticlockwise by $|X|$. Since D and E_r have the same boundary labels (i.e., labels of boundary cycles and their inverses) we can reproduce P_1 on ∂E_r by turning P_1' clockwise along ∂E_r by $|X|$. But P_1' starts at v_0 on ∂E_r just as P_1 starts at v_0 on ∂D. Hence Q_r has X^{-1} as a head, provided that $|Q_r| \geq |X|$. But X is a piece while Q_r is not, hence X is a proper subword of Q_r. We call this occurrence of P_1 on ∂E_r the *image of P_1 on ∂E_r* and denote it by P_1''. Hence $X^{-1} \in \mathcal{H}(Q_r)$. Define $Q_\rho = X$. Then we get $P_1 = Q_\rho Y = YZ$, $P_2 = ZT$. If $|P_1| = 1$ then clearly $Q \succ P_1$, hence assume that $|P_1| \geq 2$. If $\varepsilon(P_1') = -1$ then $Q_\rho Y = Z^{-1}Y$, hence by Lemma 1.2(b) $|Y| = 1$ and $Q_\rho \sim Z^{-1}Y^{-1} = P_1^{-1}$. Hence $Q_\rho \succ P_1$. So assume $\varepsilon(P_1) = 1$. Using Lemma 1.4, the first pair of equations gives

$$Q_\rho \sim P_1. \tag{2.1}$$

But P_2 and P_2' overlap, since v_1 is an inner vertex of P_2', hence $P_2 = LV$, $P_2' = KL$ and $P_1 = UK$. Applying Lemma 1.4 to the first pair of equations, if $|P_2| \geq 2$ and $\varepsilon(P_2') = 1$, then

$$P_2 \sim K, \tag{2.2}$$

186   A. Juhász

while the last equation gives

$$K \prec P_1. \tag{2.3}$$

If $|P_2| = 1$ then (2.2) and (2.3) clearly hold. Also, if $\varepsilon(P_2') = -1$ then as above (2.3) follows from Lemma 1.2(b). Now, from (2.1), (2.2) and (2.3) we get $P_1 \sim Q_\rho,\ P_2 \prec Q_\rho$.

<u>Subcase 2.</u> $v_2 \notin P_1'$ and $v_0 \in P_2'$. See Fig. 8(b). Then (2.1) above still holds, and if $\varepsilon(P_2') = 1$ then for P_2 and P_2' we get $P_2' = HP_1L$, $P_2 = LK$, where $H \in \mathcal{T}(Q)$. If $|P_2| = 1$ then clearly $Q_\rho \succ P_2$, hence assume that $|P_2| \geq 2$. Then, by Lemma 1.4, $HP_1 \sim P_2$, i.e., $QQ_\rho \succ P_2, P_1$. Also, if $\varepsilon(P_2') = -1$ then again $Q_\rho \succ P_2$ follows from Lemma 1.2(b).

<u>Subcases 3 and 4.</u> $v_2 \in P_1'$. See Fig. 8(c). Then $P_1 = XY$, $P_1' = YP_2T$ where $T \in \mathcal{H}(Q)$. As above, we may assume $|P_1| \geq 2$. Therefore, if $\varepsilon(P_1') = 1$ then by Lemma 1.4, $X \sim P_1 \sim P_2T$ and if $\varepsilon(P_1') = -1$ then by Lemma 1.2(b) $|Y| = 1$ and hence $X \sim P_1 \sim P_2T$. Hence $X \in \mathcal{H}(Q_r)$ as in Subcase 1. Thus $P_1, P_2 \prec Q_r$.

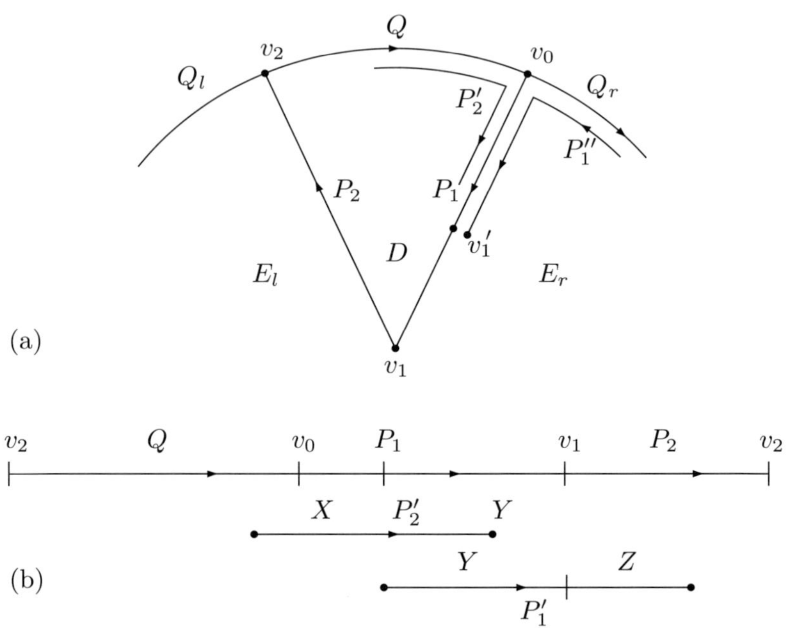

FIGURE 9

Case (2, 3). $v_1 \in P_1'$ and $v_0 \in P_2'$. See Fig. 9(a) and Fig. 9(b). We consider three main cases according as $v_1 \notin P_2'$ and $v_2 \notin P_1'$ or $v_1 \in P_2'$ or $v_1 \notin P_2$ and $v_2 \in P_1'$. In each main case we consider two cases according as $\varepsilon_2 = 1$ or $\varepsilon_2 = -1$.

Main Case 1. $v_1 \notin P_2'$ and $v_2 \notin P_1'$. We have

$$P_2' = Q_1U, P_1 = UV, P_1 = XY, P_1' = YZ \text{ and } P_2 = ZW, h(W) \neq h(Z) \tag{2.4}$$

where all subwords Q_1, U, V, X, Y, Z and W are non-empty, $Q_1 \in \mathcal{T}(Q)$ and P_1, P_2, P_1' and P_2' are reduced as written. Also, $h(W) \neq h(Z)$ because the piece-pair (P_1, P_1') is right normalized. Assume $\varepsilon_1 = -1$. Then $XY = Z^{-1}Y^{-1}$ hence by Lemma 1.2 $|Y| = 1$ and $X \sim Z \succ Y$. Hence $P_1^{-1} = yZ$ and $P_2 = ZW = y^{-1}P_1^{-1}W = y^{-1}V^{-1}U^{-1}W = Q_1U$, where $y = Y$.

Consider the equation $y^{-1}V^{-1}U^{-1}W = Q_1U$. If $U = U_0W$ then $U_0 \neq 1$ and $Q_1U_0 = y^{-1}V^{-1}W^{-1}U_0^{-1}$, hence $|U_0| = 1$ and $Q_1 \succ U, V$. Consequently, $Q_1 \succ P_1, P_2$ (by the first two equations of (2.4)). Therefore, may assume $W = Q_2U$, where $Q_1 = y^{-1}V^{-1}U^{-1}Q_2$. But then again $Q_1 \succ U_1, V, Y$, hence $Q_1 \succ P_1, P_2$. Therefore, may assume $\varepsilon_1 = 1$. Then

$$Q_1U = ZW \quad \text{and} \quad XY = YZ = UV. \tag{2.4'}$$

Observe that

$$Q_1 \in \mathcal{T}(Q) \quad \text{and} \quad X^{-1} \in \mathcal{H}(Q_r). \tag{2.4''}$$

It follows from (2.4') and Lemma 1.2 that

$$X = KL, \ Y = (KL)^kK, \ k \geq 0, \ Z = LK. \tag{2.4'''}$$

It follows from (2.4') and Lemma 1.4 that $X \succ Y, Z$ and by the second and fourth equations in (2.4) also $X \succ U, V$. Hence if we define $Q_\rho = X^{-1}$ then $Q_\rho \succ Y, Z, U, V$. Since $P_2' = Q_1U$ hence $QQ_\rho \succ Q_1, Q_\rho$ and hence $QQ_\rho \succ P_2, P_1$.

Main Case 2. $v_1 \in P_2'$. Then $P_2' = Q_1P_1U$, $P_2 = UV$, $Q_1 \in \mathcal{T}(Q)$, all expressions reduced as written. Assume that $\varepsilon_2 = -1$. Then $P_2 = U^{-1}P_1^{-1}Q_1^{-1} = UV$, hence by Lemma 1.2(b)(i) $|U| = 1$ and $P_2 \sim P_1^{-1}Q_1^{-1}$. Therefore $QQ_\rho \succ Q_1Q_\rho \succ P_1, P_2$. Finally, assume $\varepsilon_2 = 1$. Then $P_2' = Q_1P_1U$, $P_2 = UV$. Consequently, $v_2 \notin P_1'$ since $|P_2| > |P_1'|$ and hence $P_1 = XY$, $P_1' = YZ$ and $P_2 = ZW$. The equations $P_2 = UV = Q_1P_1U$ imply that $V \sim P_2 \sim Q_1P_1$, by Lemma 1.4. Also, we observe that $V^{-1} \in \mathcal{T}(Q_\ell)$. Therefore taking $Q_\lambda = V^{-1}$, we get $Q_\lambda \succ P_1^{-1}, P_2^{-1}$.

Main Case 3. $v_1 \notin P_2'$ and $v_2 \in P_1'$. Then

$$P_1 = XY, \ P_1' = YP_2Q_0, \ Q_0 \in \mathcal{H}(Q),$$
$$P_2' = Q_1U, \ P_1 = UV, \ X^{-1} \in \mathcal{H}(Q_r). \tag{2.5}$$

If $\varepsilon(P_1') = 1$ then the first two equations of (2.5) yield $X \sim P_2Q_0 \sim P_1$, hence taking $Q_\rho = X^{-1}$, gives $Q_\rho \succ P_1, P_2$. If $\varepsilon_1 = -1$ then $XY = Q_0^{-1}P_2^{-1}Y^{-1}$, hence by Lemma 1.2(b)(i) $|Y| = 1$ and $P_1 = Q_0^{-1}P_2^{-1}Y^{-1}$. Hence $P_1 = UV = Q_0^{-1}(Q_1U)^{-\varepsilon_2}Y^{-1}$. If $\varepsilon_2 = -1$ then $UV = Q_0^{-1}Q_1UY^{-1}$, hence $UVy = Q_0^{-1}Q_1U$, where $y = Y$. Consequently $Q_0^{-1}Q_1 \sim Vy \succ U$, hence $Q_0^{-1}Q_1 \succ UV = P_1 \succ P_2$ and hence $Q \succ P_1, P_2$. If $\varepsilon_2 = 1$ then $UV = Q_0^{-1}U^{-1}Q_1^{-1}Y^{-1}$. Hence $V = V_0Q_1^{-1}Y^{-1}$ and $UV_0 = Q_0^{-1}U^{-1}$. If $Q_0^{-1} = UQ_2$ then $V_0 = Q_2U^{-1}$, hence $Q_0 \succ U, V_0$ and hence $Q \succ Q_0, Q_1 \succ U, V$. Consequently $Q \succ P_1, P_2$. If $U = Q_0^{-1}U_1$ then $U_1V_0 = U_1^{-1}Q_0$, hence $|U_1| = 1$, $U \sim Q_0$ and $Q \succ U$. Also, $V_0 \sim Q_0$, hence $Q \succ V_0Q_1^{-1}Y^{-1} = V$. Therefore $Q \succ U, V$ and consequently $Q \succ P_1, P_2$. Since the equation $UV_0 = Q_0^{-1}U^{-1}$ leads either to $Q_0^{-1} = UQ_2$ or $U = Q_0^{-1}U_1$, this completes the proof of Main Case 3.

188   A. Juhász

Case (2, 4). P_2' is a subword of P_1. Assume first that v_2 is not an inner vertex of P_1'. Then

$$P_1 = XY, \ P_1' = YZ, \ P_2 = ZW \text{ and } P_1 = UP_2'V. \tag{2.6}$$

If $\varepsilon(P_1') = 1$ then the first two equations of (2.6) imply $P_1 \sim X \sim Z \succ Y$ and the last equation of (2.6) implies $P_1 \succ P_2'$. Consequently, $X \succ P_1, P_2'$.

Define $Q_\rho = X^{-1}$. Then $Q_\rho \succ P_1, P_2$. Assume now that v_2 is an inner vertex of P_1'. Assume $\varepsilon(P_1') = -1$. Then $XY = Q_0^{-1}P_2^{-1}Y^{-1} = UP_2^{\varepsilon_2}V$. By Lemma 1.2(b)(i) $|Y| = 1$ and $X = Q_0^{-1}P_2^{-1}y^{-2}$, where $y = Y$, and $P_1 = Q_0^{-1}P_2^{-1}y^{-1}$. Also, $Q_0^{-1}P_2^{-1} = UP_2^{\varepsilon_2}Vy$. Hence $|Q_0| \geq |U|$. Let $Q_0^{-1} = UQ_2^{-1}$. Then $Q_2^{-1}P_2^{-1} = P_2^{\varepsilon_2}Vy$. If $\varepsilon_2 = -1$ then $Q_2 \sim Vy \succ P_2$, hence it follows from $P_1 = Q_0^{-1}P_2^{-1}y^{-1}$ that $Q \succ Q_2$, $Q_0 \succ P_1, P_2$, i.e., $Q \succ P_1, P_2$. If $\varepsilon_2 = 1$ then $Q_2^{-1}P_2^{-1} = P_2Vy$. If $Q_2^{-1} = P_2Q_3^{-1}$ then $Q_3^{-1}P_2^{-1} = Vy$ hence $Q_2 \succ P_2, Q_3, Vy$. Consequently $Q_2, Q_0 \succ P_1$. Therefore $Q \succ P_1, P_2$. If $P_2 = Q_2^{-1}S$ then $P_2^{-1} = SVy$, hence $y^{-1}V^{-1}S^{-1} = Q_2^{-1}S$. Consequently, $|S| = 1$ and $Q_2 \sim y^{-1}V^{-1} \sim P_2$. But then $P_1 = Q_0^{-1}P_2^{-1}y^{-1}$ implies that $Q \succ P_1, P_2$. Assume therefore that $\varepsilon_1 = 1$. Then

$$P_1 = XY, \ P_1' = YP_2Q_0, \ Q_0 \neq 1, \ Q_0 \in \mathcal{H}(Q) \text{ and } P_1 = UP_2'V. \tag{2.7}$$

The first two equations of (2.7) imply $X = KL$, $P_2Q_0 = LK$ and $Y = (KL)^\alpha K$, $\alpha \geq 0$. Therefore P_2' is a subword of $P_2Q_0P_2$. It follows that $P_2 \succ Q_0$. Consequently, $Q_0 \succ P_2, P_1, L, K, X, Y$, hence $Q \succ Q_0 \succ P_1, P_2$.

Case (2, 5). Assume $v_2 \notin P_1'$. Assume first $\varepsilon(P_1') = 1$. Then $P_1 = XY$, $P_1' = YZ$ and $P_2 = ZW$. By the first two equations, via Lemma 1.4 $P_1 \sim Z$ and by the third equation $P_2 \succ Z$. Hence $P_1 \sim Z \prec P_2 \prec Q$. Consequently, $Q \succ P_2 \succ P_1$, hence $P_1, P_2 \prec Q$. Assume $\varepsilon(P_1') = -1$. Then $XY = Z^{-1}Y^{-1}$, hence $|Y| = 1$ and $X \sim Z$. Since $Q \succ P_2 \succ Z$, hence $Q \succ P_2, P_1$.

Assume now that $v_2 \in P_1'$. Then $P_1 = XY$, $P_1' = YP_2Q_0$, $Q_0 \neq 1$, $Q_0 \in \mathcal{H}(Q)$. Assume first $\varepsilon(P_1') = 1$. Then, by Lemma 1.3(a) $P_2Q_0 \succ P_1$. If $\varepsilon(P_2') = 1$ then $Q \succ P_2$, hence $Q \succ P_2Q_0 \succ P_1$, i.e., $Q \succ P_1, P_2$. Suppose $\varepsilon(P_2') = -1$. By Lemma 1.2 $X = KL$, $Y = (KL)^\alpha K$, $\alpha \geq 0$ and $P_2Q_0 = LK$. Consequently, $P_2Q_0 \succ P_1P_2$ hence $Q \succ P_1, P_2$. Assume $\varepsilon(P_1') = -1$. Then $XY = Q_0^{-1}P_2^{-1}Y^{-1}$, hence $|Y| = 1$ and $P_1 \sim Q_0^{-1}P_2^{-1}$. Consequently $P_1 \prec Q$. Hence $Q \succ P_1, P_2$.

Case (3, 3′). The cases when $v_1 \in P_1'$ or $v_1 \in P_2'$ can be dealt with in a way similar to Subcases 2 and 3 of Case $(2, 2')$. So we concentrate on the case $v_1 \notin P_1'$ and $v_1 \notin P_2'$. We have $P_1 = XU$, $P_1' = YQ_0$, $P_2 = VY$, $P_2' = Q_1X$ where $Q_0 \in \mathcal{H}(Q)$ and $Q_1 \in \mathcal{T}(Q)$. Due to the second and fourth equations, it is enough to prove $X, Y \prec Q$. If $|P_1| = 1$ or $|P_2| = 1$ then it easily follows that $P_1, P_2 \prec Q$. So assume that $|P_1| \geq 2$ and $|P_2| \geq 2$.

<u>Subcase 1.</u> $\varepsilon_1 = \varepsilon_2 = 1$. Then $Q_1X = VY$ and $XU = YQ_0$. If $X = Y$ then $Q_1 = V$ and $U = Q_0$, hence $h(U) = h(Q_0) = h(Q)$, violating the right normalization of the piece pair (P_2, P_2') (see Definition 1.5(b)). If $X = YZ$, $Z \neq 1$, then $Q_0 = ZU$, $Q_1YZ = VY$, hence $V = Q_1T$, $YZ = TY$ and, by Lemma 1.4, $Y \prec Z \sim T$.

Therefore $X = YZ$ implies $Y \prec Z \sim X \sim T$. But $Q_0 = ZU$ implies $Q_0 \succ Z$. Consequently, $Q \succ Q_0 \succ X, Y$. If $Y = XZ$, $Z \neq 1$, then $U = ZQ_0$, $Q_1 X = VXZ$. Hence $Q_1 = VT$, $TX = XZ$ and, as in previous case, $T \sim X \sim Z \prec Q_1$, hence $Y = XZ \prec Q_1$, by Lemma 1.3. Thus, $X, Y \prec Q$.

Subcase 2. $\varepsilon_1 = 1$ and $\varepsilon_2 = -1$. Then $XU = YQ_0$ and $VY = X^{-1}Q_1^{-1}$. If $X = Y$ then $U = Q_0$ and $VX = X^{-1}Q_1$. If $V = X^{-1}V_1$ then $Q_1 = V_1 X$, hence $Q_1 \succ X, Y$. If $X^{-1} = VZ$, then $Y = ZQ_1$, hence $X = ZQ_1 = V^{-1}$. Since R contains no elements of order 2, hence, by Lemma 1.2 (b), $|Z| = 1$ and the result follows.

Subcase 3. $\varepsilon_1 = -1$ and $\varepsilon_2 = 1$. This subcase dual to Subcase 2.

Subcase 4. $\varepsilon_1 = -1$ and $\varepsilon_2 = -1$. This subcase leads to $Q_0, Q_1 \succ X, Y$, by checking the cases $X = Q_0^{-1}$, $Q_0^{-1} = XQ_2$ and $X = Q_0^{-1}X$, respectively.

Case (3, 4). We have $P_2 = UV$, $P_1' = VQ_0$, $P_1 = HP_2'T$. There are four cases to check, according as $\varepsilon_1 = \pm 1$ and $\varepsilon_2 = \pm 1$. We shall check only the cases $\varepsilon_1 = 1$ and $\varepsilon_2 = \pm 1$. The other two cases are similar.

Subcase 1. $\varepsilon_1 = 1$, $\varepsilon_2 = 1$. Then $H(UV)T = VQ_0$. Hence $Q_0 = Q_0'T$ and $VQ_0' = HUV$. Consequently, $Q_0' \succ V, H, U$, hence $Q \succ P_1, P_2$.

Subcase 2. $\varepsilon_1 = 1$, $\varepsilon_2 = -1$. Then $H(V^{-1}U^{-1})T = VQ_0$, hence $Q_0 = Q_0'U^{-1}T$ and $VQ_0' = HV^{-1}$. If $V^{-1} = ZQ_0'$ then $V = HZ$, hence it follows from Lemma 1.2 (b) and the assumption that no subword of $R^* \in \mathcal{R}$ has order two, that $|Z| = 1$ and $Q_0' \succ V$. If $Q_0' = ZV^{-1}$ then $Q_0' \succ Z, V$ hence $Q_0 \succ Z, V, U, T$. Consequently, $Q \succ P_1, P_2$.

Case (3, 5'). Then $P_1' = VQ_0$, $Q_0 \in \mathcal{H}(Q)$, $P_2 = UV$, hence because P_2' is a subword of Q, $Q \succ P_2$. Consequently, $Q \succ U, V$, hence $Q \succ P_1, P_2$.

Case (4, 4'). Then P_1' is a subword of P_2 and P_2' is a subword of P_1. Without loss of generality, we may assume $|P_1| \geq 2$. Since R contains no elements of order 2, it follows that $\varepsilon_1 = \varepsilon_2$. If $\varepsilon_1 = 1$ then Q_ρ has P_1^{-1} as a head and Q_λ has P_2^{-1} as a tail, and we are done. Therefore, we may assume $\varepsilon_1 = -1$, $\varepsilon_2 = -1$. Then $|P_1| = |P_2^{-1}|$, hence $P_1 = hP_2^{-1}t$ and $P_2 = tP_1^{-1}h$, where $\alpha(h) = \alpha\big(h(P_1)\big)$ and $\alpha(t) = \alpha\big(h(P_1)\big)$. Then $P_1 P_2 = hP_2^{-1}tP_2$, hence $R = P_1 P_2 Q = hP_2^{-1}tP_2 Q$. But it is an assumption of the theorem that such word cannot occur in $\mathcal{R}$, a contradiction. Therefore, $Q_\rho \succ P_1, P_2$ and $Q_\lambda \succ P_1, P_2$.

Cases (4, 5') and (5, 5'). It follows immediately that $Q \succ P_1, P_2$.

The proposition is proved. $\qquad\square$

We close this section with the following corollary to the proposition.

Proposition 2.3. *Let M be an $\mathcal{R}$-diagram. Let P be a peak relative to a layer decomposition Λ. Let $\alpha = \partial P \cap \partial M$. Then $\Phi(\alpha)$ contains a letter from each component.*

Proof. Let $P = \langle D_1, \ldots, D_k \rangle$. If $k = 1$ then the result follows from Lemma 2.1. If $k \geq 3$ then it follows from Theorem 1.7(d) and the $T(4)$ condition that either P contains a 1-corner region or a 2-corner region D with two neighbors E_r and E_ℓ such that $\partial D \cap \partial E_r \cap \partial M$ and $\partial D \cap \partial E_\ell \cap \partial M$ are vertices with valency three and $\partial E_r \cap \partial M$ and $\partial E_\ell \cap \partial M$ are not pieces (due to the $C'(^1/_4)$ condition). In both cases the result follows by part (a) of Proposition 2.2. See Fig. 9(a), where D_1 is E_ℓ, D_2 is D and D_3 is E_r.

Finally, assume $k = 2$. See Fig. 3(b). Let $P = \langle D_1, D_2 \rangle$. Then both D_1 and D_2 are 2-corner regions, hence if we remove any of them, say D_1, then D_2 will be a 1-corner region. Therefore $\partial(M \setminus D_1) \cap \partial D_2$ contains a letter from every component, by the case $k = 1$. Consequently, if $\beta = \partial D_1 \cap \partial D_2$ and we show that if $\beta_1 = \partial D_1 \cap M$ and $\beta_2 = \partial D_2 \cap M$ then $\widehat{Q} := \Phi(\beta_1 \cup \beta_2) \succ \Phi(\beta)$, then $\widehat{Q}$ contains a letter from every component. But this follows from the "Moreover" part of Proposition 2.2, since $\Phi(\beta)$ is P_1 for D_1 and P_2^{-1} for D_2, in the notations of Proposition 2.2. Hence $\Phi(\beta) \prec \Phi(\beta_2)$ or $\Phi(\beta) \prec \Phi(\beta_1)$, respectively.

The proposition is proved. $\square$

3. Proof of the theorem

Denote by π the natural projection of F on G. Let L be a proper subset of $\{1, \ldots, n\}$ and let $\widetilde{H} = \underset{i \in L}{*}(G_i)$. Let $H = \pi\left(\widetilde{H}\right)$. Then H is a Magnus subgroup of G. Let U be a reduced word in F, $|U| \geq 2$. Without loss of generality, assume $\operatorname{Supp}(R) = \{1, \ldots, n\}$.

Our proof is based on the following proposition.

Proposition 3.1. *Let M be a connected, simply connected $\mathcal{R}$-diagram and let d be the highest valency the vertices of M have. (Thus, $d = \underset{v \in M}{\max} d(v)$.) Let μ be a boundary path of M and let ν be its complement on ∂M. (Thus, $u\mu v \nu^{-1}$ is a boundary cycle of M, where u and v are vertices.) Let $\Phi(\mu) = U = U_1 \cdots U_p$ and let $R = R_1 \cdots R_q$ in normal forms. Let $S = \left\{U_i^{\pm 1}, R_j^{\pm 1}, \ 1 \leq i \leq p, \ 1 \leq j \leq q\right\} \cup \{1\}$ and for every natural number n let S^n be the set of all the products of n elements from S. Suppose that U and $\Phi(\nu)$ are Dehn-reduced. Then each of the following holds:*

(a) *If η is a simple boundary path of M with primary endpoints then*

$$\Phi(\eta) \in \bigcup_{t=0}^{d|\eta|} S^t.$$

(b) *If $\Phi(\nu) \in \widetilde{H}$ and $|\nu| \leq c|\mu|$ for some known positive number c, then $\Phi(\nu)$ is effectively computable from S.*

(c) *If $\Phi(\nu)$ is a shortest representative of $\pi(U)$ in $\widetilde{H}$ then $\pi(U)$ has a shortest representative in F which is effectively computable from S.*

Proof. (a) Since U is Dehn-reduced, μ contains no double points, by $(*)$ in Remark 9. Consequently, M has the form like the diagram in Fig. 10(a). Now, if $M_1, \ldots, M_k$ are the connected components of $\mathrm{Int}(M)$ then the subpaths μ_i for $i = 0, \ldots, k$, which are the connected components of $M \setminus \bigcup_{i=1}^{k} \{\overline{M_i}\}$, are subpaths of μ (and ν). Hence, if e ia an edge of M (i.e., a path with primary endpoints and having length 1) then one of the following holds for e:

(i) $e \subseteq \mu_i$ for some i, $i = 0, \ldots, k$;

(ii) e is an edge of M_i, for some i, $i = 1, \ldots, k$;

(iii) $e = e_1 w e_2$, where w is a secondary vertex of M, e_1 is a terminal subpath of μ_i for some i, $i = 0, \ldots, k$ and e_2 is a half edge on ∂M_{i+1} for $i = 0, \ldots, k-1$. (See Fig. 10(b).)

(iv) $e = e_3 w' e_4$, where w' is a secondary vertex of M, e_3 is a half edge on ∂M_i and e_4 is an initial subword of μ_{i+1}, for some $i = 1, \ldots, k$. (See Fig. 10(b).)

We propose to show that in each of these cases $\Phi(e) \in \bigcup_{t=0}^{d-1} S^t$. This will clearly prove part (a).

Now, in case (i) $\Phi(e) = U_j^{\pm 1}$ for some j, $1 \leq j \leq p$, hence $\Phi(e) \in S$. In case (ii) there are two subcases to consider: either e does not contain any secondary vertex, in which case $\Phi(e) = R_j^{\pm 1}$ for some j, $1 \leq j \leq q$, hence in particular $\Phi(e) \in S$, or e contains a secondary vertex w, in which case $e = e_1 w e_2$, where e_1 and e_2 are half edges. (See Fig. 10(b).) Let $d_M(w) = d_1$. We show that $\Phi(e) \in S^{d_1 - 1}$. Since $d_M(w) = d_1$ and since w is a boundary vertex of M (since η is a boundary path of M), hence there are exactly $d_1 - 1$ regions $D_1, \ldots, D_{d_1 - 1}$, which contain w on their boundary. (See Fig. 10(c).) Let $\delta_1, \ldots, \delta_{d_1 - 1}$ be the edges of $\partial D_1, \ldots, \partial D_{d_1 - 1}$ respectively, which contain w. Then $\Phi(e) = \Phi(\delta_1) \Phi(\delta_2) \cdots \Phi(\delta_{d_1 - 1})$ is in a component G_i of F. Since $\Phi(\delta_i) = R_j^{\pm 1}$ for some j, $0 \leq j \leq q$, hence certainly $\Phi(\delta_i) \in S$, hence $\Phi(e) \in S^{d_1 - 1}$. Similar argument shows that for case (iii) $\Phi(e) \in S^{d_1 - 1}$, where $d_1 = d_M(w)$, this time $\Phi(\delta_1) = U_j^{\pm 1}$ and for case (iv) $\Phi(\delta_{d_1 - 1}) = U_j^{\pm 1}$, where $j = 1, \ldots, p$. Hence, if $\eta = \eta_1 v_1 \eta_2 \cdots v_{t-1} \eta_t$, η_i edges, then $\Phi(\eta) = \Phi(\eta_1) \Phi(\eta_2) \cdots \Phi(\eta_t) \in \bigcup_{i=0}^{td} S^i = \bigcup_{i=0}^{|\eta|d} S^i$, as required.

(b) Let $\mu = \zeta_1 [\mu] \zeta_2$ and $\nu = \theta_1 [\nu] \theta_2$ be the integral decompositions of μ and ν, respectively. (See Fig. 10(d).) ($\zeta_j = \emptyset$ or $\theta_j = \emptyset$ for $j = 1, 2$ are not excluded. This happens if u or v or both are primary vertices.) Then $\theta_1^{-1} u \zeta_1$ and $\zeta_2 v \theta_2^{-1}$ are edges of M, hence we may apply the argument of the proof of part (a) to get $\Phi(\theta_j) \in \bigcup_{i=0}^{d} S^i$ for $j = 1, 2$ and $\Phi(\zeta_j) \in \bigcup_{i=0}^{d} S^i$ for $j = 1, 2$. Hence $\Phi(\nu) \in \bigcup_{i=0}^{d|\nu|} S^i$.

Denote $\mathcal{S} = \bigcup_{i=0}^{d|\nu|} S^i$. Then $\mathcal{S}$ is a set of words computable from S since $|\nu| < c|\mu|$ and $c|\mu|$ is given. Let $\mathcal{S}_0$ be the subset of all the elements of $\mathcal{S}$ which are equal to

U in G. Since G has solvable word problem due to the condition $C'(^1/_4)\,\&\,T(4)$, $\mathcal{S}_0$ is computable from S. Finally, let $\mathcal{S}_1$ be the subset of $\mathcal{S}_0$ of the elements of $\widetilde{H}$ in $\mathcal{S}_1$. By assumption, $\mathcal{S}_1$ is non-empty, hence $\mathcal{S}_1$ can be constructed from S. (It follows from Proposition 2.3 that every word in $\widetilde{H}$ has a unique reduced expression as an element of $\widetilde{H}$. Hence in fact $\mathcal{S}_1$ consists of a single word. But we do not need this more precise result.)

(c) Like in part (b) $\Phi(\nu) \in \mathcal{S}_0$ and $\mathcal{S}_0$ can be effectively constructed from S. Choose a shortest element from $\mathcal{S}_0$, it satisfies the requirement of the proposition.

The proposition is proved. $\qquad\square$

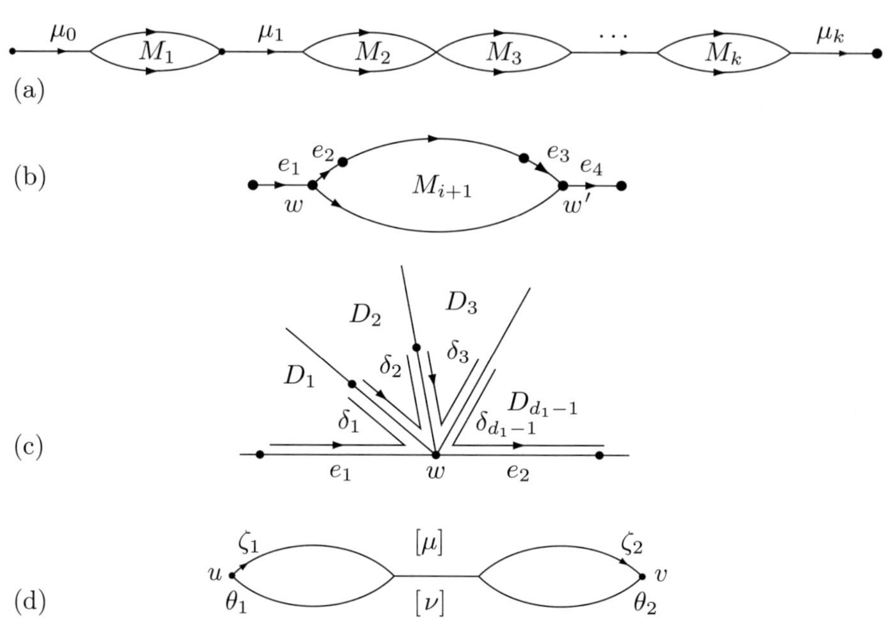

FIGURE 10

Proposition 3.1 together with Lemma 1.10 imply the following proposition.

Proposition 3.2. *Let W be a reduced word in F. Then a shortest representative U of $\overline{W} \in G$ can be obtained from W effectively.*

Proof. By Lemma 1.10 we may carry out (effectively) at most $|W|$ Dehn-reductions to get a reduced word W_1 which cannot be Dehn-reduced any further, such that $W \overset{=}{_G} W_1$. By Lemma 1.11 there is a one-layer $\mathcal{R}$-diagram M with boundary cycle $\mu\nu^{-1}$ such that $\Phi(\mu) = W_1$ and $\Phi(\nu)$ is a shortest representative of $\overline{W}$ and such that all vertices of M have valency at most four. But then by by Proposition 3.1 U can be obtained effectively.

The proposition is proved. $\qquad\square$

We start the proof of the theorem by two observations concerning U and V. First, it follows from the definition of Magnus subgroup that since $\mathrm{Supp}(R) = \{1, \ldots, n\}$ by assumption, we have

$$V \nsucc R, \quad \text{for every reduced word } V \text{ in } \widetilde{H}. \tag{3.1}$$

Next, let U be a reduced word in F. Then we can decompose U as $U = U_1 Z_1 \cdots U_k Z_k$, $k \geq 1$ where U_1 and Z_k may be empty, all the other U_i and Z_i non-empty, such that Z_i is an $\mathcal{R}$-word while U_i contains no $\mathcal{R}$-words, except perhaps the initial and the terminal letters, if U_i is not integral in U.

We divide the rest of the proof into three steps. In Step 2 we assume that M has connected interior.

Step 1. *Let V be a reduced word in $\widetilde{H}$. If M is any diagram which contains a boundary path η with $\Phi(\eta) = V$, then η cannot contain a peak, for any layer decomposition Λ of M.*

This is because if P is a peak relative to a layer decomposition Λ and $\alpha = \partial P \cap \partial M$, then by Proposition 2.3 $\Phi(\alpha) \succ R$. But since α is a subpath of η by assumption, hence $\Phi(\alpha) \prec V$ and therefore $V \succ R$, contradicting (3.1).

Step 2. *Let μ and ν be boundary paths of M with $\Phi(\mu) = U$, $\Phi(\nu) = V$ such that $u\mu v\nu^{-1}$ is a boundary cycle of M, where u and v are vertices. If μ contains no peaks then$|V| \leq 18|U||R|$ and every vertex of M has valency at most four.*

Proof. Let Λ be the layer decomposition of M with center u. Assume Λ has s layers. For $i = 1, \ldots, s-1$ let $\omega_i = \partial \mathcal{L}_i \cap \partial \mathcal{L}_{i+1}$ and for $i = 1, \ldots, s$ let $\mu_i = \partial \mathcal{L}_i \cap \mu$ and $\nu_i = \partial \mathcal{L}_i \cap \nu$. Then $\mu = \mu_1 \cdots \mu_s$, $\nu = \nu_1 \cdots \nu_s$ and by the definition of layers we have

$$1 \leq \mu_i \quad \text{and} \quad 1 \leq \nu_i. \tag{3.2}$$

Moreover, since every secondary vertex of μ (and ν) is followed by a primary vertex, hence due to (3.2)

$$\frac{1}{2} s \leq |\mu| \quad \text{and} \quad \frac{1}{2} s \leq |\nu|. \tag{3.3}$$

Suppose M has a layer $\mathcal{L}_{i+1}$ which contains at least four regions with a non-empty edge on ω_i or a vertex with valency at least five in $\mathcal{L}_{i+1}$. Then by Proposition 1.14, M has transversals α and β starting at the endpoints of ω_i, respectively, such that $M_\alpha \cap M_\beta = \emptyset$. But then it follows from Theorem 1.7 and Proposition 1.13 that M_α and M_β contain disjoint peaks P_α and P_β, respectively. Since $u\mu v\nu^{-1}$ is a boundary cycle of M, this implies that either $\partial P_\alpha \cap \partial M \subseteq \mu$ or $\partial P_\beta \cap \partial M \subseteq \nu$, where M_α is to the left of M_β. This however violates Step 1 and the assumption, that μ contains no peaks, respectively. Therefore, every vertex of M has valency at most four and every layer $\mathcal{L}_i$ of M contains at most three regions, which have a common edge with ω_i. We compute the number of regions in $\mathcal{L}_{i+1}$. Let a be the number of regions of $\mathcal{L}_{i+1}$ which have a common edge with ω_i and let b be the number of regions of $\mathcal{L}_{i+1}$ which have only a common vertex with ω_i. Then $|\mathcal{L}_{i+1}| = a + b$. Now, if $\omega_i = w_0 \theta_1 w_1 \theta_2 \cdots w_r \theta_r w_{r+1}$, where w_j vertices and θ_i

edges, then $r = a$ and each vertex contributes at most $d(w) - 3 = 1$ to b, if $w \neq w_0$ and $w \neq w_r$ and w_0 and w_r contribute at most $d(w) - 2 = 2$. Consequently, $b \leq 1 \cdot (r-1) + 2 \cdot 2 \leq r + 3$. Since $r \leq 4$, $|\mathcal{L}_{i+1}| = a + b \leq 3 + 3 + 3 = 9$.

Hence, for every layer $\mathcal{L}_i$ of Λ, we have $|\mu_i| \leq 9|R|$ and $|\nu_i| \leq 9|R|$. Together with (3.2) we get

$$1 \leq |\mu_i| \leq 9|R| \quad \text{and} \quad 1 \leq |\nu_i| \leq 9|R|. \tag{3.4}$$

But then by (3.3) and (3.4) we have $\dfrac{1}{2} s \leq |\mu| \leq 9s|R|$ and $\frac{1}{2} s \leq |\nu| \leq 9s|R|$. Consequently, $|\nu| \leq 9s|R| \leq 18|\mu||R| = 18|U||R|$, and every vertex of M has valency at most four, as required.

Step 3. *Completion of the proof.* Let U be a reduced word in F, $U \neq 1$ with $\pi(U) \neq 1$ and $\pi(U) \in H$. Then there exists a reduced word $V \in \widetilde{H}$ with $\pi(UV^{-1}) = 1$. Therefore, there exists a diagram M with boundary cycle $u\mu v \nu^{-1}$ with $\Phi(\mu) = U$ and $\Phi(\nu) = V$, where u and v vertices.

Suppose first that μ contains no peaks. Then μ has no double points, because if μ contains a double point w then w is the initial and terminal vertex of a nullhomotopic subpath μ_0 of μ, which is a boundary cycle of a simply connected, connected subdiagram of M which contains at least one region. Hence, μ_0 contains a peak, by Theorem 1.7 and Definition 1.8. It follows from Step 1 and the last argument for μ that ν has no double points. Hence M looks as depicted on Fig. 10(a). Applying Step 2 to every connected component M_i of M, it follows that $|V| \leq 18|U||R|$ and every vertex of M which is not common to μ and ν has valency at most five. Those in $\mu \cap \nu$ have a priori valency at most ten, but it follows from the analysis of transversals that, in fact, they have valency at most five. Consequently, by Proposition 3.1(b), V can be written down effectively, knowing U. This solves the Membership Problem for H, provided that μ contains no peaks.

So, we may assume that μ contains a peak. Now, due to Proposition 3.2 we may assume that $\Phi(\mu)$ is a shortest representative of $\pi(\Phi(\mu))$. But then μ cannot contain a peak, because extremal regions of peaks, Dehn reduce μ. (See Example 1.9.) A contradiction, which proves the theorem.

The theorem is proved. $\qquad\square$

Acknowledgments

I am grateful to the referee for his useful remarks which improved the presentation of the paper.

References

[D-H]   A. Duncan and J. Howie, One relator products with high-powered relators, in: G.A. Niblo and M.A. Roller (eds.), LMS Lecture Notes, No. **181**, (1993), 48–74.

[H]   J. Howie, How to generalise one-relator groups theory, in: *Ann. Math. Studies* **111** (1987), 89–115.

[J1]   A. Juhász, Small cancellation theory with a unified small cancellation condition, *J. London Math. Soc.* **2(40)** (1989), p. 57–79.

[J2]   A. Juhász, A strengthened Freiheitssatz, *Math. Proc. Camb. Phil. Soc.* **140** (2006), p. 15–35.

[J3]   A. Juhász, On quasiconvexity of Magnus subgroups in one-relator free products, in preparation.

[K-M-W] I. Kapovich, R. Wideman and A. Myasnikov, Foldings, graphs of groups and the membership problem, *J. Algebra* **125** (1997), 123–190.

[L-S]   R.C. Lyndon and P.E. Schupp, *Combinatorial Group Theory*, Springer Verlag, 1977.

[M]   W. Magnus, Das Identitätsproblem für Gruppen mit einer definierenden Relation, *Math. Ann.* **106** (1932), 295–307.

Arye Juhász
Department of Mathematics
Technion – Israel Institute of Technology
32000 Haifa, Israel

Geometric Group Theory

Trends in Mathematics, 197–224

© 2007 Birkhäuser Verlag Basel/Switzerland

Conjugacy and Centralizers for Iwip Automorphisms of Free Groups

Martin Lustig

Abstract. An automorphism α of a free group F_N of finite rank $N \geq 2$ is called *iwip* if no positive power of α maps any proper free factor F_k of F_N ($1 \leq k \leq N - 1$) to a conjugate of itself. Such automorphisms have many properties analogous to pseudo-Anosov mapping classes on surfaces. In particular, Bestvina-Handel have shown that any such α is represented by a train track map $f : \Gamma \to \Gamma$ of a graph Γ with $\pi_1 \Gamma \cong F_N$.

The goal of this paper is to give a new solution of the conjugacy problem for (outer) iwip automorphisms. We show that two train track maps $f : \Gamma \to \Gamma$ and $f' : \Gamma' \to \Gamma'$ represent iwip automorphisms that are conjugate in $\mathrm{Out}(F_N)$ if and only if there exists a map $h : \Gamma_\# \to \Gamma'$ which satisfies $f'' h = h f_\#$, where $f_\# : \Gamma_\# \to \Gamma_\#$ and $f'' : \Gamma'' \to \Gamma''$ are train track maps derived algorithmically from $f : \Gamma \to \Gamma$ and $f' : \Gamma' \to \Gamma'$ respectively, such that they represent the same pair of automorphisms. The map h maps vertices to vertices and edges to edge paths of bounded length, where the bound is derived algorithmically from f and f'.

The main ingredient of the proof, a lifting theorem of certain F_N-equivariant edge-isometric maps $i : \widetilde{\Gamma} \to T$, where T denotes the forward limit $\mathbb{R}$-tree defined by α, is a strong and useful tool in other circumstances as well.

1. Introduction

Let Γ be a connected finite graph without vertices of valence 1, with fundamental group $\pi_1 \Gamma = F_N$, a free group of finite rank $N \geq 2$. Let $f : \Gamma \to \Gamma$ be a map which maps vertices to vertices and edges to edge paths, and assume that f induces on $\pi_1 \Gamma$ an automorphism $\alpha : F_N \to F_N$.

The map f is called a *train track map* if for all edges e of Γ and all $t \in \mathbb{N}$ the map f^t is locally injective on the interior of e. Furthermore, f is *expanding* if it doesn't have periodic edges.

This article originates from the Barcelona conference.

An automorphism α of F_N is called *iwip* (for "irreducible with irreducible powers") if α^t does not map any non-trivial proper free factor of F_N to a conjugate of itself, for any $t \in \mathbb{N}$. Such automorphisms can be viewed (in more than one way) as strict analogues of pseudo-Anosov automorphisms on surfaces. Bestvina-Handel [BH92] have shown the following fundamental fact:

Every iwip automorphism α of F_N is represented
by an expanding train track map $f : \Gamma \to \Gamma$.

Indeed, in [BH92] there is an algorithm described which, given an automorphism α of F_N, either produces an expanding train track representative for α, or else explicates a non-trivial proper free factor which is invariant under some α^t with $t \geq 1$. The question, whether the automorphism given by any train track map is iwip, can also be answered algorithmically (see Remark 7.2 below).

In this paper we introduce, for any given train track map $f : \Gamma \to \Gamma$, a local modification procedure at every vertex of Γ, called *full blow-up* (see §7), which amounts to introducing the maximal number of periodic indivisible Nielsen paths (INP's) via splitting initial segments of edges, to obtain a new graph $\Gamma_\#$ and a new train track map $f_\# : \Gamma_\# \to \Gamma_\#$ which represents the same automorphism as $f : \Gamma \to \Gamma$. This full blow-up procedure is obtained algorithmically in finitely many steps.

Expanding train track representatives of iwip automorphisms have the property that for any number $m \in \mathbb{N}$ there are only finitely many (periodic or preperiodic) points which have finite orbits of order $\leq m$. Introducing all of them as new vertices, thus subdividing the original edges, does not violate the train track condition, and hence gives a new *m-subdivided* train track representative of the same automorphism. This m-subdivided train track representative is derived from the original one in finitely many steps.

We also define, for any two train track maps f and f' as above, a certain numerical invariant $L(f, f') \in \mathbb{N}$ which can be computed from the data given by f and f'.

Theorem 1.1. *The two expanding train track maps $f : \Gamma \to \Gamma$ and $f' : \Gamma' \to \Gamma'$ represent conjugate iwip automorphisms of F_N if and only if, for the fully blown-up train track map $f_\# : \Gamma_\# \to \Gamma_\#$ obtained from $f : \Gamma \to \Gamma$, and for the m-subdivided train track map $f'' : \Gamma'' \to \Gamma''$ obtained from $f' : \Gamma' \to \Gamma'$, where m is the number of vertices in $\Gamma_\#$, one has:*

There exists a map $h : \Gamma_\# \to \Gamma''$ which satisfies:

(1) h maps vertices to vertices.
(2) h maps every edge e of $\Gamma_\#$ to an edge path $h(e)$ of length $\leq L(f_\#, f'')$.
(3) For every edge e of $\Gamma_\#$ the edge paths $f''h(e)$ and $hf_\#(e)$ are equal.

There are at most finitely many such maps h, and compiling the list of all of them is a finite effective procedure.

This theorem, in a slightly more precise version (Theorem 9.1) is proved in §9. In the special case $\Gamma' = \Gamma$ and $f' = f$ it can be used to determine the centralizer of α (see §9).

Corollary 1.2. *The centralizer $C(\alpha) \subset Out(F_N)$ of any iwip automorphism α is a finite extension of the cyclic group generated by α, and a family of generators of $C(\alpha)$ can be computed explicitly.*

The proof of Theorem 1.1 is based on the fact that every iwip automorphism α of F_N has (up to uniform rescaling) precisely one α-invariant $\mathbb{R}$-tree T with stretching factor $\lambda > 1$ (see §4 below). Every train track representative $f : \Gamma \to \Gamma$ of α admits a canonical F_N-equivariant map $i : \widetilde{\Gamma} \to T$ which restricts to a homeomorphism on every edge of $\widetilde{\Gamma}$, the universal covering of Γ.

Our principal tool in the proof of the above theorem is Proposition 8.1, which states that for any second train track representative $f' : \Gamma' \to \Gamma'$ of α with canonical F_N-equivariant map $i' : \widetilde{\Gamma}' \to T$, under some additional technical assumptions, there is an F_N-equivariant map $\widetilde{h} : \widetilde{\Gamma} \to \widetilde{\Gamma}'$ which satisfies:

$$i = i'\widetilde{h}$$

This "lifting result" is also the crucial tool in the proof of the main result of [LoLu04].

Sections 2, 3, 4 and 5 are written more extensively than would be needed for a pure research paper: Most of the material discussed there is not new, except that our presentation is slightly more systematical than the previously available sources, which are also somewhat scattered in different parts of the literature. For example, we introduce (abstract) train tracks, and train track morphisms between distinct such train tracks, rather than considering just train track self-maps as done usually. The material is presented with some regard to details, and we hope that this paper can also be used by non-experts as an introduction to the subject.

The main mathematical work is done in §§6–9. What is presented there is an expanded version, for the special case of iwip automorphisms, of what has been described in [Lu01] for partial pseudo-Anosov automorphisms. The latter is a cornerstone of the complete solution of the conjugacy problem for automorphisms a free group F_N, given in [Lu00, Lu01]. It is the explicit purpose of this paper to serve as first reading, before tackling the technically more involved case of partial pseudo-Anosovs. Notice also that the special case of iwip automorphism has been treated already in [Lo96], but the methods employed there are quite different from the ones presented here.

2. Graphs, gates, turns, train tracks

Let Γ be a graph. By this we mean a connected 1-dimensional cell complex, and *we always assume* that there are no valence 1 vertices in Γ. We denote by $\mathcal{V}(\Gamma)$ the set of vertices of Γ, and by $\mathcal{E}(\Gamma)$ the set of edges of Γ. We work with the following convention: Every edge e of Γ comes with an orientation, which has been chosen randomly, mainly for notational purposes, to be able to distinguish between the initial vertex $\iota(e)$ and the terminal vertex $\tau(e)$ of e. We denote by $\overline{e}$ the edge e with reversed orientation, which gives $\iota(\overline{e}) = \tau(e)$ and $\tau(\overline{e}) = \iota(e)$. By $\overline{\mathcal{E}}(\Gamma)$ we denote the set $\{\overline{e} \mid e \in \mathcal{E}(\Gamma)\}$ which is by definition disjoint from $\mathcal{E}(\Gamma)$. To preserve symmetry we define $\overline{\overline{e}} = e$. Changing the orientation of an edge e simply amounts to exchanging corresponding elements from $\mathcal{E}(\Gamma)$ and $\overline{\mathcal{E}}(\Gamma)$. This operation does not change the graph Γ (up to graph isomorphisms as defined below).

We do not distinguish notionally between the cell complex Γ and the topological space Γ which realizes this cell complex. In particular, subdivision of edges induces a homeomorphism of graphs, but not a graph isomorphism. To be precise, a *graph isomorphism* is a homeomorphism $f : \Gamma \to \Gamma'$ which induces a bijection $f_{\mathcal{V}} : \mathcal{V}(\Gamma) \to \mathcal{V}(\Gamma')$, and, after a suitable reorientation of the edges, a bijection $f_{\mathcal{E}} : \mathcal{E}(\Gamma) \to \mathcal{E}(\Gamma')$, such that $f_{\mathcal{V}}(\tau(e)) = \tau(f_{\mathcal{E}}(e))$ and $f_{\mathcal{V}}(\iota(e)) = \iota(f_{\mathcal{E}}(e))$ for any $e \in \mathcal{E}(\Gamma)$.

We now consider, for any vertex $v \in \mathcal{V}(\Gamma)$, the set $\mathcal{E}(v)$ of all edges $e \in \mathcal{E}(\Gamma) \cup \overline{\mathcal{E}}(\Gamma)$ with initial vertex $\iota(e) = v$. A partition of $\mathcal{E}(v)$ into equivalence classes is called a *gate structure* at v, and the collection of gate structures for all $v \in \mathcal{V}(\Gamma)$ is a *gate structure* on Γ. The equivalence classes are called *gates* and denoted here by $\mathfrak{g}$ or $\mathfrak{g}_i$. Sometimes it is convenient to extend the gate structure to points which lie in the interior of an edge e: for such non-vertex points x we always define two distinct gates, each containing precisely one of the two edge segments of $e \smallsetminus \{x\}$. Notice that every graph admits a *finest* gate structure, where every gate contains precisely one edge.

Definition 2.1. A graph Γ provided with a gate structure is called a *train track*.

Gate structures occur naturally in different contexts, for example by embedding a graph into a surface or a higher dimensional manifold as branched 1-manifold: In this case every vertex inherits precisely 2 gates from this embedding. Another natural incidence are gate structures *induced* by a map of Γ to a second graph, or, more generally, by an equivariant map of the universal covering $\widetilde{\Gamma}$ of Γ to some $\mathbb{R}$-tree T, where one assumes that every edge is mapped homeomorphically to a segment. One then defines two edges of $\widetilde{\Gamma}$ with common initial vertex to belong to the same gate if and only if their images have a non-degenerate common initial segment.

Remark 2.2. The reader who is familiar with Thurston's train tracks on surfaces will notice two essential differences to the train tracks Γ as defined above:

(a) In Γ we do admit vertices with more than 2 gates, while a Thurston train track τ on a surface S has at every vertex precisely two gates.

(b) The graph Γ has fundamental group $\pi_1\Gamma$ that is isomorphic to the given free group F_N, while for a Thurston train track $\tau \subset S$ (where S has non-empty boundary) the embedding induces a homomorphisms $\pi_1\tau \to \pi_1 S \cong F_N$ which is typically non-injective.

We will see in §7 below that there is a standard procedure how to pass from the concept of train tracks presented above, implicitly introduced by Bestvina-Handel [BH92], to graphs which are closer to Thurston's train tracks in the sense of (a) and (b) above (compare also [Lu92])

Following [BH92], we define a *turn* in Γ to be a pair (e, e'), where e and e' are edges of $\mathcal{E}(\Gamma) \cup \overline{\mathcal{E}}(\Gamma)$ which have the same initial vertex $v = \iota(e) = \iota(e')$; in other words: $e, e' \in \mathcal{E}(v)$. The case $e = e'$ is included: such a turn is called *degenerate*. The vertex v is sometimes called the *initial vertex* of the turn (e, e').

For any fixed gate structure at Γ we say that the turn (e, e') is *legal* if e and e' do not belong to the same gate; otherwise the turn is called *illegal*.

An edge path γ in Γ is a sequence

$$\gamma = e_1 e_2 \ldots e_q$$

of edges $e_i \in \mathcal{E}(\Gamma) \cup \overline{\mathcal{E}}(\Gamma)$ where $\tau(e_i) = \iota(e_{i+1})$ for all $i = 1, \ldots, q-1$. Alternatively one can think of an edge path as a concatenation of subpaths,

$$\gamma = \gamma_1 \circ \gamma_2 \circ \cdots \circ \gamma_q \, ,$$

where we require that every subpath γ_i crosses over precisely one edge of $\mathcal{E}(\Gamma) \cup \overline{\mathcal{E}}(\Gamma)$. In this second view, two such edge paths are considered as "equal" if along corresponding subpaths γ_i they cross over the same edge. We call q the *simplicial length* (or simply *length*) of the above edge path γ. The case $q = 0$ is admitted: in this case γ is called *trivial*.

Unless stated explicitly otherwise, from now on we will tacitly assume that every path in Γ is an edge path in the sense just defined. Such a path γ is called *reduced* if no two subsequent subpaths γ_i, γ_{i+1} cross over the same edge, but in opposite directions. Alternatively, the path $\gamma = e_1 e_2 \ldots e_q$ is reduced if one has $e_{i+1} \neq \overline{e}_i$ for all $i = 1, \ldots, q - 1$. It is obvious that any finite edge path γ in Γ can be *reduced*, i.e., deformed by a homotopy relative its endpoints to a reduced path $[\gamma]$, which is uniquely determined by γ (up to reparametrization along the edges of Γ).

Any path of length two, $\gamma = \gamma_1 \circ \gamma_2$, determines a turn (e, e'), were γ_1 crosses over $\overline{e}$ and γ_2 crosses over e'. In this case we say that γ crosses over the turn (e, e'). More generally, any edge path γ *crosses over* a turn (e, e') if some subpath of γ of length 2 does. A path in Γ is *legal* if it crosses only over legal turns.

3. Graph morphisms and train track maps

For any two graphs Γ and Γ' we define a *graph morphism* to be a map $f : \Gamma \to \Gamma'$ which maps vertices to vertices and edges to edge paths. An edge is called *contracted* if it is mapped by f to a trivial edge path. A graph morphism with no contracted edges defines a *differential map* $Df : \mathcal{E}(\Gamma) \cup \overline{\mathcal{E}}(\Gamma) \to \mathcal{E}(\Gamma) \cup \overline{\mathcal{E}}(\Gamma)$ by defining for every edge $e \in \mathcal{E}(\Gamma) \cup \overline{\mathcal{E}}(\Gamma)$ the edge $Df(e)$ to be the first edge crossed over by the edge path $f(e)$. The map Df defines a natural map $D^2 f$ from the set of turns of Γ to the turns of Γ'.

If both, Γ and Γ', are provided with a gate structure, then we can consider a *gate structure morphism* $f : \Gamma \to \Gamma'$, i.e., a graph morphism with no contracted edges that *respects* the gate structures: The full preimage under Df of any gate of Γ', when intersected with any of the sets $\mathcal{E}(v)$, is contained in a single gate at the vertex v of Γ. Equivalently, one requires that $D^2 f$ maps legal turns to legal turns. However, for almost all purposes this notion is too weak; one rather defines:

Definition 3.1. A graph morphism $f : \Gamma \to \Gamma'$ between two train tracks Γ and Γ' is called a *train track morphism* if no edge is contracted and if

  (i) every legal turn is mapped to a legal turn, and

  (ii) every edge is mapped to a legal path.

Remark 3.2. Alternatively, one can require for a train track morphism $f : \Gamma \to \Gamma$ that

(iii) f maps legal paths to legal paths.

This condition does not require that f does not contract any edge of Γ, and, with this hypothesis, it is equivalent to (i) union (ii) of Definition 3.1. However, in this paper we will only have to deal with graph morphisms without contracted edges, which are somewhat easier to handle from an algorithmic point of view.

The most interesting case of train track morphisms is the special case where $\Gamma' = \Gamma$, i.e., f is a graph endomorphisms. Every graph endomorphism $f : \Gamma \to \Gamma$ without contracted edges defines canonically an *associated f-gate structure* on Γ, by defining a turn to be f-illegal if for some $k \in \mathbb{N}$ the f^k-image turn is degenerate. It follows directly from this definition that every graph endomorphism f of Γ is a gate structure morphism (but not necessarily a train track morphism !) with respect to the canonical f-gate structure on Γ.

Let Γ be a graph with a gate structure $\mathbf{G}$, and let $f : \Gamma \to \Gamma$ be a train track endomorphism. It is easy to see that the gate structure $\mathbf{G}$ is necessarily coarser than the canonically associated f-gate structure: Any f-illegal turn must be $\mathbf{G}$-illegal, but the converse does not necessarily hold. In case that the two gate structures coincide, we call f a *train track map*. It is easy to see that this definition is equivalent to the one given in the introduction.

Notice also that a train track map, in addition to the fact that by definition it maps legal turns to legal turns, it also has the additional property that it maps illegal turns to illegal turns. This follows directly from the definition, and it is sometimes rather useful.

A natural and useful condition on train track maps, which will be used throughout this paper, is the following:

Definition 3.3. A train track map $f : \Gamma \to \Gamma$ is called *expanding* if for every edge e of Γ there is an iterate $f^k(e)$ that has simplicial length ≥ 2. Equivalently, since f has no contracted edges, one can require that f has no periodic edges.

A path in Γ is called an *eigenpath* (or an f-*eigenpath*) if it is a subpath of $f^t(e)$, for some $t \in \mathbb{N}$ and some edge $e \in \mathcal{E}(\Gamma) \cup \overline{\mathcal{E}}(\Gamma)$. A turn crossed over by some eigenpath is called an eigenturn. Notice that every eigenpath and every eigenturn is legal, but the converse is in general not true.

To every graph endomorphism $f : \Gamma \to \Gamma$ there is canonically associated a non-negative *transition matrix* $M(f) = (m_{e,e'})_{e,e' \in \mathcal{E}(\Gamma)}$, where the coefficient $m_{e,e'}$ denotes the number of times that the edge path $f(e')$ crosses over the edge e or the inversely oriented edge $\overline{e}$ (both counted positively !).

Every edge path γ defines a vector $\vec{v}(\gamma) = (v(e))_{e \in \mathcal{E}(\Gamma)}$, where $v(e)$ denotes the number of times that the path γ crosses over the edge e or the inversely oriented edge $\overline{e}$, again both counted positively. We observe that one has always

$$\vec{v}(f(\gamma)) = M(f) \circ \vec{v}(\gamma) = \left(\sum_{e' \in \mathcal{E}(\Gamma)} m_{e,e'} v(e') \right)_{e \in \mathcal{E}(\Gamma)} \, ,$$

but in general the equality becomes an inequality if one passes from $f(\gamma)$ to the reduced path $[f(\gamma)]$. However, if f is a train track morphism and γ is legal, then one has:

$$\vec{v}([f(\gamma)]) = M(f) \circ \vec{v}(\gamma)$$

A length function $L : \mathcal{E}(\Gamma) \to \mathbb{R}_{\geq 0}$ can be expressed as *dual vector* $\vec{v}_*(L) = (L(e))_{e \in \mathcal{E}(\Gamma)}$, and the length of an edge path γ is then given by the scalar product

$$L(\gamma) = \langle \vec{v}_*(L), \vec{v}(\gamma) \rangle = \sum_{e \in \mathcal{E}(\Gamma)} L(e) v(e)$$

The transition matrix $M(f)$ of a train track map $f : \Gamma \to \Gamma$ is by definition non-negative. We consider a length function $L = L_{\vec{v}_*}$, given by a non-negative row eigenvector $\vec{v}_*$ of $M(f)$ that corresponds to some real eigenvalue $\lambda > 0$. We note that such a length function has the property that every legal path γ satisfies

$$L([f(\gamma)]) = L(f(\gamma)) = \langle \vec{v}_*, \vec{v}(f(\gamma)) \rangle = \langle \vec{v}_*, M(f) \circ \vec{v}(\gamma) \rangle$$

$$= \langle \vec{v}_* \circ M(f), \vec{v}(\gamma) \rangle = \langle \lambda \vec{v}_*, \vec{v}(\gamma) \rangle = \lambda \langle \vec{v}_*, \vec{v}(\gamma) \rangle = \lambda L(\gamma) \, ,$$

while for arbitrary paths γ' one only has

$$L([f(\gamma')]) \leq L(f(\gamma')) = \lambda L(\gamma') \, .$$

A non-negative $(m \times m)$-matrix M is called *irreducible* if for every pair of indices there is a positive power of M which has the property that the coefficient given by the index pair is non-zero (i.e., positive). Note that powers of irreducible matrices can be reducible. For irreducible matrices one has the celebrated theorem of Perron-Frobenius, which states that the (*Perron-Frobenius*) eigenvalue λ of M with maximal modulus has multiplicity 1, that it is positive real, and that the corresponding eigendirection in $\mathbb{R}^m$ is given by a (*Perron-Frobenius*) eigenvector that has positive coefficients. If the coefficients of M are integers, then it follows that $\lambda \geq 1$. If furthermore all positive powers of M are irreducible (such matrices are called *primitive*), then it follows that $\lambda > 1$, and M has a power where all coefficients are strictly bigger than 0.

Remark 3.4.

(a) If Γ_0 is a forest, i.e., a disjoint union of trees, then each of them can be contracted to give a quotient graph with less edges but same fundamental group F_N. On this quotient graph f defines a train track map that induces the same automorphism of F_N as f. This contraction of an invariant forest is an operation which in the sequel we often tacitly assume to have done. If Γ_0 is not a forest, then it defines a proper free factor of the free group $\pi_1\Gamma = F_N$ which is invariant (up to conjugation) under some positive power of the endomorphism $\alpha : F_N \to F_N$ induced by f. In this case, α cannot be an iwip automorphism (see §1).

(b) If one assumes that the train track map $f : \Gamma \to \Gamma$ is expanding (see Definition 3.3), then the above considered case of an invariant forest is excluded. Thus an expanding train track representative f of iwip automorphisms has always a transition matrix $M(f)$ which is primitive.

(c) If a train track representative $f : \Gamma \to \Gamma$ of an iwip automorphism is not expanding, then there is still a canonical Perron-Frobenius eigenvalue $\lambda > 1$ of $M(f)$, which also has multiplicity 1. Any associated non-negative Perron-Frobenius eigenvector $\vec{v}_*$ as above will then necessarily take on value 0 for all coefficients that correspond to periodic edges. It follows that the associated length function $L_{\vec{v}_*}$ automatically contracts all f-invariant forests, which justifies our tacit assumption in (a) (which amounts precisely to assuming that f is expanding).

Lemma 3.5. *Let $f : \Gamma \to \Gamma$ be a expanding train track map which induces an iwip automorphism α of $F_N = \pi_1\Gamma$. Then one has:*

(a) *For every eigenpath γ in Γ there is an edge e of Γ and an exponent $t \in \mathbb{N}$ such that $f^t(e)$ crosses twice over γ or its inverse.*

(b) *For every integer $k \geq 1$ there are only finitely many periodic points of order k, and only finitely many points which are mapped by f to a periodic point of order k.*

(c) *The set of all periodic points is dense in Γ.*

Proof. All three claims are direct consequences of Remark 3.4, which implies that for some positive power of f any edge e is mapped to an edge path which crosses (many times) over every edge of Γ. $\qquad\square$

Let us now fix a *marking isomorphism* $F_N \cong \pi_1\Gamma$. Let $f : \Gamma \to \Gamma$ be a graph morphism which induces via the fixed marking isomorphism an automorphism α of F_N: We say that f *represents* α, or that f *is a representative of* α. If $f : \Gamma \to \Gamma$ is furthermore a train track map, than it is called a *train track representative of* α.

Let $\widetilde{f} : \widetilde{\Gamma} \to \widetilde{\Gamma}$ be a lift of f to the universal covering $\widetilde{\Gamma}$ of Γ. Any other lift $\widetilde{f}' : \widetilde{\Gamma} \to \widetilde{\Gamma}$ of f is given by $\widetilde{f}' = u\widetilde{f}$, where we denote by $u : \widetilde{\Gamma} \to \widetilde{\Gamma}$ the action of $u \in F_N \cong \pi_1\Gamma$ by a deck transformation.

We say that $\widetilde{f}$ is a lift of f that α-*twistedly commutes with the F_N-action on* $\widetilde{\Gamma}$ if one has

$$\alpha(w)\widetilde{f} = \widetilde{f}w : \widetilde{\Gamma} \to \widetilde{\Gamma}$$

for all $w \in F_N$.

For any $u \in F_N$ one denotes by $\iota_u : F_N \to F_N$ the inner automorphism $\iota_u(w) = uwu^{-1}$. For the automorphism $\alpha' = \iota_u\alpha$ and the map $\widetilde{f}' = u\widetilde{f}$ as above, one has, for any $w \in F_N$, that

$$\alpha'(w)\widetilde{f}' = \iota_u(\alpha(w))u\widetilde{f} = u\alpha(w)\widetilde{f} = u\widetilde{f}w = \widetilde{f}'w : \widetilde{\Gamma} \to \widetilde{\Gamma} \, ,$$

i.e., $\widetilde{f}'$ is a lift of f' that α'-twistedly commutes with the F_N-action on $\widetilde{\Gamma}$. This gives a natural (and well-known) bijection between the different lifts of f to the universal covering $\widetilde{\Gamma}$, and the various automorphisms of F_N that define the same outer automorphism as α, which is the automorphism represented by the map $f : \Gamma \to \Gamma$.

Remark 3.6. Let $f : \Gamma \to \Gamma$ be an expanding train track representative of an iwip automorphism α of $F_N = \pi_1\Gamma$, and let $\widetilde{f}$ be the lift of f that α-twistedly commutes with the F_N-action on $\widetilde{\Gamma}$. Let $L = L_{\vec{v}_*}$ be the length function given by a Perron-Frobenius eigenvector $\vec{v}_*$ of the transition matrix $M(f)$ as above. One can lift the length function L to the universal covering $\widetilde{\Gamma}$ of Γ to define on this simplicial tree a pseudo-metric $d_{\vec{v}_*}$, where $d_{\vec{v}_*}(x, y)$ measures the length of the geodesic connecting the points $x, y \in \widetilde{\Gamma}$. As $\vec{v}_*$ has positive coefficients, the pseudo-metric $d_{\vec{v}_*}$ is actually a metric, which makes $\widetilde{\Gamma}$ into an $\mathbb{R}$-tree, with (deck transformation) action of F_N by isometries.

Then the map f can be homotoped properly into a *rigid position* so that any lift $\widetilde{f} : \widetilde{\Gamma} \to \widetilde{\Gamma}$ of f is λ-Lipschitz, where $\lambda > 1$ is the Perron-Frobenius eigenvalue of $M(f)$. In particular, any legal path γ is mapped by $\widetilde{f}$ to the path $\widetilde{f}((\gamma))$ through uniformly stretching it with factor λ and then applying an isometry.

4. Invariant $\mathbb{R}$-trees

We start this section by recalling some of the known properties of actions of a free group F_N on an $\mathbb{R}$-tree T. For details and background see [Vog02, Sha87] and the references given there.

An $\mathbb{R}$-*tree* is a metric space which is 0-hyperbolic and geodesic. Alternatively, an $\mathbb{R}$-tree is a space T with metric d where any two points $P, Q \in T$ are joined by a unique arc and this arc is isometric to the interval $[0, d(P, Q)] \subset \mathbb{R}$.

In this paper an $\mathbb{R}$-tree T always comes with a left action of F_N on T by isometries. Any isometry w of T is either *elliptic*, in which case it fixes at least one point of T, or else it is *hyperbolic*, in which case there is an *axis* $\mathrm{Ax}(w)$ in T, isometric to $\mathbb{R}$, which is w-invariant, and along which w acts as translation. The *translation length*

$$\|w\|_T = \inf\{d(P, wP) \mid P \in T\}$$

agrees in the hyperbolic case with $d(Q, wQ)$ for any point Q in $\mathrm{Ax}(w)$, while in the elliptic case it is 0.

We always assume that T is a *minimal* $\mathbb{R}$-tree, i.e., there is no non-empty F_N-invariant proper subtree of T. Another minimal $\mathbb{R}$-tree T' with isometric F_N-action is F_N-equivariantly isometric to T if and only if one has

$$\|w\|_T = \|w\|_{T'}$$

for every element $w \in F_N$. The set of such trees (or rather, of such tree actions) inherits a topology from its image in $\mathbb{R}^{F_N}$ under the map

$$T \mapsto (\|w\|_T)_{w \in F_N} \in \mathbb{R}^{F_N}.$$

A tree (or a tree action) is called *small* if any two group elements that fix pointwise a non-trivial arc in T do commute. It is called *very small* if moreover (i) the fixed set $Fix(g) \subset T$ of any elliptic element $1 \neq g \in F_N$ is a segment or a single point (i.e., "no branching"), and (ii) $Fix(g) = Fix(g^m)$ for all $g \in F_N$ and $m \geq 1$.

Let Γ be any (non-metric) graph with a marking isomorphism $\pi_1\Gamma \cong F_N$, and let $\widetilde{\Gamma}$ be its universal covering. Let $i : \widetilde{\Gamma} \to T$ be any F_N-equivariant map. Then the map i has the *bounded backtracking property (BBT)* (see [GJLL98]) if and only if for every pair of points $P, Q \in \widetilde{\Gamma}$ the i-image of the geodesic segment $[P, Q] \subset \widetilde{\Gamma}$ is contained in the C-neighborhood of the segment $[i(P), i(Q)] \subset T$, where $C \geq 0$ is an a priori constant independent of the choice of P and Q. We denote by $\mathrm{BBT}(i) \geq 0$ the smallest such constant.

It is easy to see that every $\mathbb{R}$-tree T with isometric F_N-action admits a map i as above, and that i satisfies BBT if and only if any other such map $i' : \widetilde{\Gamma}' \to T$ also satisfies BBT. Hence the property BBT is indeed a well-defined property of the tree T.

We can assume that the above map $i : \widetilde{\Gamma} \to T$ is *edge geodesic*: i maps every edge $e \subset \widetilde{\Gamma}$ homeomorphically to the geodesic segment that connects the images of the endpoints of e. One can make Γ into a metric graph by giving each edge of Γ and each of its lifts e to $\widetilde{\Gamma}$ the length of $i(e)$. Without loss of generality one can assume that the metric on every edge e is properly distributed so that i is actually *edge-isometric*, i.e., i maps every edge of $\widetilde{\Gamma}$ isometrically onto its image. In this case the inequality

$$BBT(i) \leq \mathrm{vol}(\Gamma)$$

has been proved in [GJLL98], where the *volume* $\mathrm{vol}(\Gamma)$ of Γ is the sum of the lengths of its edges.

Let $\alpha : F_N \to F_N$ be an automorphism. A non-trivial minimal $\mathbb{R}$-tree T with very small F_N-action by isometries is called α-invariant if there exists a homothety

$H : T \to T$ with stretching factor $\lambda > 0$ that α-*twistedly commutes with* the F_N-action, i.e.,

$$\alpha(w)H = Hw : T \to T$$

for every $w \in F_N$.

In this case one obtains, for every $v \in F_N$ and every $m \in \mathbb{N}$, another homothety $H' = vH^m : T \to T$ which α'-twistedly commutes with the F_N-action on T, where $\alpha' = \iota_v \circ \alpha^m : F_N \to F_N$ is the automorphisms given by composing the power α^m with the inner automorphism $\iota_v : F_N \to F_N, w \mapsto vwv^{-1}$.

Invariant $\mathbb{R}$-trees which are *expanding*, i.e., with stretching factor $\lambda > 1$, are important structural invariants of automorphisms α of F_N that have exponential growth on (some of) the conjugacy classes of F_N. In general there are finitely many such $\mathbb{R}$-trees, or at least finitely many length function simplices of such, compare [Lu01], Theorem 4.3. A particular situation occurs in the case considered in this paper, as follows for example from [LL03].

Proposition 4.1. *If α is an iwip automorphism, then there is, up to uniform rescaling of the metric, precisely one expanding α-invariant $\mathbb{R}$-tree $T = T_\alpha$. The tree T is sometimes also called the forward limit tree for α (compare Remark 5.4).* $\qquad\square$

Following [GJLL98], this α-invariant $\mathbb{R}$-tree T can be constructed from any (expanding) train track representative $f : \Gamma \to \Gamma$ of α in the following way: Let $L = L_{\vec{v}_*}$ be the length function on the edges of Γ given by a Perron-Frobenius eigenvector $\vec{v}_*$ of the transition matrix $M(f)$, with Perron-Frobenius eigenvalue $\lambda > 1$, and let $d_{\vec{v}_*}$ be the associated $\mathbb{R}$-tree metric on the universal cover $\widetilde{\Gamma}$ of Γ as in Remark 3.6. From this metric one derives a sequence of metrics d_k, for all $k \in \mathbb{N}$, by defining for any two points $x, y \in \widetilde{\Gamma}$ the distance

$$d_k(x, y) = \frac{d_{\vec{v}_*}(f^k(x), f^k(y))}{\lambda^k} .$$

We observe directly from §3 that $d_k(x, y) \geq d_{k+1}(x, y)$ for any $k \in \mathbb{N}$ and any $x, y \in \widetilde{\Gamma}$, and $d_k(x, y) = d_{k+1}(x, y)$ for any $k \in \mathbb{N}$ and points $x, y \in \widetilde{\Gamma}$ which are connected by a legal geodesic. As a consequence one obtains a non-trivial limit pseudo-metric d_∞ on $\widetilde{\Gamma}$, which has the following properties:

The metric quotient space of $\widetilde{\Gamma}$ defined by d_∞ is a minimal very small *Perron-Frobenius* $\mathbb{R}$-tree $T = T_{\vec{v}_*}$ with isometric F_N-action, inherited by the canonical F_N-action on $\widetilde{\Gamma}$ by deck transformations. The quotient map $i : \widetilde{\Gamma} \to T$ is F_N-equivariant, surjective, and 1-Lipschitz. Every legal path is mapped by i isometrically to a segment of T. The map i is thus a (generalized) train track morphism, if one provides T with the finest gate structure, see §2. (Indeed, one can show that the canonical train track structure on $\widetilde{\Gamma}$ defined by the train track map $\widetilde{f}$ is precisely equal to the train track structure induced on $\widetilde{\Gamma}$ by the finest train track structure on T, via the map i.) The map $\widetilde{f}$ induces on T a homothety $H : T \to T$ with stretching factor λ.

In particular, $T = T_{\vec{v}_*}$ is (up to rescaling the metric) canonically isometric to the unique α-invariant expanding $\mathbb{R}$-tree $T = T_\alpha$ from Proposition 4.1. It follows that the Perron-Frobenuis eigenvalue λ of $M(f)$ does not depend on the particular train track representative $f : \Gamma \to \Gamma$, but rather is an invariant of the (outer) automorphism α represented by f. We reassume these facts for future reference as follows:

Proposition 4.2. ([GJLL98]) *Let α be an iwip automorphism of F_N, and let $T = T_\alpha$ be the unique α-invariant expanding $\mathbb{R}$-tree, with associated homothety $H : T \to T$ with stretching factor $\lambda > 1$ that α-twistedly commutes with the F_N-action:*

$$\alpha(w)H = Hw : T \to T$$

for every $w \in F_N$.

Let $f : \Gamma \to \Gamma$ be any expanding train track representative of α, and let $\widetilde{f} : \widetilde{\Gamma} \to \widetilde{\Gamma}$ be the lift of f that α-twistedly commutes with the F_N-action on the universal covering $\widetilde{\Gamma}$:

$$\alpha(w)\widetilde{f} = \widetilde{f}w : \widetilde{\Gamma} \to \widetilde{\Gamma}$$

for every $w \in F_N$. Let $d_{\vec{v}_*}$ be the metric on $\widetilde{\Gamma}$ given by a Perron-Frobenius eigenvector $\vec{v}_*$ of the transition matrix $M(f)$.

Then, up to replacing $\vec{v}_*$ by a positive multiple, there is a canonical F_N-equivariant, surjective, edge-isometric, 1-Lipschitz map $i : \widetilde{\Gamma} \to T$, which maps legal paths in $\widetilde{\Gamma}$ isometrically to segments in T, and satisfies:

$$Hi = i\widetilde{f} : \widetilde{\Gamma} \to T$$

For a given $\mathbb{R}$-tree T and a train track map $f : \Gamma \to \Gamma$ with a randomly chosen Perron-Frobenius eigenvector $\vec{v}_*$ as in the above proposition, there is precisely one rescaling factor $\lambda' > 0$ such that, with respect to the rescaled metric $d_{\lambda'\vec{v}_*}$ on $\widetilde{\Gamma}$, the F_N-equivariant surjective map $i : \widetilde{\Gamma} \to T$ is edge-isometric (and hence 1-Lipschitz). This map $i : \widetilde{\Gamma} \to T$ is the called the *canonical F_N-equivariant edge-isometric map* associated to the train track representative $f : \Gamma \to \Gamma$ of the iwip automorphism α of F_N.

It is easy to see that the same construction, and hence also the statement of the above proposition, extends to train track maps $f : \Gamma \to \Gamma$ where the marking is not given by an isomorphism $\pi_1\Gamma \cong F_N$, but by a surjective homomorphisms $\pi_1\Gamma \to F_N$, given for example by adding a certain number of 2-cells to Γ to get a 2-complex Γ^2 to obtain an isomorphism $\pi_1\Gamma^2 \cong F_N$ which induces the above epimorphism $\pi_1\Gamma \to F_N$. In this case, the universal cover $\widetilde{\Gamma}$ is replaced by the 1-skeleton $\widehat{\Gamma}$ of the universal cover $\widetilde{\Gamma}^2$, and the map f is lifted to a map $\widetilde{f} : \widetilde{\Gamma}^2 \to \widetilde{\Gamma}^2$ which restricts to $\widehat{f} : \widehat{\Gamma} \to \widehat{\Gamma}$, where both α-twistedly commute with the F_N-action. The space $\widehat{\Gamma}$ is a δ-hyperbolic space in Gromov's sense, thus admitting an analogous metric limit process as described above by the metrics d_k. In this limit process, the hyperbolicity constant $\frac{\delta}{\lambda^k}$ tends to 0, thus giving again a limit tree T, and a homothety H induced by $\widehat{f}$. Thus, if α is iwip, it follows just as before that this tree T is precisely the uniquely determined limit forward tree T_α from Proposition 4.1.

5. Iterating an automorphism

Let T be an $\mathbb{R}$-tree with isometric F_N-action, and let $\alpha \in \mathrm{Aut}(F_N)$. We denote by $T' = T\alpha_*$ an isometric copy of the tree T, where we denote by $j : T \to T'$ the isometry, but with F_N-action on T' obtained from that on T by twisting with α: For any $w \in F_N$ and any $x \in T'$ define $w(x) = j\alpha(w)j^{-1}(x)$. This gives $\mathrm{Ax}_{T'}(w) = j(\mathrm{Ax}_T(\alpha(w)))$ and $\|w\|_{T'} = \|\alpha(w)\|_T$. Alternatively, the α_*-action on T can be described through replacing the *marking monomorphism* $\theta : F_N \to \mathrm{Isom}(T)$ by $\theta \circ \alpha : F_N \to \mathrm{Isom}(T)$.

For $\lambda > 0$ we denote by λT the $\mathbb{R}$-tree obtained by rescaling uniformly the metric of T by the factor λ.

Let us now assume that α is an iwip automorphism of F_N, and that T is the expanding α-invariant $\mathbb{R}$-tree as in Proposition 4.1. Let $H : T \to T$ the associated homothety with stretching factor $\lambda > 1$ that α-twistedly commutes with the F_N-action on T.

Furthermore, let Γ be a graph provided with a marking isomorphism $\theta : F_N \to \pi_1\Gamma$, and let $f : \Gamma \to \Gamma$ be a train track map that represents α. Let $\widetilde{f} : \widetilde{\Gamma} \to \widetilde{\Gamma}$ be a lift of f that α-twistedly commutes with the group action of F_N on $\widetilde{\Gamma}$. Assume furthermore that $d_{\vec{v}_*}$ is the metric on $\widetilde{\Gamma}$ given by a Perron-Frobenius eigenvector $\vec{v}_*$ of the transition matrix $M(f)$. Let $i : \widetilde{\Gamma} \to T$ be the F_N-equivariant edge-isometric map that satisfies $i\widetilde{f} = Hi$, as provided by Proposition 4.2.

For any $k \in \mathbb{Z}$ we define a new tree $\widetilde{\Gamma}_k = \frac{1}{\lambda^k}\widetilde{\Gamma}'\alpha_*^k$, where $\widetilde{\Gamma}'$ is an isometric copy of $\widetilde{\Gamma}$ with isometry $j : \widetilde{\Gamma} \to \widetilde{\Gamma}'$.

Let $\widetilde{f}_k : \widetilde{\Gamma}_k \to \widetilde{\Gamma}_k$ be the map $\widetilde{f}_k = j\widetilde{f}j^{-1}$. For any $w \in F_N$ and any $x \in \widetilde{\Gamma}_k$ one calculates $\widetilde{f}_k w(x) = j\widetilde{f}j^{-1}w(x) = j\widetilde{f}\alpha^k(w)j^{-1}(x) = j\alpha^{k+1}(w)\widetilde{f}j^{-1}(x) = \alpha^k(w)j\widetilde{f}j^{-1}(x) = \alpha^k(w)\widetilde{f}_k$, i.e., $\widetilde{f}_k$ is a map that α-twistedly commutes with the F_N-action on $\widetilde{\Gamma}_k$.

Similarly, for $i_k = H^{-k}ij^{-1} : \widetilde{\Gamma}_k \to T$, any $w \in F_N$, and any $x \in \widetilde{\Gamma}_k$ we obtain $i_k w(x) = (H^{-k}ij^{-1})(j\alpha^k(w)j^{-1})(x) = H^{-k}\alpha^k(w)ij^{-1}(x) = wH^{-k}ij^{-1}(x) = wi_k(x)$, which shows that i_k is F_N-equivariant.

Finally, we obtain $i_k\widetilde{f}_k = (H^{-k}ij^{-1})(j\widetilde{f}j^{-1}) = H^{-k}i\widetilde{f}j^{-1} = H^{-k}Hij^{-1} = HH^{-k}ij^{-1} = Hi_k$.

This proves:

Remark 5.1. For any $k \in \mathbb{Z}$, the above defined $\mathbb{R}$-tree $\widetilde{\Gamma}_k$ gives rise to a metric quotient graph $\Gamma_k = \widetilde{\Gamma}_k/F_N$, with volume $\mathrm{vol}(\Gamma_k) = \frac{1}{\lambda^k}\mathrm{vol}(\Gamma)$. The map $\widetilde{f}_k : \widetilde{\Gamma}_k \to \widetilde{\Gamma}_k$, which α-twistedly commutes with the F_N-action on $\widetilde{\Gamma}_k$, defines a train track map $f_k : \Gamma_k \to \Gamma_k$ with marking isomorphism $\theta_k = \theta \circ \alpha^k : F_N \to \pi_1\Gamma_k$, which also represents the automorphism α of F_N. The map $i_k : \widetilde{\Gamma}_k \to T$ is the F_N-equivariant edge-isometric map from Proposition 4.2.

Lemma 5.2. *For $\widetilde{\Gamma}$ and $\widetilde{\Gamma}_k$ as above, we obtain for the map $i_k : \widetilde{\Gamma}_k \to T$ the equality:*

$$BBT(i_k) = \frac{1}{\lambda^k}BBT(i)$$

Proof. By definition we have $i_k = H^{-k}ij^{-1}$, where j is an isometry, and H a homothety with stretching factor λ. Hence the desired equality follows directly from the definition of BBT. $\qquad\square$

Aside: We also observe that, for any $k \leq l$ in $\mathbb{Z}$ the map $\widetilde{f}_{l,k} := j\widetilde{f}^{l-k}j^{-1} : \widetilde{\Gamma}_k \to \widetilde{\Gamma}_l$ is 1-Lipschitz and maps every legal path isometrically to a legal path of same length. Furthermore we obtain for every $x \in \widetilde{\Gamma}$ and any $w \in F_N$ that $w\widetilde{f}_{l,k}(x) = j\alpha^l(w)j^{-1}\widetilde{f}_{l,k}(x) = j\alpha^l(w)\widetilde{f}^{l-k}j^{-1}(x) = j\widetilde{f}^{l-k}\alpha^k(w)j^{-1}(x) = \widetilde{f}_{l,k}j\alpha^k(w)j^{-1}(x) = \widetilde{f}_{l,k}w(x)$, which shows that $\widetilde{f}_{l,k}$ is F_N-equivariant.

We also note: $i_l\widetilde{f}_{l,k} = (H^{-l}ij^{-1})(j\widetilde{f}^{l-k}j^{-1}) = H^{-l}i\widetilde{f}^{l-k}j^{-1} = H^{-l}H^{l-k}ij^{-1} = H^{-k}ij^{-1} = i_k$. One also notices that, for any integers $k < l$, the map $\widetilde{f}_{l,k}$ is a train track morphism.

Remark 5.3. Let α be an iwip automorphism of F_N and $T = T_\alpha$ the α-invariant expanding $\mathbb{R}$-tree as in Proposition 4.1. Let $f : \Gamma \to \Gamma$ be a train track representative of α, and let $\vec{v}_*$ be an arbitrarily chosen Perron-Frobenius eigenvector of the transition matrix $M(f)$, with associated length function $L_{\vec{v}_*}$ on the edges of Γ. Then we have seen in Proposition 4.2 and the subsequent paragraph that, for any fixed marking isomorphism $\theta : F_N \to \pi_1\Gamma$, there exists a unique rescaling factor $\lambda' > 0$ such that with respect to the stretched eigenvector $\lambda'\vec{v}_*$ and the associated edge lengths there is a F_N-equivariant edge-isometric map $i : \widetilde{\Gamma} \to T$.

Now, using the above introduced marking changes via powers of α_*, we can achieve the same but with the additional requirement $\lambda' \in [1, \lambda)$, at the expense of replacing the marking θ by $\theta \circ \alpha_*^k$ for some $k \in \mathbb{Z}$.

To put the above definitions in prospective, we would like to conclude this section with the following remarks:

An $\mathbb{R}$-tree T with isometric F_N-actions is called *simplicial* if it arises from a graph Γ with a *marking isomorphism* $F_N \cong \pi_1\Gamma$, where the edges of Γ are given a non-negative length, which is for at least one of them strictly positive: The simplicial $\mathbb{R}$-tree T is then given by the universal covering $\widetilde{\Gamma}$, equipped with the lift of the edge lengths and with the action of F_N by deck transformations. If every edge length of Γ is non-zero, then the (simplicial) action of F_N on $\widetilde{\Gamma}$ is free. The subspace cv_N of free simplicial actions has been first considered by M. Culler and K. Vogtmann, see [CV86]. Its closure $\overline{cv}_N$ in the space of F_N-actions on $\mathbb{R}$-trees is precisely the set of all very small $\mathbb{R}$-trees as defined in §4. The *boundary* $\overline{cv}_N \smallsetminus \mathrm{cv}_N$ is denoted by $\partial\mathrm{cv}_N$. One often normalizes Γ to have volume 1, thus obtaining the subspace CV_N of cv_N, which has been named *Outer space* by P. Shalen. Alternatively, one can *projectivize* the space of tree actions: two trees T and T' are in the same equivalence class $[T]$ if they are F_N-equivariantly homothetic. This projectivization maps $\overline{cv}_N$ onto a compact space $\overline{CV}_N$, which contains a homeomorphic copy of CV_N, called the *interior*, and the projectivized image $\partial\mathrm{CV}_N$ of $\partial\mathrm{cv}_N$, called the *boundary*. Both CV_N and $\overline{CV}_N$ are contractible and finite dimensional. For more information see [Vog02, CV86].

The group $\mathrm{Out}(F_N)$ acts canonically (from the right !) on the space cv_N and on its "Thurston boundary" ∂cv_N, as well as on $CV_N \cup \partial CV_N$. This right action is defined as follows: For any $\alpha \in \mathrm{Aut}(F_N)$ and any tree $T \in \overline{cv}_N$, the length function of the image tree $T\alpha_*$ is given by

$$\|w\|_{T\alpha_*} = \|\alpha(w)\|_T\,,$$

for every $w \in F_N$.

Remark 5.4. It has been shown in [LL03] that an iwip automorphisms α of F_N induces an action α_* on $CV_N \cup \partial CV_N$ which has a North-South dynamics. The unique α-invariant expanding $\mathbb{R}$-tree $T = T_\alpha$ from Proposition 4.1, which lies on ∂CV_N, defines the attracting fixed point of this action (which explains the name "forward limit tree for α"), and the analogous unique α^{-1}-invariant expanding $\mathbb{R}$-tree $T = T_{\alpha^{-1}}$ gives the repellor. The above described trees $\widetilde{\Gamma}_k$, for $k \in \mathbb{Z}$, define (after projectivization) precisely an α_*-orbit of this action.

6. Eigensegments and their lifts

Let $\alpha : F_N \to F_N$ be an automorphism, and let T be an α-invariant expanding $\mathbb{R}$-tree, with associated homothety $H : T \to T$ with stretching factor $\lambda > 1$ that α-twistedly commutes with the F_N-action, i.e.,

$$\alpha(w)H = Hw : T \to T$$

for every $w \in F_N$.

In this case one obtains, for every $v \in F_N$ and every $m \in \mathbb{N}$, another expanding homothety $H' = vH^m : T \to T$ which α'-twistedly commutes with the F_N-action on T, where $\alpha' = \iota_v \circ \alpha^m : F_N \to F_N$ is the automorphisms given by composing the power α^m with the inner automorphism $\iota_v : F_N \to F_N, w \mapsto vwv^{-1}$.

A *ray* in T is an isometric embedding $\rho : \mathbb{R}_{>0} \to T$. For a homothety $H : T \to T$ as above an *H-eigenray* (or simply *eigenray*) is a ray ρ which satisfies $\rho(\lambda t) = H(\rho(t))$ for all $t \in \mathbb{R}_{>0}$. We say that a segment $[x, y] \subset T$ is an *eigensegment* if it is contained in some H'-eigenray, for $H' = vH^m$ as above.

Two eigenrays for H (and analogously for any H' as above) extent, for $t > 0$ converging to 0, always to the same *initial point of ρ*, which is always the (unique) fixed point of H. This fixed point, however, may not lie in T but in its metric completion $\overline{T}$; in this case there exists precisely one H-eigenray in T; such eigenrays will be called *weak*. If H has a fixed point $x \in T$, then any *direction at x*, i.e., a connected component of $T \setminus \{x\}$, contains precisely one ray which is eigenray (called a *strong eigenray*) for some $H' = vH^m$ with $vx = x$: This is a direct consequence of the index theorem of Gaboriau-Levitt [GL95], which implies that there are only finitely many $\mathrm{Stab}(x)$-orbits of directions at x. Eigensegments contained in strong eigenrays are called *strong* eigensegments.

We now assume that α is iwip, and that $f : \Gamma \to \Gamma$ is a train track representative of α, with maps $\widetilde{f} : \widetilde{\Gamma} \to \widetilde{\Gamma}$, $i : \widetilde{\Gamma} \to T$ and with distance $d_{\widetilde{v}_*}$ on $\widetilde{\Gamma}$ as in

Proposition 4.2. Recall from §3 that a path γ in Γ is called an eigenpath if it is a subpath of $f^t(e)$ for some edge e and some $t \geq 1$.

Lemma 6.1. *Every lift $\widetilde{\gamma}$ to $\widetilde{\Gamma}$ of an eigenpath γ in Γ is mapped by i to a strong eigensegment of T.*

Proof. By Lemma 3.5 (a) for any eigenpath γ in Γ there is an exponent $m \in \mathbb{N}$ and an edge e of Γ such that $f^m(e)$ crosses twice over the path γ or its inverse. Hence, for any lifts $\widetilde{f} : \widetilde{\Gamma} \to \widetilde{\Gamma}$ of f and $\widetilde{e} \subset \widetilde{\Gamma}$ of e, the path $\widetilde{f}^m(\widetilde{e})$ crosses at least twice over disjoint lifts $\widetilde{\gamma}_1$ and $\widetilde{\gamma}_2$ of γ. By Remark 3.4 every positive power of the transition matrix $M(f)$ is irreducible, and hence we can assume that $\widetilde{f}^m(\widetilde{e})$ also crosses over some lift $\widetilde{e}_1 \subset \widetilde{\Gamma}$ of e, with $\widetilde{e}_1 = w\widetilde{e}$ for some $w \in F_N$. Hence for the lift $\widetilde{f}' = \widetilde{f}^m w^{-1} = \alpha(w^{-1})\widetilde{f}^m$ of f we obtain that $\widetilde{f}^m(\widetilde{e}) = \widetilde{f}'(\widetilde{e}_1)$, and hence $\widetilde{f}'$ has a fixed point $\widetilde{q}$ on $\widetilde{e}_1$. Using the above derived equation $H'i = i\widetilde{f}'$, for $H' = \alpha(w^{-1})H^m$, we calculate for the point $Q = i(\widetilde{q}) \in T$ that $H'(Q) = H'i(\widetilde{q}) = i\widetilde{f}'(\widetilde{q}) = i(\widetilde{q}) = Q$, i.e., Q is a fixed point of H'. As $\widetilde{\gamma}_1$ and $\widetilde{\gamma}_2$ are disjoint, the point $\widetilde{q}$ can be contained in at most one of them. The other one, say $\widetilde{\gamma}_1$, must be entirely contained in a connected component of $\widetilde{\Gamma} \smallsetminus \{\widetilde{q}\}$. As by Proposition 4.2 legal paths in $\widetilde{\Gamma}$ are mapped under i isometrically to T, the subsegment $i(\widetilde{\gamma}_1)$ of the legal path $\widetilde{f}^m(\widetilde{e})$ must be contained entirely in some direction of T at Q.

It remains to observe that iterating $\widetilde{f}'$ produces a nested sequence of legal paths $\gamma'_k = \widetilde{f}'^k(\widetilde{e}_1)$, so that their images $i(\gamma'_k) = i\widetilde{f}'^k(\widetilde{e}_1) = H'^k(i(\widetilde{e}_1))$ constitute a nested sequence of segments in T, which contain Q. But then this nested sequence defines precisely two eigenrays of H' at Q, if Q is an inner point of any of the segments, or a single eigenray, if Q is a boundary point. In any case, the segment $i(\widetilde{\gamma}_1)$ is contained in one such eigenray, and hence it is a strongeigen segment of T.

For any other lift $\widetilde{\gamma}$ of γ there exists an element $u \in F_N$ with $\widetilde{\gamma} = u\widetilde{\gamma}_1$. We note that, for $\alpha' = \alpha^m \iota_{w^{-1}}$, the map $u\alpha'(u^{-1})\widetilde{f}' = u\widetilde{f}'u^{-1}$ has fixed point $u\widetilde{q}$, and that $i(\widetilde{\gamma})$ is contained in an eigenray of $u\alpha'(u^{-1})H' = uH'u^{-1}$ which starts at the fixed point $uQ \in T$, so that it also is a strong eigensegment. $\square$

Lemma 6.2.

(a) *For every strong eigensegment $[y, z] \subset T$ there exists a legal path γ in $\widetilde{\Gamma}$ that is mapped by $i : \widetilde{\Gamma} \to T$ isometrically to $[y, z]$.*

(b) *Let γ' be a second such path, and assume that y and z have distance strictly bigger than $2BBT(i)$. Then the central segments of γ and γ' of length $d(y, z) - 2BBT(i))$ agree. In particular, the central point x of $[y, z]$ has the same preimage in γ as in γ'.*

Proof. (a) As $[y, z]$ is a strong eigensegment, by definition it is contained in an eigenray ρ of H' for some $H' = vH^m$ as above, which has a fixed point $H'(Q) = Q$ in T. We consider the lift $\widetilde{f}' = v\widetilde{f}^m : \widetilde{\Gamma} \to \widetilde{\Gamma}$ which satisfies $H'i = i\widetilde{f}'$.

Let γ be any reduced path in $\widetilde{\Gamma}$ with endpoints that are mapped by i to Q and P, where $P \in T$ is a point sufficiently far out on ρ so that $[Q, P]$ contains $[y, z]$.

We now iterate $\widetilde{f'}$ on γ. Since train track morphisms map legal turns to legal turns, the number of illegal turns in these iterates will decrease and for some iterate γ' become stable. The $\widetilde{f'}$-image of each illegal turn in γ' is again an illegal turn in γ'. Now, $i(\gamma')$ is an initial segment ρ_0 of the eigenray ρ, union some extra segments, each of which connects this initial segment to the i-images of the initial vertex of some of the illegal turns on γ'. By $i\widetilde{f'} = H'i$ and the $\widetilde{f'}$-invariance of the illegal turns, each of these extra segments is contained in its H'-image, and hence they all meet the initial segment ρ_0 at the unique H-fixed point Q. We deduce that γ' is a concatenation $\gamma_0 \circ \gamma_1$, where γ_0 contains all illegal turns of γ', and the legal path γ_1 is mapped by i isometrically to the initial segment ρ_0 of ρ. Thus $i(\gamma_1)$ contains $[Q, P]$, which in turn contains $[y, z]$. This shows claim (a).

(b) We apply the property BBT to the concatenations $[\gamma' \circ \gamma_0 \circ \overline{\gamma}]$ and $[\gamma \circ \gamma_1 \circ \overline{\gamma}']$, where γ_0 is a reduced path in $\widetilde{\Gamma}$ that connects the terminal points of γ and γ', while γ_1 connects their initial points. It follows that $[\gamma' \circ \gamma_0 \circ \overline{\gamma}]$ and $[\gamma \circ \gamma_1 \circ \overline{\gamma}']$ are precisely build from the maximal boundary subpaths of γ disjoint from γ' and vice versa, and that any of them is of length smaller of equal to $BBT(i)$. $\qquad\square$

Lemma 6.3. *Let $f : \Gamma \to \Gamma$ and $f' : \Gamma' \to \Gamma'$ two train track representatives of an iwip automorphism α, and let $\widetilde{f} : \widetilde{\Gamma} \to \widetilde{\Gamma}$ and $\widetilde{f'} : \widetilde{\Gamma}' \to \widetilde{\Gamma}'$ be two lifts that α-twistedly commute with the F_N-action. Let T be the unique α-invariant expanding $\mathbb{R}$-tree as in Proposition 4.2, with canonically associated F_N-equivariant edge-isometric maps $i : \widetilde{\Gamma} \to T$ and $i' : \widetilde{\Gamma}' \to T$ that satisfy $Hi = i\widetilde{f}$ and $Hi' = \widetilde{f'}i'$. Let $h : \widetilde{\Gamma} \to \widetilde{\Gamma}'$ be an F_N-equivariant edge-isometric map, which satisfies $i'h = i$. Then the map h also satisfies:*

$$\widetilde{f'}h - h\widetilde{f}$$

Proof. We calculate $i'\widetilde{f'}h = Hi'h = Hi = i\widetilde{f} = i'h\widetilde{f}$, and conclude that for any $\widetilde{x} \in \widetilde{\Gamma}$ the points $\widetilde{f'}h(\widetilde{x})$ and $h\widetilde{f}(\widetilde{x})$ are mapped by i' to the same point $Hi(\widetilde{x})$ in T. By Lemma 3.5 (a) every point $\widetilde{x} \in \widetilde{\Gamma}$ is contained in some arbitrary long eigenpath, and thus, by Lemma 6.1, $i(\widetilde{x})$ is contained in some strong eigensegment in T of arbitrary length. Since the set of eigenrays is by definition H-invariant, the same is true for $Hi(\widetilde{x})$. By Lemma 6.2 sufficiently long such strong eigensegments have legal i'-lifts in $\widetilde{\Gamma}'$ which are well defined except possibly for boundary subsegments of length $\leq BBT(i')$. It follows that $\widetilde{f'}h(\widetilde{x})$ and $h\widetilde{f}(\widetilde{x})$ are the same point in $\widetilde{\Gamma}'$, for all $\widetilde{x} \in \widetilde{\Gamma}$. $\qquad\square$

7. Blowing up vertices of a train track

In this section we will describe in detail a procedure how to derive from a given train track an new one which is "locally finer". We first describe this procedure, subdivided into 5 steps, for general train tracks as defined in §2. In the second part of this section we review and adapt this procedure to the special case of train tracks defined by train track maps which represent iwip automorphisms of F_N.

Step 1: Let Γ_1 be a graph provided with a gate structure and with a positive length function L_1 on the edges of Γ_1. Let $s \in \mathbb{R}$ be any value which satisfies

$$0 < s < \frac{L_1(e)}{4} \, ,$$

for all edges e of Γ_1.

Step 2: We now define a new graph Γ_2 by identifying the initial segments of length s of any two edges of Γ_1 that lie in the same gate. In this process we introduce, for any vertex v of Γ_1 and any gate $\mathfrak{g}$ at v, a new edge $e(\mathfrak{g})$ of length s and a new vertex $v(\mathfrak{g}) = \tau(e(\mathfrak{g}))$. For any edge $e \in \mathcal{E}(\Gamma_1)$ contained in the gate $\mathfrak{g}$, where $\mathfrak{g}'$ is the gate which contains $\bar{e}$, the vertex $v(\mathfrak{g})$ is connected to the vertex $v(\mathfrak{g}')$ by an edge $\hat{e}$ of length $L_2(\hat{e}) = L_1(e) - 2s \, (> 0)$.

There is a canonical *folding map* $\phi : \Gamma_1 \to \Gamma_2$ which maps an edge $e \in \mathcal{E}(\Gamma_1)$ as above isometrically to the edge path $e(\mathfrak{g}) \, \hat{e} \, \bar{e}(\mathfrak{g}')$. The folding map ϕ defines an isomorphism $\phi_* : \pi_1\Gamma_1 \to \pi_1\Gamma_2$ for the fundamental groups, and it also induces a gate structure on Γ_2: Every vertex v of Γ_1 is mapped to an *old* vertex $\phi(v)$ of Γ_2, with an induced bijection between the gates at v and those at $\phi(v)$, where due to the folding of initial segments each of the gates at $\phi(v)$ contains precisely one edge. A *new* vertex $v(\mathfrak{g})$ of Γ_2, for any gate $\mathfrak{g}$ of Γ_1, has precisely two gates: One that contains the edge $\bar{e}(\mathfrak{g})$, and one which contains all other edges $\hat{e}_i$ with initial vertex $v(\mathfrak{g})$. Notice that with respect to this gate structure on Γ_2 the folding map ϕ is a train track morphism.

Step 3: We now derive from Γ_2 a new graph Γ_3^1, by considering the s-neighborhood $N_s(\phi(v)) \subset \Gamma_2$ of any old vertex $\phi(v)$ of Γ_2. The subgraph $N_s(\phi(v))$ is a star shaped tree, with center $\phi(v)$ and with edges $e(\mathfrak{g}_i)$ that have $\phi(v) = \iota(e(\mathfrak{g}_i))$ as initial vertex and the new vertex $v(\mathfrak{g}_i) = \tau(e(\mathfrak{g}_i))$ as terminal vertex. The graph Γ_3^1 is defined by replacing any of the $N_s(\phi(v))$ by the 1-skeleton $\sigma^1(v)$ of a $(m_v - 1)$-simplex $\sigma(v)$, where m_v is the number of gates at v. The vertices of $\sigma^1(v)$ are precisely the vertices $v(\mathfrak{g}_i) \in N_s(\phi(v))$, and between any (distinct) two of them, say $v(\mathfrak{g}_i)$ and $v(\mathfrak{g}_j)$, there is a *connecting edge* $e(\mathfrak{g}_i, \mathfrak{g}_j)$ in $\sigma^1(v)$ which joins $v(\mathfrak{g}_i)$ to $v(\mathfrak{g}_j)$ and has length $2s$. This defines the length function L_3 on the connecting edges of $\mathcal{E}(\Gamma_3)$; for the other edges e one sets $L_3(e) = L_2(e)$.

There is a canonical edge-isometric map $\pi : \Gamma_3^1 \to \Gamma_2$ which maps every connecting edge $e(\mathfrak{g}_i, \mathfrak{g}_j)$ to the edge path $\bar{e}(\mathfrak{g}_i)e(\mathfrak{g}_j)$. We define all edges from $\sigma^1(v)$ with initial vertex $v(\mathfrak{g}_i)$ to belong to a common new gate at $v(\mathfrak{g}_i)$. As a consequence, we see that every vertex of Γ_3^1 has precisely two gates. Note that with respect to these gates the map π is a train track morphisms, and that every legal path in Γ_2 has, up to boundary segments of length $\leq s$, precisely one legal preimage path in Γ_3^1 under the map π.

One can add to Γ_3^1 the 2-cells of each of the simplices $\sigma(v)$. More generally, we check for cycles (= reduced loops which run over any vertex at most once) in $\sigma^1(v)$, and to each of them we glue a 2-cell along its boundary to Γ_3^1, to get a 2-complex $\Gamma_3^2 = \Gamma_3$. The map π extends easily to these 2-cells, where each such 2-cell has as image the star shaped subgraph of $N_s(\phi(v))$ which is given by the π-image of its boundary. This gives an induced isomorphism $\pi_* : \pi_1\Gamma_3 \to \pi_1\Gamma_2$.

Step 4: We now derive from Γ_3 a new 2-complex Γ_4 by deleting some of the connecting edges $e(\mathfrak{g}_i, \mathfrak{g}_j)$, and all 2-cells with boundary subpath crossing over $e(\mathfrak{g}_i, \mathfrak{g}_j)$. Clearly, with respect to the inherited gate structure on Γ_4, the inclusion $e : \Gamma_4 \to \Gamma_3$ is a train track morphism, and it is edge-isometric with respect to the length function L_4 which is simply the restriction of L_3 to $\mathcal{E}(\Gamma_4) \subset \mathcal{E}(\Gamma_3)$.

Note that it may well happen that after the above deletions, the left-over subgraph $\sigma^*(v)$ of $\sigma^1(v)$ is no longer connected. In this case the inclusion $e : \Gamma_4 \subset \Gamma_3$ induces, for each connected component Γ_4' of Γ_4, a map

$$e_* : \pi_1\Gamma_4' \to \pi_1\Gamma_3\,,$$

which has as image a (possibly trivial) proper free factor of $\pi_1\Gamma_3$. Below we will suppose that $\pi_1\Gamma_4$ is isomorphic to $\pi_1\Gamma_3$, which excludes the case that $\sigma^*(v)$ is not connected.

Step 5: The fifth step in our construction consists in *blowing down* every cycle χ of any of the subgraphs $\sigma^*(v)$ of Γ_4, and to replace it by a star shaped tree that is isomorphic to the corresponding subgraph $\pi(\chi)$ of $N_s(\phi(v))$. Of course, the identifications of the "half-edges" of $\sigma^*(v)$ induced by all these quotient maps $\chi \to \pi(\chi)$ may lead to identifications of half-edges which are not contained in a common cycle: However, it is important to note that the quotient of this blowing down operation is in general not a subgraph of $N_s(\phi(v))$, but rather a collection of (not necessarily disjoint) such subgraphs N_k, as well as of those edges e_j of $\sigma^*(v)$ which are not contained in any cycle of $\sigma^*(v)$. Here any two of the N_k or e_j can only meet in a single vertex. The reader may note that each such subgraph N_k or edge e_j corresponds precisely to a connected component of the space obtained from $\sigma^*(v)$ after removing all vertices which separate (so called *cut points*). The cut points describe precisely the locus where any two of the N_k or e_j will meet.

Clearly this quotient of the 1-skeleton Γ_4^1 of the 2-complex Γ_4 is a graph Γ_5, and the induced quotient map $q' : \Gamma_4^1 \to \Gamma_5$ is edge-isometric with respect to the length function L_5 on Γ_5 inherited canonically from L_4. Furthermore, the map q' extends canonically to all 2-cells of Γ_4, by contracting every 2-cell with boundary χ to the tree $\pi(\chi)$, to give a quotient map $q : \Gamma_4 \to \Gamma_5$ which induces an isomorphism $q_* : \pi_1\Gamma_4 \to \pi_1\Gamma_5$. If we give each of the subgraphs N_k of $N_s(\phi(v))$ the gate structure inherited by that on $N_s(\phi(v))$ (i.e., in each gate at the central vertex of N_k there is precisely one edge), then the map q is easily seen to be a train track morphism.

Below we will adapt the above described 5 step blow-up procedure to the case of train track representatives of iwip automorphisms. Before doing so, we note the above described procedure admits naturally the following generalizations:

Modification A. In Step 2, one does not necessarily have to perform the described folding procedure at all vertices of Γ: It suffices to perform this "initial segments folding" at those gates which have at their initial vertex 3 or more gates. In the subsequent steps of the blow-up procedure, "old" vertices will thus mean vertices which have 3 or more gates. Furthermore, the length of the folded initial segments does not have to be uniformly equal to the given parameter $s > 0$, but may well be larger, say, for example between s and $2s$: It is enough to assure that for every edge $e \in \mathcal{E}(\Gamma_1)$ the left-over edge $\hat{e}$ has positive length.

Modification B. In Step 3, if the length of unique edge in any of the gates at an old vertex $\phi(v)$ varies (for example between s and $2s$), then the length of the connecting edges in the simplices $\sigma^1(v)$ will take on corresponding lengths, so that the canonical quotient map $\pi : \Gamma_3^1 \to \Gamma_2$ is edge-isometric.

Let now $f : \Gamma \to \Gamma$ be an expanding train track map which represents an iwip automorphism α of F_N. We now perform the above 5 blow-up steps, adapted to the given situation in the following way:

For *Step* 1, we set $\Gamma_1 = \Gamma$ and $f_1 = f$, and we assume that the length function L_1 on the edges of Γ_1 is given by a Perron-Frobenius eigenvector $\vec{v}_{1,*}$ of the transition matrix $M(f_1)$ as in §3. Of course, the gate structure considered on Γ_1 is the canonical one defined by the graph morphism f_1, see §3.

For *Step* 2, we will derive now from f_1 a new train track map $f_2 : \Gamma_2 \to \Gamma_2$ as in Modification A above, by iterative subdivision and folding of edges, in the following way:

If e and e' are edges with common initial vertex, and with edge paths $f(e)$ and $f(e')$ which have their first edge in common, one first subdivides Γ further by pulling back vertices via f, so that $f(e)$ and $f(e')$ have at least the first three edges (with respect to the cell structure after subdivision) in common. This is possible by Lemma 3.5 (c).

One then folds the initial segments of e and e' corresponding to the first common edge in $f(e)$ and $f(e')$. Clearly one obtains from f an induced map on this quotient graph. As in this folding process all vertices obtained from subdivision of edges have after the folding still only two gates, this process strictly decreases the total number of edges with initial vertex that has 3 or more gates. Thus we can iterate this procedure, and obtain, after possibly subdividing edges once more, a graph Γ_2, an induced train track map $f_2 : \Gamma_2 \to \Gamma_2$, and an edge-isometric folding map $\phi : \Gamma_1 \to \Gamma_2$ with $f_2\phi = \phi f_1$, which satisfy:

(a) At every vertex with 3 or more gates, there is precisely one edge in every gate.

(b) At every vertex with 2 gates, but with more than two adjacent edges, for any two edges e and e' in the same gate, there is a power f^m of f such that $f^m(e)$

and $f^m(e')$ agree along initial subpaths of at least half the total length (of either).

It is easy to see that the gates of Γ_2 inherited via ϕ from Γ_1 give precisely the canonical gate structure defined by f_2. Furthermore, the folding map ϕ defines an induced length function L_2 on Γ_2 which is easily seen to come also from a positive eigenvector $\vec{v}_{2,*}$ of the transition matrix $M(f_2)$.

We now perform *Step* 3 of the blow-up procedure, with the above Modification B. As the map π lifts legal paths in Γ_2 (essentially) uniquely to legal paths in Γ_3, there is a canonical way to lift the map $f_2\pi : \Gamma_3 \to \Gamma_2$ via π to a train track map $f_3 : \Gamma_3 \to \Gamma_3$. In the subsequent *Step* 4 we erase precisely those connecting edges of Γ_3^1 which do not correspond to eigenturns of the train track map f_2. This defines an f_3-invariant subgraph Γ_4^1 of Γ_3^1, and the restriction f_4 of f_3 to Γ_4^1 has again an irreducible transition matrix (which was not necessarily true for f_3). Finally, we perform *Step* 5 to blow-down all 2-cells from Γ_4, to get a graph Γ_5 and a train track map f_5. Note that the map f_5 inherits the above properties (a) and (b) from f_2

We observe that the maps π, e, and q from Steps 3, 4 and 5 of the blow-up procedure satisfy $f_2\pi = \pi f_3$, $f_3 e = e f_4$ and $f_5 q = q f_4$. Furthermore, one sees easily that the gates of each of the Γ_k inherited via the maps π, e or q from Γ_{k-1} give precisely the canonical gate structure defined by the train track map f_k. In addition, any of the maps π, e and q defines an induced length function L_k on Γ_k which is easily seen to come also from a positive eigenvector $\vec{v}_{k,*}$ of the transition matrix $M(f_k)$. Note that this is also true for f_3, despite of the above observation that $M(f_3)$ may be reducible.

Furthermore, the above maps ϕ, π and q induce on the fundamental groups homomorphisms ϕ_*, π_* and q_* which are easily seen to be isomorphisms. The map e induces a homomorphism e_* which also must be an isomorphism, as otherwise it defines an α-invariant proper free factor of F_N, which contradicts the definition of an iwip automorphism. Hence each of the $f_k : \Gamma_k \to \Gamma_k$, for $k = 1, \ldots, 5$, is a train track representative for α.

It is easy to see that if one iterates the blow-up procedure, i.e., one performs again Steps 3, 4 and 5 to the train track map $f_5 : \Gamma_5 \to \Gamma_5$ obtained already in Step 5, then one obtains in Step 3 a 2-complex Γ_3^+ and a train track map f_3^+, in Step 4 one obtains a subcomplex $\Gamma_4^+ \subset \Gamma_3^+$ without cut point on any of the $\sigma^*(v)$, as well as a restriction f_4^+ of f_3^+, and in Step 5 one just reproduce again the train track map $f_5 : \Gamma_5 \to \Gamma_5$. For $s > 0$ we define a train track map $f_\# : \Gamma_\# \to \Gamma_\#$ to be *fully s-blown-up* if it satisfies the conditions (a) and (b) above, if every edge has length $\geq 2s$, and if performing Steps 3–5 as above reproduces the map $f_\#$. The corresponding 2-complex $\Gamma_4^+ =: \Gamma_*$ and the train track map $f_4^+ = f_* : \Gamma_* \to \Gamma_*$ are called *the associated 2-gate s-blow-up*, with *canonical quotient map* $q : \Gamma_* \to \Gamma_\#$, which satisfies $f_\# q = q f_*$.

Remark 7.1. We note the following important consequence of the above definition: The 2-gate s-blow-up $f_* : \Gamma_* \to \Gamma_*$ has the property that any two edges e, e', which have the same initial vertex $\iota(e) = \iota(e')$ and belong to the same gate at this vertex, have an initial segment of length $\geq s$ which will be identified by some power f_*^m of f_* (by property (b) above). As a consequence, any lift to the universal covering $\widetilde{\Gamma}_*$ of such segments will be identified by the lift $\widetilde{f}_*^m$ of f_*^m, and hence also by the canonically associated F_N-equivariant edge-isometric map $i_* : \widetilde{\Gamma}_* \to T$ as given by Proposition 4.2. This follows directly from the equality $Hi_* = i_*\widetilde{f}_*$ given there (compare also the subsequent discussion).

Remark 7.2. In Step 4 of the above algorithm we have argued that, if the automorphism α represented by the given train track map $f : \Gamma \to \Gamma$, and hence also by $f_3 : \Gamma_3 \to \Gamma_3$, is iwip, then passing to the subcomplex $\Gamma_4 \subset \Gamma_3$ and to the corresponding restriction f_4 of f_3 gives an "eigenturn" 2-complex with $\pi_1\Gamma_4 = F_N$ on which f_4 also induces α.

In fact, the following stronger fact is true (a proof will be given elsewhere): An automorphism α is iwip if and only if

(1) some train track representant $f : \Gamma \to \Gamma$ of α has irreducible transition matrix $M(f)$, and

(2) performing the blow-up Steps 1 - 4 on $f : \Gamma \to \Gamma$ yields a 2-complex Γ_4 which is connected and where the canonical embedding $\Gamma_4 \subset \Gamma_3$ induces an isomorphism on the fundamental groups.

Hence, in the algorithmic process presented here we are implicitly checking whether a given train track map represents an iwip automorphisms or not.

8. The lifting theorem

The goal of this section is to prove the following lifting theorem, which is the main tool for the algorithm presented in the next section.

Proposition 8.1. *Let $\alpha \in \mathrm{Aut}(F_N)$ be iwip, and let T be its forward limit tree. Let $f_\# : \Gamma_\# \to \Gamma_\#$ be a fully s-blown-up train track representative of α, for some constant $s > 0$, and let $f' : \Gamma' \to \Gamma'$ be a second train track representative of α, with canonically associated F_N-equivariant edge-isometric maps $i_\# : \widetilde{\Gamma}_\# \to T$ and $i' : \widetilde{\Gamma}' \to T$. Assume*

$$BBT(i') < s.$$

Then there exists an F_N-equivariant edge-isometric map $\widetilde{h}_\# : \widetilde{\Gamma}_\# \to \widetilde{\Gamma}'$ which satisfies

$$i_\# = i'\widetilde{h}_\# \qquad and \qquad \widetilde{f}'\widetilde{h}_\# = \widetilde{h}_\#\widetilde{f}_\# ,$$

where $\widetilde{f}_\# : \widetilde{\Gamma}_\# \to \widetilde{\Gamma}_\#$ and $\widetilde{f}' : \widetilde{\Gamma}' \to \widetilde{\Gamma}'$ are the lifts of $f_\#$ and f' respectively which α-twistedly commute with the F_N-action. In particular, on the quotient graphs $\Gamma_\# = \widetilde{\Gamma}_\#/F_N$ and $\Gamma' = \widetilde{\Gamma}'/F_N$ the induced map $h : \Gamma_\# \to \Gamma'$ satisfies

$$f'h = hf_\# .$$

Proof. We consider the canonical 2-gate s-blow-up $f_* : \Gamma_* \to \Gamma_*$ associated to $f_\# : \Gamma_\# \to \Gamma_\#$, as defined at the end of §7, with canonical quotient map $q : \Gamma_* \to \Gamma_\#$ that satisfies $f_\# q = q f_*$.

For any point $x \in \Gamma_*$, let $N(x)$ denote the s-neighborhood of x. We notice that, by definition of an s-blow-up in §7, the minimal length of the edges of Γ_* is bigger or equal to $2s$. Hence $N(x)$ can contain at most one vertex v of Γ_*.

It follows that any connected component $N(\widetilde{x})$ of the full lift of $N(x)$ to the universal covering $\widetilde{\Gamma}_*$ is homeomorphic to $N(x)$ and hence contains precisely one preimage $\widetilde{x}$ of x. From Remark 7.1 it follows that $i_*(N(\widetilde{x}))$ is a geodesic segment in T of length $2s$ centered at $i_*(\widetilde{x})$.

As every vertex of Γ_* has precisely two gates, and since the transition matrix $M(f_*)$ is irreducible (see §7), it follows from Lemma 3.5 (a) that some eigenpath will cross over $\widetilde{x}$ and hence intersect $N(\widetilde{x})$ in a legal path of length $2s$. We obtain from Lemma 6.1 that the geodesic segment $i(N(\widetilde{x}))$ belongs to some strong eigenray.

We now use the hypothesis $BBT(i') < s$: We use Lemma 6.2 (a) to lift the strong eigensegment $i(N(\widetilde{x}))$ via i' to a legal path γ in $\widetilde{\Gamma}'$. From Lemma 6.2 (b) we obtain on γ an i'-lift $\widetilde{x}'$ of $i(\widetilde{x})$ which is independent of the special choice of the legal path γ, as long as $i'(\gamma)$ is equal to the segment $i(N(\widetilde{x}))$. We define a map $\widetilde{h}_* : \widetilde{\Gamma}_* \to \widetilde{\Gamma}'$ by setting $\widetilde{h}_*(\widetilde{x}) = \widetilde{x}'$.

In the last paragraph it is not just the i'-lift $\widetilde{x}'$, but also a small open neighborhood of $\widetilde{x}'$ on γ of length $\geq 2s - BBT(i')$, which is independent on the particular choice of the path γ. It follows that in the corresponding neighborhood of $\widetilde{x}$ the map $\widetilde{h}_*$ is given by first applying i_* and then lifting the image segment isometrically via i' to γ. This shows that the map $\widetilde{h}_*$ is continuous. Indeed, this argument shows that $\widetilde{h}_*$ is isometric on eigenpaths, and thus in particular edge-isometric.

Furthermore, by construction $\widetilde{h}_*$ is F_N-equivariant and it satisfies $i_* = i'\widetilde{h}_*$. Hence we can apply Lemma 8.2 below to obtain an F_N-equivariant edge-isometric map $\widetilde{h}_\# : \widetilde{\Gamma}_\# \to \widetilde{\Gamma}'$ which satisfies $\widetilde{h}_* = \widetilde{h}_\# \widetilde{q} : \widetilde{\Gamma}_* \to \widetilde{\Gamma}'$, where $\widetilde{q} : \widetilde{\Gamma}_* \to \widetilde{\Gamma}_\#$ is the F_N-equivariant lift of the map $q : \Gamma_* \to \Gamma_\#$.

Finally, we apply Lemma 6.3 to obtain the equality $\widetilde{f}' \widetilde{h}_\# = \widetilde{h}_\# \widetilde{f}_\#$, where $\widetilde{f}_\# : \widetilde{\Gamma}_\# \to \widetilde{\Gamma}_\#$ is the lift of $f_\# : \Gamma_\# \to \Gamma_\#$ that α-twistedly commutes with the F_N-action on $\widetilde{\Gamma}_\#$. By the F_N-equivariance of $\widetilde{h}_\#$ this gives the desired equation $h f_\# = f' h : \Gamma_\# \to \Gamma'$. $\qquad\square$

In order to complete the last proof, we still need to show the following:

Lemma 8.2. *Let $f_\# : \Gamma_\# \to \Gamma_\#$ be a fully blown-up train track representative of an iwip automorphism α of F_N, and let $f_* : \Gamma_* \to \Gamma_*$ be the canonically associated 2-gate blow-up, with canonical quotient map $q : \Gamma_* \to \Gamma_\#$ that satisfies $f_\# q = q f_*$. Let Γ' be a second train track representative of α, let T be the forward limit tree for α, with canonically associated F_N-equivariant edge-isometric maps $i_* : \widetilde{\Gamma}_* \to T$, $i_\# : \widetilde{\Gamma}_\# \to T$ and $i' : \widetilde{\Gamma}' \to T$ which satisfy $i_\# \widetilde{q} = i_*$, where $\widetilde{q} : \widetilde{\Gamma}_* \to \widetilde{\Gamma}_\#$ is* *the F_N-equivariant lift of the above map $q : \Gamma_* \to \Gamma_\#$. Assume furthermore that there exists an F_N-equivariant edge-isometric map $\widetilde{h}_* : \widetilde{\Gamma}_* \to \widetilde{\Gamma}'$ which satisfies $i_* = i'\widetilde{h}_*$.*

Then there exists an F_N-equivariant edge-isometric map $\widetilde{h}_\# : \widetilde{\Gamma}_\# \to \widetilde{\Gamma}'$ which satisfies $\widetilde{h}_* = \widetilde{h}_\#\widetilde{q} : \widetilde{\Gamma}_* \to \widetilde{\Gamma}'$.

Proof. The boundary path γ of every 2-cell of $\widetilde{\Gamma}_*$ is mapped by i_* to a star shaped finite tree $i_*(\gamma) \subset T$ with center $i_\#(v)$, where v is the vertex in $\widetilde{\Gamma}_\#$ that gives rise to the subgraph $\sigma^*(v) \subset \widetilde{\Gamma}_*$ which contains γ (compare Step 4 of the blow-up procedure from the last section). As $\widetilde{\Gamma}'$ is also a tree, the image $\widetilde{h}_*(\gamma)$ must also be a graph, and from $i_* = i'\widetilde{h}_*$ and the fact that i_* and $\widetilde{h}_*$ are edge-isometric, one deduces that i' maps $\widetilde{h}_*(\gamma)$ isometrically to $i_*(\gamma)$. However, precisely the same argument applies to the map $\widetilde{q} : \widetilde{\Gamma}_* \to \widetilde{\Gamma}_\#$, so that $i_\#$ also maps $\widetilde{q}(\gamma)$ isometrically to $i_*(\gamma)$. But then $i'^{-1}i_\#$ defines an isometry $\widetilde{h}_\#$ on $\widetilde{q}(\gamma)$ which satisfies $\widetilde{h}_* = \widetilde{h}_\#\widetilde{q}$. Since $\widetilde{q}$ is precisely the quotient defined by such paths γ, this shows that $\widetilde{h}_*$ induces on all of $\widetilde{\Gamma}_\#$ an F_N-equivariant edge-isometric map $\widetilde{h}_\# : \widetilde{\Gamma}_\# \to \widetilde{\Gamma}'$ which satisfies $\widetilde{h}_* = \widetilde{h}_\#\widetilde{q} : \widetilde{\Gamma}_* \to \widetilde{\Gamma}'$. $\square$

We observe, from material presented in this section, that for many train track representatives $f' : \Gamma' \to \Gamma'$ of α, the statement of the last lemma (and hence also that of Proposition 8.1) stays valid if the full blow-up $f_\# : \Gamma_\# \to \Gamma_\#$ is is replaced by a further quotient: Blowing up a vertex amounts to introducing locally *(periodic) indivisible Nielsen paths (INP's)*, see [BH92]. Thus it suffices to control the INP's of $f' : \Gamma' \to \Gamma'$ and to partially blow-up the preimage train track in order to introduce the corresponding INP's. Here the correspondence is given through the α-invariant tree T, where *blow-up classes* of train track representatives can be defined by mimicking turns at a vertex of $\widetilde{\Gamma}$ by pairs of directions at the corresponding branch point of T. For more detail see [LoLu04].

We now draw the main conclusion from Proposition 8.1 with respect to the conjugacy problem for iwip automorphisms:

Corollary 8.3. *Let $\alpha \in Aut(F_N)$ be iwip, with stretching factor $\lambda > 1$. Let $f_\# : \Gamma_\# \to \Gamma_\#$ be a fully s-blown-up metric train track representative of α, for some constant $s > 0$. Let $f' : \Gamma' \to \Gamma'$ be a second train track map which represents a conjugate α' of the given automorphism α.*

(a) *Then there exists a map (not a graph morphism!) $h : \Gamma_\# \to \Gamma'$ which satisfies*

$$f'h = hf_\# \, .$$

(b) *There exists a Perron-Frobenius length function $L_{\vec{v}'}$ on the edges of Γ' such that the volume of Γ' is contained in the interval $[\frac{s}{\lambda}, s) \subset \mathbb{R}$, with the property that h maps every edge e of $\Gamma_\#$ via a local isometry to a path in Γ' of same length as e.*

Proof. Let $\vec{v}_*$ be the Perron-Frobenius eigenvector of $M(f_\#)$ that determines the edge lengths of $\Gamma_\#$, and let $T = T_{\vec{v}_*}$ be the Perron-Frobenius tree defined by the limit process described after Proposition 4.1. By this proposition T is also the (up to rescaling) uniquely determined forward limit $\mathbb{R}$-tree for α, which has expansion factor λ. Let $i_\# : \widetilde{\Gamma}_\# \to T$ be the canonically associated F_N-equivariant edge-isometric map.

If $f' : \Gamma' \to \Gamma'$ is a train track representative of a conjugate of α, then it suffices to apply a marking change (using the conjugator) to transform it into a train track representative of α. We then pick a random Perron-Frobenius eigenvector $\vec{v}''_*$ of $M(f')$ to define a length function on the edges of Γ', and rescale it by a factor $\lambda' > 0$ as discussed after Proposition 4.2, to obtain an F_N-equivariant edge-isometric map $i' : \widetilde{\Gamma}' \to T$. We then apply a second marking change of Γ' through iteration of the induced map α_*, to replace $f' : \Gamma' \to \Gamma'$ by $f'_k : \Gamma'_k \to \Gamma'_k$ for some $k \in \mathbb{Z}$ as in Remark 5.1 and Lemma 5.2, in order to achieve that the volume of Γ'_k is contained in $[\frac{s}{\lambda}, s) \subset \mathbb{R}$. As a consequence one has, for the canonically associated F_N-equivariant edge-isometric map $i'_k : \widetilde{\Gamma}'_k \to T$, that $BBT(i_k) < s$. We can then apply Proposition 8.1, to $f'_k : \Gamma'_k \to \Gamma'_k$ instead of $f' : \Gamma' \to \Gamma'$. As these two train track maps differ only by a marking change, we obtain from Proposition 8.1 directly the desired map h. $\qquad\square$

9. The algorithm

Let $f : \Gamma \to \Gamma$ and $f' : \Gamma' \to \Gamma'$ be two train track maps. We assume that their fundamental groups have the same rank, and that both represent iwip automorphisms. The latter can be checked algorithmically, see Remark 7.2. We describe now a finite algorithmic procedure which decides whether the automorphisms represented by f and f' are conjugate or not in $\mathrm{Out}(F_N)$.

We first iterate the map f sufficiently often, to find all f-invariant forests (= the set of periodically mapped edges of Γ, compare Remark 3.4), and on the canonical quotient graph, still denoted Γ, we compute the canonical gate structure defined by f on Γ. Note that one can always determine, through finitely many iterations of f, which of the turns of Γ are eigenturns, and which are not. The same is done with f'.

We then apply to $f : \Gamma \to \Gamma$ the 5 step blow-up procedure described in §7 for train tracks with canonical gate structure given by a train track map, to obtain a fully s-blown-up train track map $f_\# : \Gamma_\# \to \Gamma_\#$, for some $s > 0$ as in §7. We compute the transition matrix, which must be irreducible; otherwise the original map f did not represent an iwip automorphism. We then calculate approximations for the Perron-Frobenius eigenvalue λ and also for a Perron-Frobenius eigenvector $\vec{v}_*$ of $M(f_\#)$. This enables us to calculate uniform lower and upper bounds $l_\# < L_{\vec{v}_*}(e) < L_\#$ to the length $L_{\vec{v}_*}(e)$ of any edge e of $\Gamma_\#$. If necessary we decrease $l_\#$ further to obtain $s \geq \frac{l_\#}{2}$.

If f and f' represent conjugate iwip automorphisms, we can now apply Corollary 8.3 to get a map $h : \Gamma_\# \to \Gamma'$. However, to obtain a combinatorial procedure, it is convenient to transform the map h obtained by Corollary 8.3 into a graph morphism, i.e., subdividing Γ' so that the h-image of every vertex is a vertex:

Let m be the largest cardinality of any $f_\#$-orbit of vertices in $\Gamma_\#$. We now subdivide Γ' so that every periodic point of order $\leq m$ becomes a vertex, and also any of its $(f')^m$-preimage points. By Lemma 3.5 (b) there are finitely many such subdivision points, and, as f' is expanding, they can easily be found algorithmically (they are in 1-1 correspondence with the occurrence of any edge e or $\bar{e}$ in the image path $(f')^{m!}(e)$, for all $e \in \mathcal{E}(\Gamma')$).

Let Γ'' denote the subdivided graph, and let $f'' : \Gamma'' \to \Gamma''$ be (topologically) the same map as f'. As for $f_\#$, we also compute from the transition matrix $M(f'')$ approximations $\widehat{\lambda} \geq \lambda''$ of the Perron-Frobenius eigenvalue λ'' and also for some Perron-Frobenius eigenvector $\vec{v}''_*$ of $M(f'')$, as well as uniform lower and upper bounds $l'' < L_{\vec{v}''_*}(e) < L''$ to the length of any edge e of Γ''.

We now compute an upper bound $L(f_\#, f'') \in \mathbb{N}$ to the simplicial length of the edge path $h(e)$ in Γ'', for any edge e of $\Gamma_\#$ (where h is the map obtained above from Corollary 8.3). Let $v > 0$ be the number $\#\mathcal{E}(\Gamma'')L''$, which is an upper bound to the volume of Γ''. Thus rescaling the edge length vector $\vec{v}''_*$ by a factor $\rho := \frac{l_\#}{2v\widehat{\lambda}}$ gives a new graph Γ''' which is combinatorially the same as Γ'', with length vector $\vec{v}'''_*$. It satisfies $\mathrm{vol}(\Gamma''') \leq \rho v = \frac{l_\#}{2\widehat{\lambda}} \leq \frac{s}{\widehat{\lambda}} \leq \frac{s}{\lambda''}$, so that we can deduce that the vector $\vec{v}'''_*$ is smaller or equal than the edge length vector on Γ'' obtained from Corollary 8.3 (under the name of $\vec{v}'_*$). Recall from Corollary 8.3 that for every edge e of $\Gamma_\#$ the lengths of e and $h(e)$ (measured with respect to $\vec{v}'_*$) agree, so that both are bounded above by $L_\#$. As a consequence, the simplicial length in Γ'' of any edge path $h(e)$ is bounded above by $\frac{L_\#}{\rho l''}$. We thus pose $L(f_\#, f'')$ to be the smallest integer bigger or equal to

$$\frac{L_\#}{\rho l''} = \frac{L_\# 2v\widehat{\lambda}}{l_\# l''} = 2\frac{L_\#}{l_\#}\frac{L''}{l''}\widehat{\lambda}\#\mathcal{E}(\Gamma'').$$

We then look for maps $h_i : \Gamma_\# \to \Gamma''$ which satisfy:

(1) h_i maps vertices to vertices.

(2) h_i maps every edge e of $\Gamma_\#$ to an edge path $h_i(e)$ in Γ'' which has simplicial length smaller or equal than $L(f_\#, f'')$.

(3) For every edge e of $\Gamma_\#$ the edge paths $f'' h_i(e)$ and $h_i f_\#(e)$ agree.

Theorem 9.1. *The two maps $f : \Gamma \to \Gamma$ and $f' : \Gamma' \to \Gamma'$ represent conjugate outer automorphisms of F_N if and only if there exists a map h_i as above. There are at most finitely many such maps, and compiling the list of all of them is a finite effective procedure.*

Proof. Modifying the graphs Γ and Γ' as described above, to get the fully blown-up map $f_\# : \Gamma_\# \to \Gamma_\#$ and the m-subdivision $f'' : \Gamma'' \to \Gamma''$ is clearly a finite procedure. As all data describing the two graphs and the two graph maps are finite, it is a finite procedure to calculate the bound $L(f_\#, f'')$ and to compile the complete list of maps h_i with the above conditions (1)–(3).

If there exists at least one such h_i, then the two maps represent conjugate outer automorphisms, where the conjugator is given through the marking isomorphisms and the map on the fundamental groups induced by h_i.

The fact, that the existence of such a map h_i is necessary for a conjugacy between the two outer automorphisms, follows directly from Corollary 8.3 and the above definition of the bound $L(f_\#, f'')$. This proves the claim. $\square$

We now prove that the above derived finite list of graph morphisms h_i can in a special case used to determine the centralizer of α:

Corollary 9.2. *The centralizer $C(\alpha) \subset Out(F_N)$ of any iwip automorphism α is a finite extension of the cyclic group generated by α, and a family of generators of $C(\alpha)$ can be computed explicitly.*

Proof. Let β be an automorphism of F_N which defines an outer automorphism $\widehat{\beta} \in C(\alpha)$. Let $f : \Gamma \to \Gamma$ be a train track representative of α, and let $f' : \Gamma' \to \Gamma'$ be obtained from f by a change of the marking of Γ via application of β_*. Since $\widehat{\beta} \in C(\alpha)$, it follows that $f' : \Gamma' \to \Gamma'$ is also a train track representative of α. It follows now directly from the proof of Corollary 8.3 that for some suitable $k \in \mathbb{Z}$ the change of marking of Γ by $\beta \circ \alpha^k$ is realized by one of the maps h_i described above. Hence $C(\alpha)$ is generated by α and by the finitely many outer automorphisms of F_N induced by any of the maps h_i obtained as above, for the special case $\Gamma' = \Gamma$ and $f' = f$, with respect to the canonical identifications $\pi_1\Gamma = \pi_1\Gamma_\#$ and $\pi_1\Gamma' = \pi_1\Gamma''$. $\square$

Acknowledgements

The author would like to thank Martin Bridson, Jérôme Los and Karen Vogtmann for providing the proper motivation to write this paper.

References

[BH92]   M. Bestvina and M. Handel. Train tracks for automorphisms of the free group. *Annals of Math.* **135**, 1–51, 1992.

[CV86]   M. Culler and K. Vogtmann. Moduli of graphs and automorphisms of free groups. *Invent. Math.* **84**, 91–119, 1986.

[GJLL98]   D. Gaboriau, A. Jäger, G. Levitt, and M. Lustig. An index for counting fixed points of automorphisms of free groups. *Duke Math. J.* **93**, 425–452, 1998.

[GL95]   D. Gaboriau and G. Levitt. The rank of actions on **R**-trees. *Ann. Sci. École Norm. Sup.* **28**, 549–570, 1995.

[LL03]   G. Levitt and M. Lustig. Irreducible automorphisms of F_n have north-south dynamics on compactified outer space. *J. Inst. Math. Jussieu* **2**, 59–72, 2003.

[Lo96]   J. Los. On the conjugacy problem for automorphisms of free groups. With an addendum. *Toplogy* **53**, 779–808, 1996.

[LoLu04]   J. Los and M. Lustig. The set of train track representatives of an irreducible free group automorphism is contractible. Preprint 2004. (http://www.crm.es/Publications/Preprints04.htm)

[Lu92]   M. Lustig. *Automorphismen von freien Gruppen.* Habilitationsschrift, Bochum 1992.

[Lu00]   M. Lustig. Structure and conjugacy for automorphisms of free groups I. *MPI-Bonn preprint series* 2000, No. 241 (http://www.mpim-bonn.mpg.de/preprints)

[Lu01]   M. Lustig. Structure and conjugacy for automorphisms of free groups II. *MPI-Bonn preprint series* 2001, No. 4 (http://www.mpim-bonn.mpg.de/preprints)

[Sha87]   P. Shalen. Dendrology of groups: an introduction. In *Essays in group theory*, Math. Sci. Res. Inst. Publ. **8**, 265–319. Springer, New York, 1987.

[Vog02]   K. Vogtmann. Automorphisms of free groups and outer space. *Geom. Dedicata* **94**, 1–31, 2002.

Martin Lustig
Mathématiques (LATP)
Université Paul Cézanne – Aix-Marseille III
Av. Escadrille Normandie-Niémen
F-13397 Marseille 20, France
e-mail: `Martin.Lustig@univ-cezanne.fr`

Geometric Group Theory
Trends in Mathematics, 225–253
© 2007 Birkhäuser Verlag Basel/Switzerland

Algebraic Extensions in Free Groups

Alexei Miasnikov, Enric Ventura and Pascal Weil

Abstract. The aim of this paper is to unify the points of view of three recent and independent papers (Ventura 1997, Margolis, Sapir and Weil 2001 and Kapovich and Miasnikov 2002), where similar modern versions of a 1951 theorem of Takahasi were given. We develop a theory of algebraic extensions for free groups, highlighting the analogies and differences with respect to the corresponding classical field-theoretic notions, and we discuss in detail the notion of algebraic closure. We apply that theory to the study and the computation of certain algebraic properties of subgroups (*e.g.*, being malnormal, pure, inert or compressed, being closed in certain profinite topologies) and the corresponding closure operators. We also analyze the closure of a subgroup under the addition of solutions of certain sets of equations.

1. Introduction

A well-known result by Nielsen and Schreier states that all subgroups of a free group F are free. A non-specialist in group theory could be tempted to guess from this pleasant result that the lattice of subgroups of F is simple, and easy to understand. This is however far from being the case, and a closer look quickly reveals the classical fact that inclusions do not respect rank. In fact, the free group of countably infinite rank appears many times as a subgroup of the free group of rank 2. There are also many examples of subgroups H, K of F such that the rank of $H \cap K$ is greater than the ranks of H and K. These are just a few indications that the lattice of subgroups of F is not easy.

Although the lattice of subgroups of free groups was already studied by earlier authors, Serre and Stallings in their seminal 1977 and 1983 papers [14, 16], introduced a powerful new technique, that has since turned out to be extremely useful in this line of research. It consists in thinking of F as the fundamental group of a bouquet of circles R, and of subgroups of F as covering spaces of R, i.e., some special types of graphs. With this idea in mind, one can understand and prove many properties of the lattice of subgroups of F using graph theory. These

---

This article originates from the Barcelona conference.

techniques are also very useful to solve algorithmic problems and to effectively compute invariants concerning subgroups of F.

The present paper offers a contribution in this direction, by analyzing a tool (an invariant associated to a given subgroup $H \leq F$) which is suggested by a 1951 theorem of Takahasi [17] (see Section 2.3). The algorithmic constructions involved in the computation of this invariant actually appeared in recent years, in three completely independent papers [20], [11] and [7], where the same notion was invented in independent ways. In chronological order, we refer:

- to the *fringe of a subgroup*, constructed in 1997 by Ventura (see [20]), and applied to the study of maximal rank fixed subgroups of automorphisms of free groups;
- to the *overgroups* of a subgroup, constructed in 2001 by Margolis, Sapir and Weil (see [11]), and applied to improve an algorithm of Ribes-Zaleskii for computing the pro-p topological closure of a finitely generated subgroup of a free group, among other applications; and
- to the *algebraic extensions* constructed in 2002 by Kapovich and Miasnikov (see [7]), in the context of a paper where the authors surveyed, clarified and extended the list of Stallings graphical techniques.

Turner also used the same notion, restricted to the case of cyclic subgroups, in his paper [18] (again, independently) when trying to find examples of test elements for the free group.

The terminology and the notation used in the above mentioned papers are different, but the basic concept – that of algebraic extension for free groups – is the same. Although aimed at different applications, the underlying basic result in these three papers is a modern version of an old theorem by Takahasi [17]. It states that, for every finitely generated subgroup H of a free group F, there exist finitely many subgroups $H_0, \ldots, H_n$ canonically associated to H, such that every subgroup of F containing H is a free multiple of H_i for some $i = 0, \ldots, n$. The original proof was combinatorial, while the proof provided in [20], [11] and [7] (which is the same up to technical details) is graphical, algorithmic, simpler and more natural.

The aim of this paper is to unify the points of view in [20], [11] and [7], and to systematize the study of the concept of algebraic extensions in free groups. We show how algebraic extensions intervene in the computation of certain abstract closure properties for subgroups, sometimes making these properties decidable. This was the idea behind the application of algebraic extensions to the study of profinite topological closures in [11], but it can be applied in other contexts. In particular, we extend the discussion of the notions of pure closure, malnormal closure, inert closure, etc (a discussion that was initiated in [7]).

A particularly interesting application concerns the property of being closed under the addition of the solutions of certain sets of equations. In this case, new results are obtained, and in particular one can show that the rank of the closure of a subgroup H is at most equal to $\mathrm{rk}(H)$.

The paper is organized as follows.

In Section 2, we remind the readers of the fundamentals of the representation of finitely generated subgroups of a free group F by finite labeled graphs. This method, which was initiated by Serre and Stallings at the end of the 1970s, quickly became one of the major tools of the combinatorial theory of free groups. This leads us to the short, algorithmic proof of Takahasi's theorem discussed above (see Section 2.3).

Section 3 introduces algebraic extensions, essentially as follows: the algebraic extensions of a finitely generated subgroup H are the minimum family that can be associated to H by Takahasi's theorem. We also discuss the analogies that arise between this notion of algebraic extensions and classical field-theoretic notions, and we discuss in detail the corresponding notion of algebraic closure.

Section 4 is devoted to the applications of algebraic extensions. We show that whenever an abstract property of subgroups of free groups is closed under free products and finite intersections, then every finitely generated subgroup H admits a unique closure with respect to this property, which is finitely generated and is one of the algebraic extensions of H. Examples of such properties include malnormality, purity or inertness, as well as the property of being closed for certain profinite topologies. In a number of interesting situations, this leads to simple decidability results. Equations over a subgroup, or rather the property of being closed under the addition of solutions of certain sets of equations, provide another interesting example of such an abstract property of subgroups, which we discuss in Section 4.4.

Finally, in Section 5, we collect the open questions and conjectures suggested by previous sections.

2. Preliminaries

Throughout this paper, A is a finite non-empty set and $F(A)$ (or simply F if no confusion may arise) is the free group on A.

In the algorithmic or computational statements on subgroups of free groups, we tacitly assume that the free group F is given together with a basis A, that the elements of F are expressed as words over A, and that finitely generated subgroups of F are given to us by finite sets of generators, and hence by finite sets of words.

2.1. Representation of subgroups of free groups

In his 1983 paper [16], Stallings showed how many of the algorithmic constructions introduced in the first half of the 20th century to handle finitely generated subgroups of free groups, can be clarified and simplified by adopting a graph-theoretic language. This method has been used since then in a vast array of articles, including work by the co-authors of this paper.

The fundamental notion is the existence of a natural, algorithmically simple one-to-one correspondence between subgroups of the free group F with basis A, and certain A-labeled graphs – mapping finitely generated subgroups to finite graphs and vice versa. This is nothing else than a particular case of the more general covering theory for topological spaces, particularized to graphs and free groups. We briefly describe this correspondence in the rest of this subsection. More detailed expositions can be found in the literature: see Stallings [16] or [20, 7] for a graph-oriented version, and see one of [4, 22, 11, 15] for a more combinatorial-oriented version, written in the language of automata theory.

By an *A-labeled graph* Γ we understand a directed graph (allowing loops and multiple edges) with a designated vertex written 1, and in which each edge is labeled by a letter of A. We say that Γ is *reduced* if it is connected (more precisely, the underlying undirected graph is connected), if distinct edges with the same origin (resp. with the same end vertex) have distinct labels, and if every vertex $v \neq 1$ is adjacent to at least two different edges.

In an A-labeled graph, we consider paths, where we are allowed to travel backwards along edges. The *label* of such a path p is the word obtained by reading consecutively the labels of the edges crossed by p, reading a^{-1} whenever an edge labeled $a \in A$ is crossed backwards. The path p is called *reduced* if it does not cross twice consecutively the same edge, once in one direction and then in the other. Note that if Γ is reduced then every reduced path is labeled by a reduced word in $F(A)$.

The subgroup of $F(A)$ associated with a reduced A-labeled graph Γ is the set of (reduced) words, which label reduced paths in Γ from the designated vertex 1 back to itself. One can show that every subgroup of $F(A)$ arises in this fashion, in a unique way. That is, for each subgroup H of $F(A)$, there exists a unique reduced A-labeled graph, written $\Gamma_A(H)$, whose set of labels of reduced closed paths at 1 is exactly H.

Moreover, if the subgroup H is given together with a finite set of generators $\{h_1, \ldots, h_r\}$ (where the h_i are non-empty reduced words over the alphabet $A \sqcup A^{-1}$), then one can effectively construct $\Gamma_A(H)$, proceeding as follows. First, one constructs r subdivided circles around a common distinguished vertex 1, each labeled by one of the h_i (and following the above convention: an inverse letter, say a^{-1} with $a \in A$, in a word h_i gives rise to an a-labeled edge in the reverse direction on the corresponding circle). If h_i has length n_i, then the corresponding circle has n_i edges and $n_i - 1$ vertices, in addition to the vertex 1. Then, we iteratively identify identically labeled pairs of edges starting (resp. ending) at the same vertex. One shows that this process terminates, that it does not matter in which order identifications take place, and that the resulting A-labeled graph is reduced and equal to $\Gamma_A(H)$. In particular, it does not depend on the choice of a set of generators of H. Also, this shows that $\Gamma_A(H)$ is finite if and only if H is finitely generated (see one of [16, 20, 4, 22, 11, 7, 15] for more details).

Example 2.1. Let $A = \{a, b, c\}$. The above procedure applied to the subgroup $H = \langle aba^{-1}, aca^{-1} \rangle$ of $F(A)$ is represented in Figure 1, where the last graph is $\Gamma_A(H)$. $\qquad\square$

Let Γ and Δ be reduced A-labeled graphs as above. A mapping φ from the vertex set of Γ to the vertex set of Δ (we write $\varphi \colon \Gamma \to \Delta$) is a *morphism of*

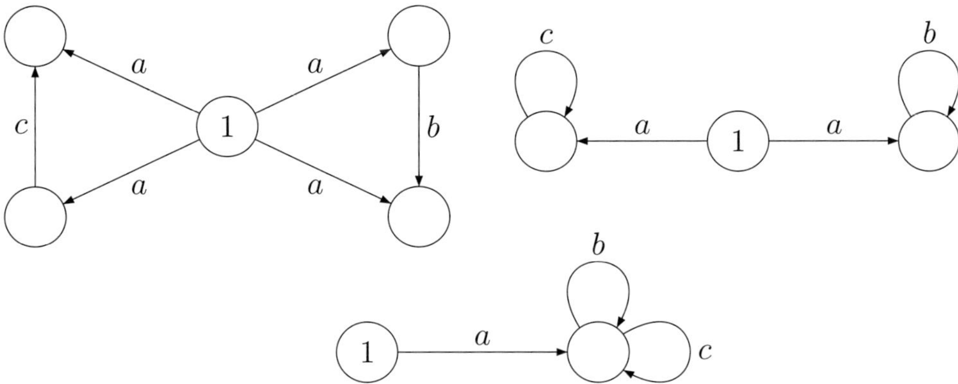

FIGURE 1. Computing the representation of $H = \langle aba^{-1}, aca^{-1}\rangle$

reduced (A)-labeled graphs if it maps the designated vertex of Γ to the designated
vertex of Δ and if, for each $a \in A$, whenever Γ has an a-labeled edge e from vertex
u to vertex v, then Δ has an a-labeled edge f from vertex $\varphi(u)$ to vertex $\varphi(v)$.
The edge f is uniquely defined since Δ is reduced. We then extend the domain
and range of φ to the edge sets of the two graphs, by letting $\varphi(e) = f$.

Note that such a morphism of reduced A-labeled graphs is necessarily locally
injective (an *immersion* in [16]), in the following sense: for each vertex v of Γ,
distinct edges starting (resp. ending) at v have distinct images. Further following
[16], we say that the morphism $\varphi\colon \Gamma \to \Delta$ is a *cover* if it is locally bijective, that
is, if the following holds: for each vertex v of Γ, each edge of Δ starting (resp.
ending) at $\varphi(v)$ is the image under φ of an edge of Γ starting (resp. ending) at v.

The graph with a single vertex, called 1, and with one a-labeled loop for each
$a \in A$ is called the *bouquet of A circles*. It is a reduced graph, equal to $\Gamma_A(F(A))$,
and every reduced graph admits a trivial morphism into it. One can show that a
subgroup H of $F(A)$ has finite index if and only if this natural morphism from
$\Gamma_A(H)$ to the bouquet of A circles is a cover, and in that case, the index of H
in $F(A)$ is the number of vertices of $\Gamma_A(H)$. In particular, it is easily decidable
whether a finitely generated subgroup of $F(A)$ has finite index.

This graph-theoretic representation of subgroups of free groups leads to many
more algorithmic results, some of which are discussed at length in this paper. We
will use some well-known facts (see [16]). If H is a finitely generated subgroup of
$F(A)$, then the rank of H is given by the formula

$$\mathrm{rk}(H) = E - V + 1,$$

where E (resp. V) is the number of edges (resp. vertices) in $\Gamma_A(H)$. A more precise
result shows how each spanning tree in $\Gamma_A(H)$ (a subtree of the graph $\Gamma_A(H)$ which
contains every vertex) determines a basis of H. It is also interesting to note that if
H and K are finitely generated subgroups of $F(A)$, then $\Gamma_A(H \cap K)$ can be easily

constructed from $\Gamma_A(H)$ and $\Gamma_A(K)$: one first considers the A-labeled graph whose vertices are pairs (u, v) consisting of a vertex u of $\Gamma_A(H)$ and a vertex v of $\Gamma_A(K)$, with an a-labeled edge from (u, v) to (u', v') if and only if there are a-labeled edges from u to u' in $\Gamma_A(H)$ and from v to v' in $\Gamma_A(K)$. Finally, one considers the connected component of vertex $(1, 1)$ in this product, and we repeatedly remove the vertices of valence 1, other than the distinguished vertex $(1, 1)$ itself, to make it a reduced A-labeled graph (this is the pull-back construction in [16]).

To conclude this section, it is very important to observe that if we change the ambient basis of F from A to B, we may radically modify the labeled graph associated with a subgroup H of F, see Example 2.2 below. In fact, a clearer understanding of the transformation from $\Gamma_A(H)$ to $\Gamma_B(H)$ (put otherwise: of the action of the automorphism group of $F(A)$ on the A-labeled reduced graphs) is one of the challenges of the field.

Example 2.2. Let F be the free group with basis $A = \{a, b, c\}$, and let $H = \langle ab, acba \rangle$. Note that $B = \{a', b', c'\}$ is also a basis of F, where $a' = a$, $b' = ab$ and $c' = acba$. The graphs $\Gamma_A(H)$ and $\Gamma_B(H)$ are depicted in Figure 2. $\qquad\square$

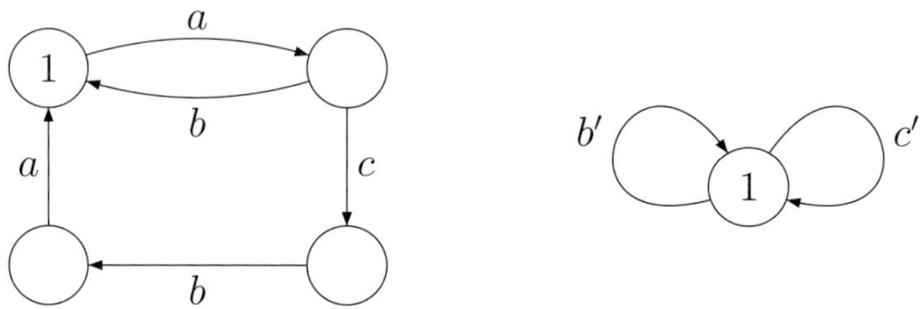

FIGURE 2. The graphs $\Gamma_A(H)$ and $\Gamma_B(H)$

2.2. Subgroups of subgroups

A pair of free groups $H \leq K$ is called an *extension* of free groups. If $H \leq M \leq K$ are free groups then $H \leq M$ will be referred to as a *sub-extension* of $H \leq K$.

If $H \leq K$ is an extension of free groups, we use the following shorthand notation: $H \leq_{\mathsf{fg}} K$ means that H is finitely generated; $H \leq_{\mathsf{fi}} K$ means that H has finite index in K; and $H \leq_{\mathsf{ff}} K$ means that H is a free factor of K.

Extensions can be characterized by means of the labeled graphs associated with subgroups as in Section 2.1. We first note the following simple result (see [11, Proposition 2.4] or [7, Section 4]).

Lemma 2.3. *Let H, K be subgroups of a free group F with basis A. Then $H \leq K$ if and only if there exists a morphism of labeled graphs $\varphi_{H,K}$ from $\Gamma_A(H)$ to $\Gamma_A(K)$. If it exists, this morphism is unique.*

Given an extension $H \leq K$ between subgroups of the free group with basis A, certain properties of the resulting morphism $\varphi_{H,K}$ have a natural translation on the relation between H and K. For instance, it is not difficult to verify that $\varphi_{H,K}$ is a covering if and only if H has finite index in K (and the index is the cardinality of each fibre). This generalizes the characterization of finite index subgroups of $F(A)$ given in the previous section.

If $\varphi_{H,K}$ is one-to-one (and that is, if and only if it is one-to-one on vertices), then H is a free factor of K. Unfortunately, the converse is far from holding since each non-cyclic free group has infinitely many free factors. Furthermore, given K, the particular collection of free factors $H \leq_{\mathrm{ff}} K$ such that $\varphi_{H,K}$ is one-to-one heavily depends on the ambient basis.

We recall here, for further reference, the following well-known properties of free factors (see [8] or [9]).

Lemma 2.4. *Let H, K, L, $(H_i)_{i\in I}$ and $(K_i)_{i\in I}$ be subgroups of a free group F.*

(i) *If $H \leq_{\mathrm{ff}} K \leq_{\mathrm{ff}} L$, then $H \leq_{\mathrm{ff}} L$.*
(ii) *If $H_i \leq_{\mathrm{ff}} K_i$ for each $i \in I$, then $\bigcap_i H_i \leq_{\mathrm{ff}} \bigcap_i K_i$.*

In particular, if H is a free factor of each K_i, then H is a free factor of their intersection; and an intersection of free factors of K is again a free factor of K.

Finally, in the situation $H \leq K$, we say that K is an *A-principal overgroup* of H if $\varphi_{H,K}$ is onto (both on vertices and on edges). We refer to the set of all *A*-principal overgroups of H as the *A-fringe* of H, denoted $\mathcal{O}_A(H)$. As seen later, this set strongly depends on A. The *A*-fringe of H is finite whenever H is finitely generated.

Principal overgroups were first considered under the name of *overgroups* in [11] (see [22] as well). They also appeared later as *principal quotients* in [7], and their first introduction is in the earlier [20], where $\mathcal{O}_A(H)$ was called the *fringe* of H, its *orla* in catalan. We shall use the phrase *principal overgroup* (to stress the fact that not every K containing H is a principal overgroup of H) and *fringe*, omitting the reference to the basis A when there is no risk of confusion. Both *orla* and *overgroup* justify the notation $\mathcal{O}_A(H)$.

Given a finitely generated subgroup $H \leq F(A)$, the fringe $\mathcal{O}_A(H)$ is computable: it suffices to compute $\Gamma_A(H)$, and to consider each equivalence relation $\sim$ on the set of vertices of $\Gamma_A(H)$. Say that such an equivalence relation $\sim$ is a *congruence* (with respect to the labeled graph structure of $\Gamma_A(H)$) if, whenever $p \sim q$ and there are a-labeled edges from p to p' and from q to q' (resp. from p' to p and from q' to q), then $p' \sim q'$. Then each congruence gives rise to a surjective morphism from $\Gamma_A(H)$ onto $\Gamma_A(H)/\sim$, and hence to a principal overgroup K of H such that $\Gamma_A(K) = \Gamma_A(H)/\sim$. Moreover, each principal overgroup $K \in \mathcal{O}_A(H)$ arises in this fashion. At the time of writing, a computer program is being developed with the purpose, among others, of efficiently computing the fringe of a finitely generated subgroup of a free group.

Example 2.5. Let F be the free group with basis $A = \{a, b, c\}$, and let $H = \langle ab, acba \rangle \leq F$ (the graph $\Gamma_A(H)$ was constructed in Example 2.2). Successively identifying pairs of vertices of $\Gamma_A(H)$ and reducing the resulting A-labeled graph in all possible ways, one concludes that $\Gamma_A(H)$ has six congruences, whose corresponding quotient graphs are depicted in Figure 3.

Thus the A-fringe of H consists on $\mathcal{O}_A(H) = \{H_0, H_1, H_2, H_3, H_4, H_5\}$, where $H_0 = H$, $H_1 = \langle ab, ac, ba \rangle$, $H_2 = \langle ab, a^2, acba^{-1} \rangle$, $H_3 = \langle ab, ac, ab^{-1}, a^2 \rangle$, $H_4 = \langle ab, aca, acba \rangle$ and $H_5 = \langle a, b, c \rangle = F(A)$.

However, with respect to the basis $B = \{a, ab, acba\}$ of F, the graph $\Gamma_B(H)$ has a single vertex, and hence the B-fringe of H is much simpler, $\mathcal{O}_B(H) = \{H\}$. $\square$

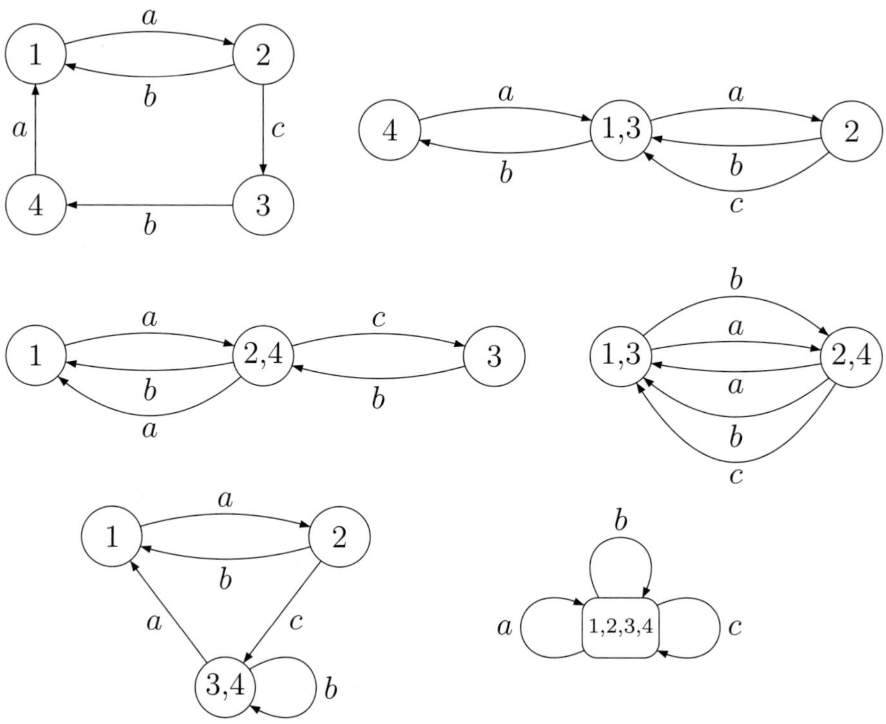

FIGURE 3. The six quotients of $\Gamma_A(\langle ab, acba \rangle)$

Finally we observe that, if $H \leq_{\mathrm{fi}} F(A)$, then $\mathcal{O}_A(H)$ consists of all the extensions of H. Indeed, suppose that $H \leq K \leq F(A)$ and $H \leq_{\mathrm{fi}} F(A)$. Since $\Gamma_A(H)$ is a cover of the bouquet of A circles (that is, each vertex of $\Gamma_A(H)$ is the origin and the end of an a-labeled edge for each $a \in A$), the range of $\varphi_{H,K}$ is also a cover of the bouquet of A circles. It follows that $\varphi_{H,K}$ is onto, since $\Gamma_A(K)$ is connected, and so $K \in \mathcal{O}_A(H)$. In particular, if $H \leq_{\mathrm{fi}} F(A)$, then $\mathcal{O}_A(H)$ does not depend on A, in contrast with what happens in general.

2.3. Takahasi's theorem

Of particular interest to our discussion is the following 1951 result by Takahasi (see [9, Section 2.4, Exercise 8], [17, Theorem 2] or [20, Theorem 1.7]).

Theorem 2.6 (Takahasi). *Let $F(A)$ be the free group on A and $H \leq F(A)$ a finitely generated subgroup. Then, there exists a finite computable collection of extensions of H, say $H = H_0, H_1, \ldots, H_n \leq F(A)$ such that every extension K of H, $H \leq K \leq F(A)$, is a free multiple of one of the H_i.*

The original proof, due to M. Takahasi was combinatorial, using words and their lengths with respect to different sets of generators. The geometrical apparatus described in this section leads to a clear, concise and natural proof, which was discovered independently by Ventura in [20] and by Kapovich and Miasnikov in [7]. Margolis, Sapir and Weil, also independently considered the same construction in [11] for a slightly different purpose. Finally, we note that Turner considered a similar construction in the case of cyclic subgroups, in his work about test words [18]. We now give this proof of Takahasi's theorem.

Proof. Let K be an extension of H, and let $\varphi_{H,K} \colon \Gamma_A(H) \to \Gamma_A(K)$ be the resulting graph morphism. Note that the image of $\varphi_{H,K}$ is a reduced subgraph of $\Gamma_A(K)$, and let $L_{H,K}$ be the subgroup of $F(A)$ such that $\Gamma_A(L_{H,K}) = \varphi_{H,K}(\Gamma_A(H))$. By definition, $L_{H,K}$ is an A-principal overgroup of H and, by construction, $\Gamma_A(L_{H,K})$ is a subgraph of $\Gamma_A(K)$, which implies $L_{H,K} \leq_{\mathrm{ff}} K$ (see Section 2.2). It follows immediately that the A-fringe of H, $\mathcal{O}_A(H)$, satisfies the required conditions. $\square$

Thus, for a given $H \leq_{\mathrm{fg}} F(A)$, the A-principal overgroups of H form one possible collection of extensions that satisfy the requirements of Takahasi's theorem, let us say, a *Takahasi family for H*. This is certainly not the only one: firstly, we may add arbitrary subgroups to a Takahasi family; secondly, we observe that the statement of the theorem does not depend on the ambient basis, so if B is another basis of $F(A)$, then $\mathcal{O}_B(H)$ forms a Takahasi family for H as well. There does however exist a minimum Takahasi family for H (see Proposition 3.7 below), which in particular does not depend on the ambient basis. The main object of this paper is a discussion of this family, which is introduced in the next section.

3. Algebraic extensions

The notion of algebraic extension discussed in this paper was first introduced by Kapovich and Miasnikov [7]. It seems to be mostly of interest for finitely generated subgroups, but many definitions and results hold in general and we avoid restricting ourselves to finitely generated subgroups until that becomes necessary.

3.1. Definitions

Let $H \leq K$ be an extension of free groups and let $x \in K$. We say that x is K-*algebraic over* H if every free factor of K containing H, $H \leq L \leq_{\mathrm{ff}} K$, satisfies

$x \in L$. Otherwise (i.e., if there exists $H \leq L \leq_{\text{ff}} K$ such that $x \notin L$) we say that x is K-*transcendental over* H.

Example 3.1. If $H \leq K$, then every element $x \in H$ is obviously K-algebraic over H.

Every element $x \in K$ is K-algebraic over $\langle x^n \rangle$, for each integer $n \neq 0$. In fact, it is straightforward to verify that if x^n lies in a free factor L of K, then so does x.

If x is primitive in K (that is, if $\langle x \rangle \leq_{\text{ff}} K$), then every element of $K \setminus \langle x \rangle$ is K-transcendental over the subgroup $\langle x \rangle$.

The notion of algebraicity over H is relative to K. For example, in $F = F(a,b)$, a^2 is $\langle a^2, b^2 \rangle$-transcendental over $H = \langle a^2 b^2 \rangle$ since $a^2 b^2$ is primitive in $\langle a^2, b^2 \rangle$. However, a^2 is F-algebraic over H because no proper free factor of F contains $a^2 b^2$. $\qquad\square$

The following is a trivial but useful observation.

Fact 3.2. *Let $H \leq K$ be an extension of free groups, and let $x, y \in K$.*

   (i) *If x, y are K-algebraic over H then so are x^{-1} and xy.*

   (ii) *If x, y are K-transcendental over H then so is x^{-1} (but not in general xy).*

We say that an extension of free groups $H \leq K$ is *algebraic*, and we write $H \leq_{\text{alg}} K$, if every element of K is K-algebraic over H. It is called *purely transcendental* if every element of K is either in H or is K-transcendental over H. Naturally, there are extensions that are neither algebraic nor purely transcendental. These concepts were originally introduced in [7], and the following propositions further describe their properties.

Proposition 3.3. *Let $H \leq K$ be an extension of free groups. The following are equivalent:*

   (a) *H is contained in no proper free factor of K;*

   (b) *$H \leq_{\text{alg}} K$, that is, every $x \in K$ is K-algebraic over H;*

   (c) *there exists $X \subseteq K$ such that $K = \langle H \cup X \rangle$ and every $x \in X$ is K-algebraic over H (furthermore, if K is finitely generated, one may choose X to be finite).*

Proof. (b) follows from (a) by definition. If (b) holds, then (c) holds with X any system of generators for K. Finally, (a) follows from (c) in view of Fact 3.2 (i). $\quad\square$

Proposition 3.4. *Let $H \leq K$ be an extension of free groups. The following are equivalent:*

   (a) *H is a free factor of K,*

   (b) *$H \leq K$ is purely transcendental, that is, every $x \in K \setminus H$ is K-transcendental over H.*

Proof. (a) implies (b) by definition. To prove the converse, let M be the intersection of all the free factors of K containing H. By Lemma 2.4, M is a free factor of K containing H, and (b) implies that $M = H$. $\quad\square$

Example 3.5. It is easily verified (say, using Example 3.1) that if $1 \neq x \in F$ and $n \neq 0$, we have $\langle x^n \rangle \leq_{\mathsf{alg}} \langle x \rangle$.

By Proposition 3.4, an extension of the form $\langle x \rangle \leq F$ is purely transcendental if and only if x is a primitive element of F. Moreover, if F has rank two, then $\langle x \rangle \leq F$ is algebraic if and only if x is not a power of a primitive element of F.

Assuming again that F has rank two, $H \leq_{\mathsf{alg}} F$ for every non-cyclic subgroup H. Indeed, every proper free factor of F is cyclic and hence cannot contain H. $\square$

We denote by $\mathsf{AE}(H)$ the set of algebraic extensions of H, and we observe that, in contrast with the definition of principal overgroups, this set does not depend on the choice of an ambient basis. This same observation can be expressed as follows.

Fact 3.6. *Let $H \leq K \leq F$ be extensions of free groups and let $\varphi \in \mathsf{Aut}(F)$. Then $H \leq_{\mathsf{alg}} K$ if and only if $\varphi(H) \leq_{\mathsf{alg}} \varphi(K)$.*

We can now express the connection between algebraic extensions and Takahasi's theorem.

Proposition 3.7. *Let $H \leq_{\mathsf{fg}} F(A)$ be an extension of free groups. Then we have:*

  (i) $\mathsf{AE}(H) \subseteq \mathcal{O}_A(H)$;
 (ii) $\mathsf{AE}(H)$ *is finite (i.e., H admits only a finite number of algebraic extensions)*;
(iii) $\mathsf{AE}(H)$ *is the set of $\leq_{\mathsf{ff}}$-minimal elements of every Takahasi family for H (see Section 2.3)*;
(iv) $\mathsf{AE}(H)$ *is the minimum Takahasi family for H.*

Proof. Let K be an algebraic extension of H. The proof of Takahasi's theorem shows that K is a free multiple of some principal overgroup $L \in \mathcal{O}_A(H)$. Then, Proposition 3.3 implies that $L = K$ proving (i). Statement (ii) follows immediately.

Let $\mathcal{L}$ be a Takahasi family for H and let $K \subset \mathsf{AE}(H)$. By definition of $\mathcal{L}$, there exists a subgroup $L \in \mathcal{L}$ such that $H \leq L \leq_{\mathsf{ff}} K$. By Proposition 3.3, it follows that $L = K$, so $K \in \mathcal{L}$. Thus $\mathsf{AE}(H)$ is contained in every Takahasi family for H. For the same reason, K is $\leq_{\mathsf{ff}}$-minimal in $\mathcal{L}$.

Now suppose that $K \in \mathcal{L}$ is $\leq_{\mathsf{ff}}$-minimal in $\mathcal{L}$, and let M be an extension of H such that $H \leq M \leq_{\mathsf{ff}} K$. By definition of a Takahasi family, there exists $L \in \mathcal{L}$ such that $H \leq L \leq_{\mathsf{ff}} M$, so $L \leq_{\mathsf{ff}} K$. Since K is $\leq_{\mathsf{ff}}$-minimal in $\mathcal{L}$, it follows that $L = K$, so $M = K$. Hence, $H \leq_{\mathsf{alg}} K$ concluding the proof of (iii).

Finally, it is immediate that the $\leq_{\mathsf{ff}}$-minimal elements of a Takahasi family for H again form a Takahasi family. Statement (iv) follows directly. $\square$

Example 3.8. If $H \leq_{\mathsf{fi}} K$, then $H \leq_{\mathsf{alg}} K$. This follows immediately from the observation that a proper free factor of K has infinite index.

It follows that, if $H \leq_{\mathsf{fi}} F(A)$, then $\mathsf{AE}(H) = \mathcal{O}_A(H)$ is equal to the set of all extensions of H. Indeed, we have already observed at the end of Section 2.2 that every extension of H is an A-principal overgroup of H, and since H has finite index in each of its extensions, it is algebraic in each. $\square$

Proposition 3.7 shows that $\mathsf{AE}(H)$ is contained in $\mathcal{O}_A(H)$ for each ambient basis A. We conjecture that $\mathsf{AE}(H)$ is in fact equal to the intersection of the sets $\mathcal{O}_A(H)$, when A runs over all the bases of F. Example 3.8 shows that the conjecture holds if H has finite index. It also holds if $H \leq_{\mathsf{ff}} F$, since in that case, $\mathsf{AE}(H) = \{H\}$, and F admits a basis B relative to which $\Gamma_B(H)$ is a graph with a single vertex.

We conclude with a simple but important statement.

Proposition 3.9. *Let $F(A)$ be the free group on A and $H \leq_{\mathsf{fg}} F(A)$. The set $\mathsf{AE}(H)$ is computable.*

Proof. Since every algebraic extension of H is in $\mathcal{O}_A(H)$, it suffices to compute $\mathcal{O}_A(H)$ and then, for each pair of distinct elements $K, L \in \mathcal{O}_A(H)$, to decide whether $L \leq_{\mathsf{ff}} K$: $\mathsf{AE}(H)$ consists of the principal overgroups of H that do not contain another principal overgroup as a free factor.

In order to conclude, we observe that deciding whether $L \leq_{\mathsf{ff}} K$ can be done, for example, using the first part of the classical Whitehead's algorithm. More precisely, Whitehead's algorithm (see [8, Proposition 4.25]) shows how to decide whether a tuple of elements, say $u = (u_1, \ldots, u_r)$, of a free group K can be mapped to another tuple $v = (v_1, \ldots, v_r)$ by some automorphism of K. The first part of this algorithm reduces the sum of the length of the images of the u_i to its minimal possible value. And it is easy to verify that this minimal total length is exactly r if and only if $\{u_1, \ldots, u_r\}$ freely generates a free factor of K. $\qquad\square$

We point out here that two alternative algorithms were recently proposed for the second part of the proof of Proposition 3.9, with several advantages. Silva and Weil [15] gave an algorithm completely based on graphical tools, while Roig, Ventura and Weil [13] gave another one improving significantly the complexity of the previous ones.

Remark 3.10. The terminology adopted for the concepts developed in this section is motivated by an analogy with the theory of field extensions. More precisely, if an element $x \in K$ is K-transcendental over H, then H is a free factor of $\langle H, x \rangle$ and $\langle H, x \rangle = H * \langle x \rangle$ (see Proposition 3.13 below). This is similar to the field-theoretic definition of transcendental elements: an element x is transcendental over H if and only if the field extension of H generated by x is isomorphic to the field of rational fractions $H(X)$.

However, the analogy is not perfect and in particular, the converse does not hold. For instance, a^2 is $\langle a, b \rangle$-algebraic over $\langle a^2 b^2 \rangle$ (see Example 3.1), but $\langle a^2 b^2, a^2 \rangle = \langle a^2 b^2 \rangle * \langle a^2 \rangle$. This stems from the fact, noticed earlier, that the notion of an element x being K-algebraic over H, depends on K and not just on x.

It is natural to ask whether the analogy also extends to the definition of algebraic elements: in other words, is there a natural analogue in this context for the notion of roots of a polynomial with coefficients in H? The discussion of equations in Section 4.4 offers some insight into this question. $\qquad\square$

3.2. Composition of extensions

We now consider compositions of extensions. Some of the results in the following proposition come from [7]. We restate and extend them here with simpler proofs. We also include in the statement well-known facts (the primed statements), in order to emphasize the dual properties of algebraic and purely transcendental extensions.

Proposition 3.11. *Let $H \leq K$ be an extension of free groups, and let $H \leq K_i \leq K$ be two sub-extensions, $i = 1, 2$.*

   (i) *If $H \leq_{\mathsf{alg}} K_1 \leq_{\mathsf{alg}} K$ then $H \leq_{\mathsf{alg}} K$.*
   (i') *If $H \leq_{\mathsf{ff}} K_1 \leq_{\mathsf{ff}} K$ then $H \leq_{\mathsf{ff}} K$.*
   (ii) *If $H \leq_{\mathsf{alg}} K$ then $K_1 \leq_{\mathsf{alg}} K$, while $H \leq K_1$ need not be algebraic.*
   (ii') *If $H \leq_{\mathsf{ff}} K$ then $H \leq_{\mathsf{ff}} K_1$, while $K_1 \leq K$ need not be purely transcendental.*
   (iii) *If $H \leq_{\mathsf{alg}} K_1$ and $H \leq_{\mathsf{alg}} K_2$ then $H \leq_{\mathsf{alg}} \langle K_1 \cup K_2 \rangle$, while $H \leq K_1 \cap K_2$ need not be algebraic.*
   (iii') *If $H \leq_{\mathsf{ff}} K_1$ and $H \leq_{\mathsf{ff}} K_2$ then $H \leq_{\mathsf{ff}} K_1 \cap K_2$, while $H \leq \langle K_1 \cup K_2 \rangle$ need not be purely transcendental.*

Proof. Statement (i') and the positive parts of statements (ii') and (iii') can be found in Lemma 2.4. The free group F on $\{a, b\}$ already contains counterexamples for the converse statements in (ii') and (iii'): for the first one, we have $\langle a \rangle \leq_{\mathsf{ff}} F$ while $\langle a \rangle \leq_{\mathsf{ff}} \langle a, b^2 \rangle \leq_{\mathsf{alg}} F$ (see Example 3.5). And for the second one, we have $\langle [a, b] \rangle \leq_{\mathsf{ff}} \langle a, [a, b] \rangle$ and $\langle [a, b] \rangle \leq_{\mathsf{ff}} \langle b, [a, b] \rangle$, whereas $\langle [a, b] \rangle \leq_{\mathsf{alg}} \langle a, [a, b], b \rangle = F$.

Now assume that $H \leq_{\mathsf{alg}} K_1 \leq_{\mathsf{alg}} K$ and let L be a free factor of K containing H. Then, $L \cap K_1$ is a free factor of K_1 containing H by Lemma 2.4. Since $H \leq K_1$ is algebraic, we deduce that $L \cap K_1 = K_1$, and hence $K_1 \leq L$. But $K_1 \leq_{\mathsf{alg}} K$, so $L = K$. Thus, the extension $H \leq K$ is algebraic, which proves (i).

The first part of (ii) is clear. A counterexample for the second part in $F = F(a, b)$ is as follows: we have $\langle [a, b] \rangle \leq_{\mathsf{ff}} \langle a, [a, b] \rangle \leq F$, while $\langle [a, b] \rangle \leq_{\mathsf{alg}} F$ by Example 3.5.

Suppose now that $H \leq_{\mathsf{alg}} K_1$ and $H \leq_{\mathsf{alg}} K_2$, and let L be a free factor of $\langle K_1 \cup K_2 \rangle$ containing H. Then Lemma 2.4 shows that, for $i = 1, 2$, $L \cap K_i \leq_{\mathsf{ff}} K_i$ containing H. Since $H \leq_{\mathsf{alg}} K_i$, we deduce that $L \cap K_i = K_i$ and hence, $K_i \leq L$. Thus, $L = \langle K_1 \cup K_2 \rangle$, and the extension $H \leq \langle K_1 \cup K_2 \rangle$ is algebraic, thus proving the positive part of (iii).

Finally, to conclude the proof of (iii), it suffices to exhibit subgroups H, K_1, K_2 such that $H \leq_{\mathsf{alg}} K_i$ ($i = 1, 2$) but $H \leq_{\mathsf{ff}} K_1 \cap K_2$. Again in $F(a, b)$ take, for example, $K_1 = \langle a^2, b^2 \rangle$ and $K_2 = \langle a^3, b^3 \rangle$, whose intersection is $K_1 \cap K_2 = \langle a^6, b^6 \rangle$. Letting $H = \langle a^6 b^6 \rangle$, we have $H \leq_{\mathsf{ff}} K_1 \cap K_2$ but $H \leq_{\mathsf{alg}} K_1$ and $H \leq_{\mathsf{alg}} K_2$. $\square$

To close this section, let us note another natural property of algebraic extensions, which slightly generalizes a result of Kapovich and Miasnikov [7].

Proposition 3.12. *Let F be a free group. If $H_i \leq_{\mathsf{alg}} K_i \leq F$ ($i \in I$), then $\langle \bigcup_i H_i \rangle \leq_{\mathsf{alg}} \langle \bigcup_i K_i \rangle$. The converse holds if $\langle \bigcup_i K_i \rangle = *_i K_i$.*

Proof. Suppose that $\langle \bigcup_i H_i \rangle \leq L \leq_{\mathrm{ff}} \langle \bigcup_i K_i \rangle$. Let $j \in I$. By Lemma 2.4, we have $L \cap K_j \leq_{\mathrm{ff}} \langle \bigcup_i K_i \rangle \cap K_j = K_j$. Moreover, $H_j \leq L \cap K_j$, so $K_j = L \cap K_j$ since $H_j \leq_{\mathrm{alg}} K_j$, and hence $K_j \subseteq L$. This holds for each $j \in I$, so $L = \langle \bigcup_i K_i \rangle$ and we have shown that $\langle \bigcup_i H_i \rangle \leq_{\mathrm{alg}} \langle \bigcup_i K_i \rangle$.

For the converse, suppose that $\langle \bigcup_i K_i \rangle = *_i K_i$. It follows that $\langle \bigcup_i H_i \rangle = *_i H_i$. Now we assume that $*_i H_i \leq_{\mathrm{alg}} *_i K_i$. Let $j \in I$. If $H_j \leq L \leq_{\mathrm{ff}} K_j$, then $*_i H_i \leq L * *_{i \neq j} K_i \leq_{\mathrm{ff}} *_i K_i$ and hence $L * *_{i \neq j} K_i = *_i K_i$. Taking the projection onto K_j, it follows that $L = K_j$. Thus $H_j \leq_{\mathrm{alg}} K_j$ for each $j \in I$. $\qquad\square$

Note that the converse of Proposition 3.12 does not hold in general, as can be seen from the counterexample provided in the proof of Proposition 3.11 (iii′).

3.3. Elementary extensions

We say that an extension of free groups $H \leq K$ is *elementary* if $K = \langle H, x \rangle$ for some $x \in K$. Elementary extensions turn out to be either algebraic or purely transcendental, as we now see.

Proposition 3.13. *Let $H \leq F$ be an extension of free groups and let $x \in F$. Let also X be a new letter, not in F. The following are equivalent:*

  (a) *the morphism $H * \langle X \rangle \to F$ acting as the identity over H and sending X to x is injective;*
  (b) *H is a proper free factor of $\langle H, x \rangle$;*
  (c) *H is contained in a proper free factor of $\langle H, x \rangle$.*

If, in addition, H is finitely generated, then these are further equivalent to:

  (d) $\mathrm{rk}(\langle H, x \rangle) = \mathrm{rk}(H) + 1$;
  (e) $\mathrm{rk}(\langle H, x \rangle) > \mathrm{rk}(H)$.

Proof. It is immediately clear that (a) implies (b), and (b) implies (c).

At this point, let us assume that H has finite rank. It is immediate that $\mathrm{rk}(\langle H, x \rangle) \leq \mathrm{rk}(H) + 1$, so (b) implies (d) and (d) and (e) are equivalent. Now consider the morphism from $H * \langle X \rangle$ to $\langle H, x \rangle$ mapping H identically to itself, and X to x. This morphism is surjective by construction, and if $\mathrm{rk}(\langle H, x \rangle) = \mathrm{rk}(H * \langle X \rangle)$, then it is injective by the Hopfian property of finitely generated free groups. That is, (d) implies (a). Thus we have shown that if H has finite rank, then statements (a), (b), (d) and (e) are equivalent. It only remains to prove that (c) implies (a).

We now return to the general case, where H may have infinite rank, and we assume that (c) holds, that is, $\langle H, x \rangle = K * L$ for some $L \neq 1$ and $H \leq K$. We have $\langle H, x \rangle \leq \langle K, x \rangle$, and hence $\langle K, x \rangle = \langle H, x \rangle = K * L$. Moreover, $x \notin K$ and we let $x = k_0 \ell_1 k_1 \cdots \ell_r k_r$ be the normal form of x in the free product $K * L$.

Let M be a finitely generated free factor of K containing the k_i, and let N be such that $K = M * N$. First we observe that

$$\langle M, x \rangle \leq M * L \leq_{\mathrm{ff}} K * L = M * N * L = \langle K, x \rangle = \langle M, N, x \rangle.$$

It follows that N is a free complement of $\langle M, x \rangle$ in $\langle K, x \rangle$, that is, $\langle K, x \rangle = \langle M, x \rangle * N$.

Next we note that $M \leq_{\mathrm{ff}} K \leq_{\mathrm{ff}} \langle K, x \rangle$, so $M \leq_{\mathrm{ff}} \langle K, x \rangle$ and hence $M \leq_{\mathrm{ff}} \langle M, x \rangle$. Since $x \notin K$, M is a finitely generated, proper free factor of $\langle M, x \rangle$, and we already know that this implies that the morphism from $M * \langle X \rangle$ to F mapping M identically to itself and mapping X to x, is injective. Since N is a free complement of the range of this morphism in $\langle H, x \rangle$, and also a free complement of M in K, it follows that the natural map from $K * \langle X \rangle$ to F mapping X to x is injective. Its restriction to $H * \langle X \rangle$ is therefore injective, and statement (a) holds, which completes the proof. $\qquad\square$

Proposition 3.13 immediately translates into the following.

Corollary 3.14. *Let F be a free group and $H \leq K$ be an elementary extension of subgroups of F. Then, either $H \leq_{\mathrm{alg}} K$ or $H \leq_{\mathrm{ff}} K$. Furthermore, if H is finitely generated then $\mathrm{rk}(K) \leq \mathrm{rk}(H) + 1$ with equality if and only if $H \leq_{\mathrm{ff}} K$.*

Let us say that an extension $H \leq K$ is *e-algebraic*, written $H \leq_{\mathrm{ealg}} K$, if it splits as a finite composition of algebraic, elementary extensions, $H \leq_{\mathrm{alg}} H_1 \leq_{\mathrm{alg}} \cdots \leq_{\mathrm{alg}} H_k = K$. Then Proposition 3.13 yields the following.

Corollary 3.15. *Let H be a finitely generated subgroup of a free group F and let $H \leq_{\mathrm{ealg}} K$ be an e-algebraic extension. Then $\mathrm{rk}(K) \leq \mathrm{rk}(H)$.*

Obviously, every extension $H \leq K$ with K finitely generated, splits into a composition of elementary extensions, but an algebraic extension $H \leq_{\mathrm{alg}} K$ cannot always be split into a composition of algebraic elementary extensions. In view of Corollary 3.15, this is the case for the algebraic extension $\langle [a, b] \rangle \leq_{\mathrm{alg}} F(a, b)$. Thus, $H \leq_{\mathrm{alg}} K$ does not imply $H \leq_{\mathrm{ealg}} K$.

3.4. Algebraic closure of a subgroup

If $H \leq K$ is an extension of free groups, there exists a greatest algebraic extension of H inside K. This can be deduced from Proposition 3.12, but the following theorem is a more precise statement.

Theorem 3.16. *Let $H \leq L \leq K$ be extensions of free groups. The following are equivalent.*

(a) *$H \leq_{\mathrm{alg}} L \leq_{\mathrm{ff}} K$.*
(b) *L is the intersection of the free factors of K containing H.*
(c) *L is the set of elements of K that are K-algebraic over H.*
(d) *L is the greatest algebraic extension of H contained in K.*

In this case, the subgroup L is uniquely determined by H and K.

Proof. Let $x \in K$. By definition, x is K-algebraic over H if and only if x sits in every free factor of K containing H. This is exactly the equivalence of statements (b) and (c). The equivalence of (c) and (d) is a direct consequence of the fact that the elements that are K-algebraic over H form a subgroup (Fact 3.2). Thus statements (b), (c) and (d) are equivalent.

Now let L be defined as in (b). By (d), $H \leq_{\mathsf{alg}} L$. And by Lemma 2.4 (ii), $L \leq_{\mathsf{ff}} K$. This proves (b) implies (a).

Finally, let us assume that $H \leq_{\mathsf{alg}} L \leq_{\mathsf{ff}} K$. Let M be such that $H \leq M \leq_{\mathsf{ff}} K$. Then $L \cap M \leq_{\mathsf{ff}} L$ by Lemma 2.4 (ii). But we also have $H \leq L \cap M \leq L$ and $H \leq_{\mathsf{alg}} L$. It follows that $L \cap M = L$, that is $L \leq M$, and (b) follows. This concludes the proof. $\qquad\square$

Remark 3.17. It is interesting to compare Theorem 3.16 with M. Hall's Theorem, stating that every finitely generated subgroup $H \leq F$ is a free factor of a subgroup M of finite index in F. In other words, one can split the extension $H \leq F$ in two parts, $H \leq_{\mathsf{ff}} M \leq_{\mathsf{fi}} F$, the first being purely transcendental, and the second being finite index (and hence, algebraic). Note that the intermediate subgroup M is not unique in general. Theorem 3.16 yields a "dual" splitting of the extension $H \leq F$, where the order between the transcendental and the algebraic parts is switched around, and with the additional nice property that the intermediate extension is now uniquely determined by $H \leq F$. $\qquad\square$

Let $H \leq K$ be an extension of free groups. The subgroup L characterized in Theorem 3.16 is called the *K-algebraic closure* of H, denoted $\mathsf{cl}_K(H)$. It is natural to consider the extremal situations, where $\mathsf{cl}_K(H) = H$ (we say that H is *K-algebraically closed*) and where $\mathsf{cl}_K(H) = K$ (we say that H is *K-algebraically dense*). Of course, these situations coincide with $H \leq K$ being purely transcendental and algebraic, respectively.

Fact 3.18. *Let $H \leq K$ be an extension of free groups. Then,*

　(i) *H is K-algebraically closed if and only if $H \leq_{\mathsf{ff}} K$,*
　(ii) *H is K-algebraically dense if and only if $H \leq_{\mathsf{alg}} K$.*

As established in the following proposition, maximal proper retracts of a finitely generated free group K are good examples of extremal subgroups, i.e., subgroups of K that are either K-algebraically closed or K-algebraically dense. Recall that a subgroup $H \leq K$ is a *retract* of K if the identity $\mathsf{id} \colon H \to H$ extends to a homomorphism $K \to H$, called a *retraction* (see [9] for a general description of retracts of finitely generated free groups); in particular, free factors of K are retracts of K. Note that if H is a retract of K then $\mathsf{rk}(H) \leq \mathsf{rk}(K)$. Moreover, if K is finitely generated, the Hopfian property of finitely generated free groups shows that K is the unique retract of K with rank equal to $\mathsf{rk}(K)$. So, if H is a proper retract of K then $\mathsf{rk}(H) < \mathsf{rk}(K)$.

We also say that H is *compressed in* K (see [5]) if $\mathsf{rk}(H) \leq \mathsf{rk}(L)$ for each $H \leq L \leq K$. By restricting a retraction to L, it is clear that every retract of K (and, in particular, every free factor of K) is compressed in K.

Proposition 3.19. *Let K be a finitely generated free group. A maximal proper compressed subgroup (resp. a maximal proper retract) H of K is either K-algebraically* *dense, or K-algebraically closed. In the latter case, H is in fact a free factor of K, of rank $rk(K) - 1$.*

Proof. The algebraic closure $\mathsf{cl}_K(H)$ is a free factor of K, and hence it is also a retract and a compressed subgroup. By definition of H, either $\mathsf{cl}_K(H) = K$, and H is K-algebraically dense; or $\mathsf{cl}_K(H) = H$ and H is K-algebraically closed and a free factor. Maximality then implies the announced rank property. $\qquad\square$

We now discuss the behavior of the algebraic closure operator.

Proposition 3.20. *Let $H_i \leq K$, $i = 1, 2$, be two extensions of free groups. Then, $\mathsf{cl}_K(H_1 \cap H_2) \leq_{\mathsf{ff}} \mathsf{cl}_K(H_1) \cap \mathsf{cl}_K(H_2)$, and the equality is not true in general.*

Proof. By Theorem 3.16, $\mathsf{cl}_K(H_i)$ is a free factor of K containing H_i, so $\mathsf{cl}_K(H_1) \cap \mathsf{cl}_K(H_2)$ is a free factor of K containing $H_1 \cap H_2$ (Lemma 2.4). Again by Theorem 3.16, $\mathsf{cl}_K(H_1 \cap H_2)$ is a free factor of $\mathsf{cl}_K(H_1) \cap \mathsf{cl}_K(H_2)$.

A counterexample for the reverse inclusion is as follows: let $K = F(a, b)$, $H_1 = \langle [a, b] \rangle$ and $H_2 = \langle [a, b^{-1}] \rangle$. Both these subgroups are K-algebraically dense (see Example 3.5) and their intersection is trivial. $\qquad\square$

Proposition 3.21. *Let $K_i \leq K$, $i = 1, 2$, be two extensions of free groups and let $H \leq K_1 \cap K_2$. Then, $\mathsf{cl}_{K_1 \cap K_2}(H) \leq_{\mathsf{ff}} \mathsf{cl}_{K_1}(H) \cap \mathsf{cl}_{K_2}(H)$, and the equality is not true in general.*

Proof. By Theorem 3.16, $\mathsf{cl}_{K_i}(H)$ is a free factor of K_i containing H, so $\mathsf{cl}_{K_1}(H) \cap \mathsf{cl}_{K_2}(H)$ is a free factor of $K_1 \cap K_2$ containing H (Lemma 2.4). Again by Theorem 3.16, $\mathsf{cl}_{K_1 \cap K_2}(H)$ is a free factor of $\mathsf{cl}_{K_1}(H) \cap \mathsf{cl}_{K_2}(H)$.

The following is a counter-example for the converse inclusion. Let $K = \langle a, b, c \rangle$ be a free group of rank 3, let $H = \langle [a, b], [a, c] \rangle$, $K_1 = \langle a, b, [a, c] \rangle$ and $K_2 = \langle a, c, [a, b] \rangle$. One can verify that $K_1 \cap K_2 = \langle a, [a, b], [a, c] \rangle$, so $\mathsf{cl}_{K_1 \cap K_2}(H) = H$. On the other hand, $H \leq_{\mathsf{alg}} K_i$ by Example 3.5 and Proposition 3.12, so $\mathsf{cl}_{K_i}(H) = K_i$ and $\mathsf{cl}_{K_1}(H) \cap \mathsf{cl}_{K_2}(H) = K_1 \cap K_2 \neq \mathsf{cl}_{K_1 \cap K_2}(H)$. $\qquad\square$

Remark 3.22. If $H \leq K_1 \leq K_2$, Proposition 3.21 shows that $\mathsf{cl}_{K_1}(H) \leq \mathsf{cl}_{K_2}(H)$. If in addition $K_1 \leq_{\mathsf{ff}} K_2$, Proposition 3.16 shows that $\mathsf{cl}_{K_1}(H) = \mathsf{cl}_{K_2}(H)$. However, in general, even the inclusion $\mathsf{cl}_{K_1}(H) \leq K_1 \cap \mathsf{cl}_{K_2}(H)$ may be strict, as the following counterexample shows.

Let $K_2 = \langle a, b \rangle$ be a free group of rank 2, and let $H = \langle [a, b] \rangle$ and $K_1 = \langle a, [a, b] \rangle$. Then $H \leq_{\mathsf{ff}} K_1 \leq_{\mathsf{alg}} F$ and $H \leq_{\mathsf{alg}} F$. So, $\mathsf{cl}_{K_1}(H) = H$ is properly contained in $K_1 \cap \mathsf{cl}_F(H) = K_1 \cap F = K_1$. $\qquad\square$

Finally, let us consider e-algebraic extensions. There too, there exists a greatest e-algebraic extension, at least for finitely generated subgroups. We first prove the following technical lemma.

Lemma 3.23. *Let $H \leq K \leq F$ be extensions of free groups and let $x \in F$. If $H \leq_{\mathsf{alg}} \langle H, x \rangle$, then $K \leq_{\mathsf{alg}} \langle K, x \rangle$.*

242   A. Miasnikov, E. Ventura and P. Weil

Proof. Assume $H \leq_{\mathsf{alg}} \langle H, x \rangle$. If $K \leq \langle K, x \rangle$ is not algebraic, then $x \notin K$ and $K \leq_{\mathsf{ff}} \langle K, x \rangle$ by Proposition 3.13. It follows that $H \leq \langle H, x \rangle \cap K \leq_{\mathsf{ff}} \langle H, x \rangle \cap \langle K, x \rangle = \langle H, x \rangle$, which forces either $\langle H, x \rangle \cap K = H$ or $\langle H, x \rangle \cap K = \langle H, x \rangle$. The first possibility implies $H = \langle H, x \rangle \cap K \leq_{\mathsf{ff}} \langle H, x \rangle$ contradicting the hypothesis, while the second possibility contradicts $x \notin K$. $\qquad\square$

Corollary 3.24. *Let $H \leq F$ be an extension of free groups and let $H \leq_{\mathsf{ealg}} K_i \leq F$ ($i = 1, \ldots, n$) be a finite family of e-algebraic extensions of H. Then $K_i \leq_{\mathsf{ealg}} \langle \bigcup_j K_j \rangle$ for each i.*

In particular, if H is finitely generated, then H admits a greatest e-algebraic extension in F.

Proof. It suffices to prove the first statement for $n = 2$. Let us assume that $H = H_0 \leq_{\mathsf{alg}} H_1 \leq_{\mathsf{alg}} \cdots \leq_{\mathsf{alg}} H_p = K_1$ and that $x_1, \ldots, x_p$ are such that $H_i = \langle H_{i-1}, x_i \rangle$ for each $1 \leq i \leq p$. Then a repeated application of Lemma 3.23 shows that $K_2 \leq_{\mathsf{ealg}} \langle K_2, x_1, \ldots, x_p \rangle = \langle K_1 \cup K_2 \rangle$.

If $H \leq_{\mathsf{fg}} F$, H has finitely many algebraic extensions, and among them finitely many e-algebraic extensions. The join of these extensions is again an e-algebraic extension and this concludes the proof. $\qquad\square$

The greatest e-algebraic extension of a subgroup $H \leq F$, whose existence is asserted in Corollary 3.24, is called its *e-algebraic closure*. We say that H is *e-algebraically closed* if it is equal to its e-algebraic closure. Proposition 3.13 immediately implies the following characterization.

Corollary 3.25. *Let $H \leq F$ be an extension of free groups. Then H is e-algebraically closed if and only if $\langle H, x \rangle = H * \langle x \rangle$ for each $x \notin H$.*

Example 3.26. Let $x \in F$ be an element of a free group not being a proper power. Then, for every $y \in F$, either $\langle x \rangle = \langle x, y \rangle$ or $\mathsf{rk}(\langle x, y \rangle) = 2$. In other words, maximal cyclic subgroups of free groups are e-algebraically closed.

A subgroup $H \leq F$ is said to be *strictly compressed* if $\mathsf{rk}(H) < \mathsf{rk}(K)$ for each proper extension $H < K \leq F$. It is immediate that strictly compressed subgroups form a natural class of e-algebraically closed subgroups.

By Example 3.5, we know that if F has rank two, then $\langle x \rangle \leq F$ is algebraic if and only if x is not a power of a primitive element of F. Hence, situations like $H = \langle [a, b] \rangle < \langle a, b \rangle$ are examples of algebraic extensions where the base group H is e-algebraically closed. This is a behavior significantly different from what happens in field theory. $\qquad\square$

Corollary 3.27. *Let $H \leq F(A)$ be an extension of free groups. If H is finitely generated, it is decidable whether H is e-algebraically closed.*

Proof. Let $x \notin H$, viewed as a reduced word on the alphabet A, let p be the longest prefix of x labeling a path starting at the designated vertex 1 in $\Gamma_A(H)$, and let s be the longest suffix of x labeling a path to 1 in $\Gamma_A(H)$. We denote by $1 \cdot p$ and $1 \cdot s^{-1}$ the end vertices of these two paths.

First assume that the sum of the length of p and s is less than the length of x, that is, if $x = pys$ for some non-empty word y. Then $\Gamma_A(\langle H, x \rangle)$ is obtained from $\Gamma_A(H)$ by gluing a path (made of new vertices and new edges) from $1 \cdot p$ to $1 \cdot s^{-1}$, labeled y. In particular, $\mathsf{rk}(\langle H, x \rangle) = \mathsf{rk}(H) + 1$.

We now assume that the sum of the lengths of p and s is greater than or equal to the length of x, and we let t be the longest suffix of p which is also a prefix of s. That is, $p = p't$, $s = ts'$ and $x = p'ts'$. Let $1 \cdot p'$ be the end vertex of the path starting at 1 and labeled p' in $\Gamma_A(H)$. If $1 \cdot p' = 1 \cdot s^{-1}$, then $x = p's$ labels in fact a loop at 1, that is, $x \in H$, a contradiction. So the labeled graph $\Gamma_A(\langle H, x \rangle)$ is the quotient of $\Gamma_A(H)$ by the congruence generated by the pair $(1 \cdot p', 1 \cdot s^{-1})$ (see the end of Section 2.2).

Thus, in view of Corollary 3.25, H is e-algebraically closed if and only if the following holds: for each pair of distinct vertices (v, w) in $\Gamma_A(H)$, the subgroup represented by the quotient of $\Gamma_A(H)$ by the congruence generated by (v, w) has rank at most $\mathsf{rk}(H)$. This is decidable, and concludes the proof. $\square$

4. Abstract properties of subgroups

Let F be a free group. An abstract *property* of subgroups of F is a set $\mathcal{P}$ of subgroups of F containing at least the total group F itself. For simplicity, if $H \in \mathcal{P}$, we will say that the subgroup H *satisfies property* $\mathcal{P}$.

We say that the property $\mathcal{P}$ is *(finite) intersection closed* if the intersection of any (finite) family of subgroups of F satisfying $\mathcal{P}$ also satisfies $\mathcal{P}$, and that it is *free factor closed* if every free factor of a subgroup of F satisfying $\mathcal{P}$ also satisfies $\mathcal{P}$. Finally, we say that the property $\mathcal{P}$ is *decidable* if there exists an algorithm to decide whether a given finitely generated subgroup $H \leq F(A)$ satisfies $\mathcal{P}$.

4.1. $\mathcal{P}$-closure of a subgroup

Let F be a free group, $\mathcal{P}$ be an abstract property of subgroups of F, and let $H \leq F$. If there exists a unique minimal subgroup of F satisfying $\mathcal{P}$ and containing H, it is called the $\mathcal{P}$-*closure* of H, denoted by $\mathsf{cl}_{\mathcal{P}}(H)$; in this situation, we say that H admits a *well-defined* $\mathcal{P}$-closure.

Proposition 4.1. *Let F be a free group and let $\mathcal{P}$ be an abstract property of subgroups of F.*

(i) *If $\mathcal{P}$ is intersection closed, then every subgroup $H \leq F$ admits a well-defined $\mathcal{P}$-closure.*

(ii) *If $\mathcal{P}$ is finite intersection closed and free factor closed then every finitely generated subgroup $H \leq_{\mathsf{fg}} F$ admits a well-defined $\mathcal{P}$-closure.*

(iii) *If $\mathcal{P}$-closures are well defined and $\mathcal{P}$ is free factor closed, then for every subgroup $H \leq F$, we have $H \leq_{\mathsf{alg}} \mathsf{cl}_{\mathcal{P}}(H)$. In particular, if H is finitely generated, then so is $\mathsf{cl}_{\mathcal{P}}(H)$.*

Proof. Statement (i) is immediate: it suffices to consider the intersection of all the extensions of H satisfying $\mathcal{P}$ (there is at least one, namely F itself).

If $\mathcal{P}$ is only finite intersection closed, but is also free factor closed, we use Theorem 3.16: since every extension of a finitely generated subgroup H is a free multiple of an algebraic extension of H, then every extension of H in $\mathcal{P}$ contains an algebraic extension of H in $\mathcal{P}$. It follows that the intersection of all extensions of H in $\mathcal{P}$ is equal to the intersection of the algebraic extensions of H in $\mathcal{P}$. But the latter intersection is finite, and hence it satisfies $\mathcal{P}$ as well, which concludes the proof of (ii).

Finally, if $\mathcal{P}$ is free factor closed, then H is not contained in any proper free factor of its $\mathcal{P}$-closure, that is, $H \leq_{\mathsf{alg}} \mathsf{cl}_{\mathcal{P}}(H)$. $\qquad\square$

It would be interesting to produce an example of an abstract property $\mathcal{P}$ that is closed under free factors and finite intersections, not closed under intersections, and non-trivial for finitely generated subgroups (note that the property *to be finitely generated* satisfies the required closure and non-closure properties, but it is trivial for finitely generated subgroups).

Remark 4.2. It is well known that the property of being normal in F is closed under intersections and not under free factors, and that given a subgroup $H \leq F$, the normal closure of H is well defined, but is not in general finitely generated, even if H is. $\qquad\square$

Proposition 4.3. *Let $\mathcal{P}$ be an abstract property for subgroups of $F(A)$ for which $\mathcal{P}$-closures are well defined. If $\mathcal{P}$-closures of finitely generated subgroups of $F(A)$ are computable, then $\mathcal{P}$ is decidable. The converse holds if, additionally, $\mathcal{P}$ is free factor closed.*

Proof. Let us assume that $\mathcal{P}$-closures are computable. Then, in order to decide whether a given $H \leq_{\mathsf{fg}} F(A)$ satisfies $\mathcal{P}$, it suffices to compute $\mathsf{cl}_{\mathcal{P}}(H)$, and to verify whether $H = \mathsf{cl}_{\mathcal{P}}(H)$.

Conversely, suppose that $\mathcal{P}$ is free factor closed and decidable. Then, given $H \leq_{\mathsf{fg}} F(A)$, one can compute the set $\mathsf{AE}(H)$, check which algebraic extensions of H satisfy $\mathcal{P}$ and identify the minimal one(s). By Proposition 4.1, only one of them is minimal, and that one must be $\mathsf{cl}_{\mathcal{P}}(H)$. $\qquad\square$

Remark 4.4. Proposition 4.1 states that every property of subgroups that is closed under (finite) intersections and under free factors yields a well-defined closure operator for (finitely generated) subgroups of F, that can be obtained by looking exclusively at algebraic extensions.

A form of converse holds too: if $K \leq_{\mathsf{fg}} F$, let $\mathcal{P}_K$ be the following property. A subgroup L satisfies $\mathcal{P}_K$ if and only if L is a free factor of an extension of K. Clearly, F satisfies this property, and one can verify that $\mathcal{P}_K$ is intersection and free factor closed. Moreover, one can use Proposition 3.16 to verify that the $\mathcal{P}_K$-closure of a subgroup $H \leq K$ is exactly the K-algebraic closure of H. In particular, for every algebraic extension $H \leq_{\mathsf{alg}} K$, K is the $\mathcal{P}$-closure of H for some well-chosen property $\mathcal{P}$. $\qquad\square$

4.2. Some algebraic properties

Let us recall the definition of certain properties of subgroups, that have been discussed in the literature. Let $H \leq F$ be an extension of free groups. We say that H is

- *malnormal* if $H^g \cap H = 1$ for all $g \in F \setminus H$;
- *pure* if $x^n \in H$, $n \neq 0$ implies $x \in H$ (this property is also called being closed under radical, or being isolated);
- *p-pure* (for a prime p) if $x^n \in H$, $(n,p) = 1$ implies $x \in H$;

The following results on malnormal and pure closure were first shown in [7, Section 13]. The proof given here, while not fundamentally different, is simpler and more general. Corollary 4.14 below gives further properties of these closures.

Proposition 4.5. *Let $F(A)$ be a free group. The properties (of subgroups) defined by malnormal, pure, p-pure (p a prime), retract and e-algebraically closed subgroups are intersection and free factor closed, and decidable for finitely generated subgroups.*

For each of these properties $\mathcal{P}$, each subgroup $H \leq F(A)$ admits a well-defined $\mathcal{P}$-closure $cl_{\mathcal{P}}(H)$, which is an algebraic extension of H. Finally, if $H \leq_{fg} F(A)$, the $\mathcal{P}$-closure of H has finite rank and is computable.

Proof. The closure under intersections and free factors of malnormality is immediate from the definition. The decidability of malnormality was established in [1], with a simple algorithm given in [7, Corollary 9.11].

The closure under intersections of the properties of purity and p-purity is immediate. Now, assume that K is pure, $H \leq_{ff} K$, and let x be such that $x^n \in H$ with $n \neq 0$. Since K is pure, we have $x \in K$, and we simply need to show that a free factor of a free group F is pure, which was established in Example 3.1 above. Thus purity is free factor closed. The proof of the same property for p-purity is identical. The decidability of purity and p-purity was proved in [3, 4].

It is shown in [2, Lemma 18] that an arbitrary intersection of retracts of F is again a retract of F. Moreover, it follows from the definition of retracts that a retract of a retract is a retract, and that a free factor is a retract. Thus the property of being a retract of F is free factor closed. The decidability of this property was established by Turner, but as no proof seems to have been published, we give his in Proposition 4.6 below.

Suppose that $H \leq_{ff} K \leq F$, K is e-algebraically closed and $x \notin H$. If $x \notin K$, then $\langle K, x \rangle = K * \langle x \rangle$, so $\langle H, x \rangle = H * \langle x \rangle$. If $x \in K \setminus H$, then we have that H is a free factor of $\langle H, x \rangle \leq K$ and so, by Proposition 3.13, we also conclude that $\langle H, x \rangle = H * \langle x \rangle$. Thus the property of being e-algebraically closed is closed under free factors. Next, let $(H_i)_{i \in I}$ be a family of e-algebraically closed subgroups, let $H = \bigcap_i H_i$, and let $x \notin H$. There exists $i \in I$ such that $x \notin H_i$, so $\langle H_i, x \rangle = H_i * \langle x \rangle$. Using Lemma 3.23, we conclude that $\langle H, x \rangle = H * \langle x \rangle$. Thus the property of being e-algebraically closed is also closed under intersections. Finally, this property is decidable by Corollary 3.27.

The last part of the statement follows from Proposition 4.1. $\qquad\square$

As announced in the proof of Proposition 4.5, we prove the decidability of retracts, that was established by Turner [19].

Proposition 4.6. *Let $H \leq F(A)$ be an extension of finitely generated free groups. It is decidable whether H is a retract of $F(A)$.*

Proof. (Turner) Suppose that $A = \{a_1, \ldots, a_n\}$ and let $u_1, \ldots, u_r$ be a basis of H. Then H is a retract of $F(A)$ if and only if there exist $x_1, \ldots, x_n \in H$ such that the endomorphism φ of $F(A)$ defined by $\varphi(a_i) = x_i$ maps H identically to itself. That is, if $u_i(x_1, \ldots, x_n) = u_i$ for $i = 1, \ldots, r$. This can be expressed in terms of systems of equations.

Let e_i be the word on alphabet $\{X_1, \ldots, X_n\}$ obtained from the word u_i (on alphabet A) by substituting X_j for a_j for each j. Then H is a retract of F if and only if the system of equations $e_i(X_1, \ldots, X_n) = u_i$, $i = 1, \ldots, r$ (where u_i are viewed as constants in H) admits a solution in H. This is decidable by Makanin's algorithm [10] (note that the form of the system (i.e., the words e_i) depends on the way H is embedded in F, but once this form is established, the system itself is entirely set within H, so Makanin's algorithm works, applied to this system over H). $\qquad\square$

Let $H \leq F$ be an extension of free groups. Recall that H is *compressed* if $\mathsf{rk}(H) \leq \mathsf{rk}(K)$ for every $K \leq F$ containing H (see Section 3.4), and say that H is *inert* if $\mathsf{rk}(H \cap K) \leq \mathsf{rk}(K)$ for every $K \leq F$. Both these properties were introduced by Dicks and Ventura [5] in the context of the study of subgroups of free groups that are fixed by sets of endomorphisms or automorphisms (see also [21]).

It is clear that an inert or compressed subgroup is finitely generated, with rank at most $\mathsf{rk}(F)$. It is also clear that inert subgroups (and retracts) of F are compressed. On the other hand, we do not know whether all compressed subgroups are inert, nor whether retracts are inert (both these facts are conjectured in [21] and related to other conjectures about fixed subgroups in free groups).

Proposition 4.7. *Let F be a free group. The properties of inertness and compressedness are closed under free factors. In addition, inertness is closed under intersections.*

Each subgroup $H \leq F$ admits an inert closure, which is an algebraic extension of H.

Proof. The closure of inertness under intersections is shown in [5, Corollary I.4.13]. Free factors of F are trivially inert. Moreover, if $H \leq K \leq F$, H is inert in K and K is inert in F, then H is inert in F. So inertness is also closed under free factors.

Now suppose that $H = L * M \leq F$ is compressed, and let $L \leq K \leq F$. Since $H \leq \langle K, M \rangle$, we have

$$\mathsf{rk}(L) + \mathsf{rk}(M) = \mathsf{rk}(H) \leq \mathsf{rk}(\langle K, M \rangle) \leq \mathsf{rk}(K) + \mathsf{rk}(M).$$

It follows that $\mathsf{rk}(L) \leq \mathsf{rk}(K)$, and hence L is compressed. Thus, compressedness is closed under free factors.

The last statement is a direct application of Proposition 4.1. $\qquad\square$

Note that, even though a finitely generated subgroup H admits an inert closure, which is one of its (finitely many) algebraic extensions of H, we do not know how to compute this closure, nor how to decide whether a subgroup is inert.

It is not known either whether compressedness is closed under intersections, or even finite intersections, so we don't know whether each subgroup admits a *compressed closure*. However it is decidable whether a finitely generated subgroup of F is compressed [20]. Indeed if $H \leq_{\mathsf{fg}} F$, then H is compressed if and only if $\mathsf{rk}(H) \leq \mathsf{rk}(K)$ for every algebraic extension $H \leq_{\mathsf{alg}} K \leq F$, which reduces the verification to a finite number of rank comparisons, for $K \in \mathsf{AE}(H)$, say.

4.3. On certain topological closures

Let $\mathcal{T}$ be a topology on a free group F. The abstract property of subgroups consisting of the subgroups that are closed in $\mathcal{T}$ is trivially closed under intersections. This property becomes more interesting when the topology is related to the algebraic structure of F. This is the case of the pro-$\mathbf{V}$ topologies that we analyze now.

A *pseudovariety of groups* $\mathbf{V}$ is a class of finite groups that is closed under taking subgroups, quotients and finite direct products. $\mathbf{V}$ is called *non-trivial* if it contains some non-trivial finite group. Additionally, if for every short exact sequence of finite groups, $1 \rightarrow G_1 \rightarrow G_2 \rightarrow G_3 \rightarrow 1$, with G_1 and G_3 in $\mathbf{V}$, one always has $G_2 \in \mathbf{V}$, we say that $\mathbf{V}$ is *extension-closed*.

For every non-trivial pseudovariety of groups $\mathbf{V}$, the *pro-$\mathbf{V}$* topology on a free group F is the initial topology of the collection of morphisms from F into groups in $\mathbf{V}$, or equivalently, the topology for which the normal subgroups N such that $F/N \in \mathbf{V}$ form a basis of neighborhoods of the unit. We refer the readers to [11, 22] for a survey of results concerning these topologies with regard to finitely generated subgroups of free groups. In particular, Ribes and Zalesskiĭ showed that if $\mathbf{V}$ is extension-closed then every free factor of a closed subgroup is closed [12]. The following observation then follows from Proposition 4.1.

Fact 4.8. *Let $\mathbf{V}$ be a non-trivial extension-closed pseudovariety of groups. Then the pro-$\mathbf{V}$ closure of a finitely generated subgroup H is finitely generated, and an algebraic extension of H.*

In the case of the pro-p topology (p is a prime and the pseudovariety $\mathbf{V}$ is that of finite p-groups, which is closed under extensions), Ribes and Zalesskiĭ [12] showed that one can compute the closure of a given finitely generated subgroup of $F(A)$. A polynomial time algorithm was later given by Margolis, Sapir and Weil [11], based on the finiteness of the number of principal overgroups of H, that is, essentially on the spirit of Fact 4.8. Moreover, they showed that one can simultaneously compute the pro-p closures of H, for all primes p, using the fact that they are all algebraic extensions, and hence that they take only finitely many values. This was also used to show the computability of the pro-nilpotent closure of a finitely generated subgroup: even though the pseudovariety of finite nilpotent groups is not closed under extensions, it still holds that the pro-nilpotent closure of a finitely generated subgroup is finitely generated and computable.

At this point, several remarks are in order. First, Ribes and Zalesskiĭ [12] proved that if $\mathbf{V}$ is extension-closed and if $\bar{H}$ is the pro-$\mathbf{V}$ closure of H, then $\mathsf{rk}(\bar{H}) \leq \mathsf{rk}(H)$. The proof of this fact can be reduced to dimension considerations in appropriate vector spaces. This proof does not seem related with the idea of e-algebraic extensions, which also lowers the rank (Corollary 3.15).

Next, not every algebraic extension arises as a pro-$\mathbf{V}$ closure for some $\mathbf{V}$. This is clear if $H \leq_{\mathsf{alg}} K$ and $\mathsf{rk}(K) > \mathsf{rk}(H)$ by the result of Ribes and Zalesskiĭ cited above, but rank is not the only obstacle. Consider indeed $H = \langle a, bab^{-1} \rangle \leq F(a,b)$. Then $H \leq_{\mathsf{alg}} F$ (Example 3.5) and $\mathsf{AE}(H) = \mathcal{O}_A(H) = \{H, F\}$. We now verify that H is $\mathbf{V}$-closed for each non-trivial extension-closed pseudovariety $\mathbf{V}$, so F is never the $\mathbf{V}$-closure of H. Since $\mathbf{V}$ is non-trivial, the cyclic p-element group $C_p = \langle c \mid c^p \rangle$ sits in $\mathbf{V}$ for some prime p. Let $\varphi_p \colon F \to C_p$ be the morphism defined by $\varphi_p(a) = 1$ and $\varphi_p(b) = c$, and let $N_p = \ker \varphi_p$. Then $H \leq N_p$ and N_p is $\mathbf{V}$-closed, so H is not topologically dense in F. Since the $\mathbf{V}$-closure of H is in $\mathsf{AE}(H)$, it follows that H is closed in the pro-$\mathbf{V}$ topology.

Solvable groups form an extension-closed pseudovariety, so the above results apply to it: in particular, given an extension $H \leq_{\mathsf{fg}} F(A)$, we can compute a finite list of candidates for being the pro-solvable closure of H, namely $\mathsf{AE}(H)$ (or even this list, restricted to the extensions of rank at most $\mathsf{rk}(H)$). However, it is a wide open problem to compute this closure.

Finally, let us consider the (uncountable) collection of extension-closed pseudovarieties of finite groups $\mathbf{V}$ as above. For each finitely generated subgroup $H \leq F$, the pro-$\mathbf{V}$ closures of H are among the (finitely many) algebraic extensions of H, so each finitely generated subgroup H naturally induces a finite index equivalence relation on the collection of the $\mathbf{V}$'s. It would be interesting to investigate the properties of these equivalence relations. In particular, the intersection of these equivalence relations, as H runs over all the (countably many) finitely generated subgroups of $F(a,b)$, has countably many classes, so there are pseudovarieties $\mathbf{V}$ that are indistinguishable in this way.

4.4. Equations over a subgroup

In this section we use equations over free groups to define abstract properties of subgroups. Let $H \leq F$ be an extension of free groups. A *(one variable) H-equation* (or *equation over H*) is an element $e = e(X)$ of the free group $H * \langle X \rangle$, where X is a new free letter, called the *variable*. An element $x \in F$ is a *solution of* $e(X)$ if $e(x) = 1$ in F (technically: if the morphism $H * \langle X \rangle \to F$ mapping H identically to itself and X to x, maps e to 1).

Example 4.9. If $H = \langle a^2 \rangle$, the H-equation $e(X) = Xa^2X^{-1}a^{-2}$ admits a as a solution. So does the H-equation X^2a^{-2}.

If e does not involve X, that is, $e \in H$, then e has no solution unless it is the *trivial* equation $e = 1$, in which case every element of F is a solution. $\square$

We immediately observe the following.

Lemma 4.10. *Let $H \leq F$ be an extension of free groups and let $x \in F$. The element x is a solution of some non-trivial H-equation if and only if the elementary extension $H \leq \langle H, x \rangle$ is algebraic.*

Proof. Let X be a new free generator and let $\varphi \colon H * \langle X \rangle \to F$ be the morphism that maps H identically to itself and X to x. By definition, x is a solution of some non-trivial equation over H if and only if φ is not injective, and we conclude by Proposition 3.13 and Corollary 3.14 that this is equivalent to $H \leq_{\mathsf{alg}} \langle H, x \rangle$. $\square$

In order to make this natural definition of equations independent on the choice of the subgroup H, we consider a countable set $X, Y_1, Y_2, \ldots$ of variables and we call *equation* any element e of the free group on these variables. If $H \leq F$ is an extension of free groups, a *particularization of e over H* is the H-equation $e(X, h_1, h_2, \ldots)$ obtained by substituting elements $h_1, h_2, \ldots \in H$ for the variables $Y_1, Y_2, \ldots$ (and having X as variable).

A *solution of the equation e over H* is a solution of some non-trivial particularization of e over H, that is, an element $x \in F$ such that, for some $h_1, h_2, \ldots \in H$, $e(X, h_1, h_2, \ldots) \neq 1$ but $e(x, h_1, h_2, \ldots) = 1$. (Note that even when X occurs in e, some particularizations of e over H can be trivial.)

Let $\mathcal{E}$ be an arbitrary set of equations. We say that a subgroup $H \leq F$ is *$\mathcal{E}$-closed* if H contains every solution over H of every equation in $\mathcal{E}$. Note that, when looking for solutions, the set $\mathcal{E}$ is not considered as a system of equations, but as a set of mutually unrelated equations. In particular, a larger set $\mathcal{E}$ yields a larger set of solutions.

Proposition 4.11. *Let F be a free group and let $\mathcal{E}$ be a set of equations. Then the property of being $\mathcal{E}$-closed is closed under intersections and under free factors.*

Proof. The closure under intersections follows directly from the definition. Now assume that $K \leq F$ is $\mathcal{E}$-closed and let $H \leq_{\mathsf{ff}} K$. Let x be a solution of an equation of $\mathcal{E}$ over H. Then x is also a solution over K, and hence $x \in K$. Now, by Lemma 4.10, $H \leq_{\mathsf{alg}} \langle H, x \rangle \leq K$. This contradicts $H \leq_{\mathsf{ff}} K$ unless $H = \langle H, x \rangle$, and hence $x \in H$. $\square$

Corollary 4.12. *Let $H \leq F$ and let $\mathcal{E}$ be a set of equations. There exists a least $\mathcal{E}$-closed extension of H, denoted by $cl_{\mathcal{E}}(H)$ and called the $\mathcal{E}$-closure of H. Moreover, $H \leq_{\mathsf{alg}} cl_{\mathcal{E}}(H)$.*

If in addition H is finitely generated, then $H \leq_{\mathsf{ealg}} cl_{\mathcal{E}}(H)$, $rk(cl_{\mathcal{E}}(H)) \leq rk(H)$ and there exists a finite subset $\mathcal{E}_0$ of $\mathcal{E}$ such that $cl_{\mathcal{E}_0}(H) = cl_{\mathcal{E}}(H)$.

Proof. Propositions 4.1 and 4.11 directly prove the first part of the statement.

We now suppose that $H \leq_{\mathsf{fg}} F$ and we let $H_0 = H$ and suppose that we have constructed distinct extensions $H_0 \leq_{\mathsf{ealg}} H_1 \leq_{\mathsf{ealg}} \cdots \leq_{\mathsf{ealg}} H_n$ $(n \geq 0)$, elements $x_1, \ldots, x_n \in F$, and equations $e_1, \ldots, e_n \in \mathcal{E}$ such that $H_i = \langle H_{i-1}, x_i \rangle$ and x_i is a solution of e_i over H_{i-1}. If H_n is not $\mathcal{E}$-closed, then there exists an equation $e_{n+1} \in \mathcal{E}$, and an element $x_{n+1} \notin H_n$ such that x_{n+1} is a solution of a non-trivial particularization of e_{n+1} over H_n. Then $H_{n+1} = \langle H_n, x_{n+1} \rangle$ is a proper elementary algebraic extension of H_n by Lemma 4.10. Since H has only a finite number of algebraic extensions, this construction must stop, that is, for some n, H_n is $\mathcal{E}$-closed. It follows easily that H_n is the $\mathcal{E}$-closure of H, whose existence was already established. In particular $H \leq_{\mathrm{ealg}} \mathrm{cl}_{\mathcal{E}}(H)$, and $\mathrm{rk}(\mathrm{cl}_{\mathcal{E}}(H)) \leq \mathrm{rk}(H)$ by Corollary 3.15.

Finally, let $\mathcal{E}_0 = \{e_1, \ldots, e_n\}$. Any $\mathcal{E}$-closed subgroup is also $\mathcal{E}_0$-closed, and the $\mathcal{E}_0$-closure of H must contain $x_1, \ldots, x_n$. Thus $\mathrm{cl}_{\mathcal{E}}(H) = \mathrm{cl}_{\mathcal{E}_0}(H)$. $\square$

We conclude with the observation that some of the properties discussed in Section 4.2 can be expressed in terms of equations. Let p be a prime number and let

$$\begin{aligned}
\mathcal{E}_{\mathrm{mal}} &= \{X^{-1}Y_1 X Y_2\}, \\
\mathcal{E}_p &= \{X^n Y_1 \mid (n,p) = 1\}, \\
\mathcal{E}_{\mathbb{Z}} &= \{X^n Y_1 \mid n \neq 0\} = \bigcup_p \mathcal{E}_p, \\
\mathcal{E}_{\mathrm{com}} &= \{X^{-1}Y_1^{-1} X Y_1\}.
\end{aligned}$$

Proposition 4.13. *Let $H \leq F$ be an extension of free groups. The subgroup H is*

(i) *malnormal if and only if it is $\mathcal{E}_{\mathrm{mal}}$-closed;*
(ii) *p-pure if and only if it is $\mathcal{E}_p$-closed;*
(iii) *pure if and only if it is $\mathcal{E}_{\mathbb{Z}}$-closed, and if and only if it is $\mathcal{E}_{\mathrm{com}}$-closed.*

Proof. H is $\mathcal{E}_{\mathrm{mal}}$-closed if and only if, for all $h_1, h_2 \in H$, not simultaneously trivial, every solution of the equation $X^{-1}h_1 X h_2 = 1$ belongs to H. That is, if and only if $x^{-1}Hx \cap H \neq 1$ implies $x \in H$. This is precisely the malnormality property for H. This proves (i).

H is $\mathcal{E}_p$-closed if and only if H contains the nth roots of every one of its elements, for all n such that $(n,p) = 1$. Again, this is exactly the definition of p-purity, showing (ii).

Similarly, H is $\mathcal{E}_{\mathbb{Z}}$-closed if and only if H is pure. Finally, we recall that two elements x and y in F commute if and only if they are powers of a common $z \in F$. Thus the subgroup generated by H and all the roots of its elements is exactly the $\mathcal{E}_{\mathrm{com}}$-closure of H. $\square$

Corollary 4.12 immediately implies the following.

Corollary 4.14. *Let $H \leq_{\mathrm{fg}} F$ and let K be the malnormal (resp. pure, p-pure) closure of H. Then $H \leq_{\mathrm{ealg}} K$ and $\mathrm{rk}(K) \leq \mathrm{rk}(H)$.*

5. Some open questions

To conclude this paper, we would like to draw the readers' attention to a few of the questions it raises.

(1) We believe that the algebraic extensions of a finitely generated subgroup $H \leq_{\mathsf{fg}} F$ are precisely the extensions which occur as principal overgroups of H for every choice of an ambient basis. That is, we conjecture that $\mathsf{AE}(H) = \bigcap_A \mathcal{O}_A(H)$, where A runs over all the bases of F. As noticed in Section 3.1, this is the case when $H \leq_{\mathsf{fi}} F$ or $H \leq_{\mathsf{ff}} F$, but nothing is known in general.

(2) With reference to Corollary 3.15, we would like to find an algebraic extension $H \leq_{\mathsf{alg}} K$ of finitely generated groups, where $\mathsf{rk}(K) \leq \mathsf{rk}(H)$, yet the extension is not e-algebraic. It would be appropriate to look for such an extension where H is e-algebraically closed in K, that is, $\langle H, x \rangle = H * \langle x \rangle$ for each $x \in K \setminus H$ (Corollary 3.25).

(3) Even though a finitely generated subgroup H admits an inert closure, which is one of the finitely many (computable) algebraic extensions of H, we do not know how to compute this closure. Equivalently, it would be interesting to find an algorithm to decide whether a subgroup is inert (see Section 4.2).

(4) It is not known whether an intersection, even a finite intersection, of (strictly) compressed subgroups is again (strictly) compressed. In other words, does a finitely generated subgroup admit a (*strictly*) *compressed closure*? If the answer was affirmative, then these closures would be computable, as indicated in Section 4.2.

(5) As pointed out in Section 4.3, we know that if $\mathbf{V}$ is a non-trivial extension-closed pseudovariety of groups and $H \leq_{\mathsf{fg}} F$, then $\bar{H}$, the pro-$\mathbf{V}$ closure of H, is an algebraic extension of H with rank at most $\mathsf{rk}(H)$. However the known proof of this fact does not rely on the notion of e-algebraic extensions. We would like to find an example of such a subgroup H and a pseudovariety $\mathbf{V}$ such that the extension $H \leq \bar{H}$ is not e-algebraic – or alternately to give a new proof of Ribes and Zalesskiĭ's result (that in this situation, $\mathsf{rk}(\bar{H}) \leq \mathsf{rk}(H)$), by showing that $H \leq_{\mathsf{ealg}} \bar{H}$.

(6) As indicated at the end of Section 4.3, it would be interesting to find and investigate explicit examples of pseudovarieties $\mathbf{V}_1$ and $\mathbf{V}_2$, such that the pro-$\mathbf{V}_1$ and pro-$\mathbf{V}_2$ closures of H do coincide, for every $H \leq_{\mathsf{fg}} F$. As argued above, there are uncountably many such pairs that are indistinguishable by means of closures of finitely generated subgroups.

(7) Finally, Corollary 4.12 shows that for every set of equations $\mathcal{E}$ and every $H \leq_{\mathsf{fg}} F$, there exists a finite subset $\mathcal{E}_0 \subseteq \mathcal{E}$ such that $\mathsf{cl}_{\mathcal{E}_0}(H) = \mathsf{cl}_{\mathcal{E}}(H)$. Is it true that such a finite set always exists satisfying the previous equality for all finitely generated subgroups of F at the same time (showing a kind of Noetherian behavior)?

Acknowledgements

E. Ventura thanks the DGI of the Spanish government (grant BFM2003-06613) and the Generalitat de Catalunya (grant ACI-013) for their support. P. Weil acknowledges support from the European Science Foundation program AUTOMATHA. E. Ventura and P. Weil wish to thank the Mathematics Department of the University of Nebraska (Lincoln), where they were invited Professors when part of this work was developed. All three authors gratefully acknowledge the support of the Centre de Recerca Matemàtica (Barcelona) for their warm hospitality during different periods of the academic year 2004–2005, while part of this paper was written.

References

[1] G. Baumslag, A. Miasnikov and V. Remeslennikov, Malnormality is decidable in free groups, *Internat. J. Algebra Comput.*, **9** (1999), no. 6, 687–692.

[2] G.M. Bergman, Supports of derivations, free factorizations and ranks of fixed subgroups in free groups, *Trans. Amer. Math. Soc.*, **351** (1999), 1531–1550.

[3] J.-C. Birget, S. Margolis, J. Meakin, P. Weil. PSPACE-completeness of certain algorithmic problems on the subgroups of free groups, in *ICALP 94* (S. Abiteboul, E. Shamir éd.), Lecture Notes in Computer Science **820** (Springer, 1994) 274–285.

[4] J.-C. Birget, S. Margolis, J. Meakin, P. Weil. PSPACE-completeness of certain algorithmic problems on the subgroups of free groups, *Theoretical Computer Science* **242** (2000) 247–281.

[5] W. Dicks, E. Ventura, The group fixed by a family of injective endomorphism of a free group, *Contemp. Math.*, **195** (1996), 1–81.

[6] S. Gersten, On Whitehead's algorithm, *Bull. Am. Math. Soc.*, **10** (1984), 281–284.

[7] I. Kapovich and A. Miasnikov, Stallings Foldings and Subgroups of Free Groups, *J. Algebra*, **248**, 2 (2002), 608–668.

[8] R. Lyndon and P. Schupp, *Combinatorial group theory*, Springer, (1977, reprinted 2001).

[9] W. Magnus, A. Karras and D. Solitar, *Combinatorial group theory*, Dover Publications, New York, (1976).

[10] G.S. Makanin. Equations in a free group, *Izvestiya Akad. Nauk SSSR* **46** (1982), 1199–1273 (in Russian). (English translation: *Math. USSR Izvestiya* **21** (1983), 483–546.)

[11] S. Margolis, M. Sapir and P. Weil, Closed subgroups in pro-V topologies and the extension problems for inverse automata, *Internat. J. Algebra Comput.* **11**, 4 (2001), 405–445.

[12] L. Ribes and P.A. Zalesskiĭ, The pro-p topology of a free group and algorithmic problems in semigroups, *Internat. J. Algebra Comput.* **4** (1994) 359–374.

[13] A. Roig, E. Ventura, P. Weil, On the complexity of the Whitehead minimization problem, *Internat. J. Algebra Comput.*, to appear.

[14] J.-P. Serre, *Arbres, amalgames, SL_2*, Astérisque **46**, Soc. Math. France, (1977). English translation: *Trees*, Springer Monographs in Mathematics, Springer, (2003).

[15] P. Silva and P. Weil, On an algorithm to decide whether a free group is a free factor of another, *Theoretical Informatics and Applications*, to appear.

[16] J.R. Stallings, Topology of finite graphs, *Inventiones Math.* **71** (1983), 551–565.

[17] M. Takahasi, Note on chain conditions in free groups, *Osaka Math. Journal* **3**, 2 (1951), 221–225.

[18] E.C. Turner, Test words for automorphisms of free groups, *Bull. London Math. Soc.*, **28** (1996), 255–263.

[19] E.C. Turner, private communication, 2005.

[20] E. Ventura, On fixed subgroups of maximal rank, *Comm. Algebra*, **25** (1997), 3361–3375.

[21] E. Ventura, Fixed subgroups in free groups: a survey, *Contemp. Math.*, **296** (2002), 231–255.

[22] P. Weil. Computing closures of finitely generated subgroups of the free group, in *Algorithmic problems in groups and semigroups* (J.-C. Birget, S. Margolis, J. Meakin, M. Sapir éds.), Birkhäuser, 2000, 289–307.

Alexei Miasnikov
Dept of Mathematics and Statistics
McGill University
Burnside Hall, room 915
805 Sherbrooke West
Montréal, Quebec, Canada, H3A 2K6

and

Dept of Mathematics and Computer Science
City University of New York
e-mail: **alexeim@math.mcgill.ca**

Enric Ventura
EPSEM, Universitat Politècnica de Catalunya
Av. Bases de Manresa 61-73
08242 Manresa (Barcelona), Catalonia

and

Centre de Recerca Matemàtica
Bellaterra (Barcelona), Catalonia
e-mail: **enric.ventura@upc.edu**

Pascal Weil
LaBRI
Université de Bordeaux and CNRS
351 cours de la Libération
F-33400 Talence   France
e-mail: **pascal.weil@labri.fr**

Progress in Mathematics (PM)

Edited by
Hyman Bass, University of Michigan, USA
Joseph Oesterlé, Institut Henri Poincaré, Université Paris VI, France
Alan Weinstein, University of California, Berkeley, USA

Progress in Mathematics is a series of books intended for professional mathematicians and scientists, encompassing all areas of pure mathematics. This distinguished series, which began in 1979, includes research level monographs, polished notes arising from seminars or lecture series, graduate level textbooks, and proceedings of focused and refereed conferences. It is designed as a vehicle for reporting ongoing research as well as expositions of particular subject areas.

PM 260: Holzapfel, R.-P. / Uludağ, A.M. / Yoshida, M. (Eds.)
Arithmetic and Geometry Around Hypergeometric Functions (2007)
ISBN 978-3-7643-8283-4

PM 259: Capogna, L. / Danielli, D. / Pauls, S.D. / Tyson, J.T.
An Introduction to the Heisenberg Group and the Sub-Riemannian Isoperimetric Problem (2007)
ISBN 978-3-7643-8132-5

PM 258: Gan, W.T. / Kudla, S.S. / Tschinkel, Y. (Eds.)
Eisenstein Series and Applications (due 2008)
ISBN 978-0-8176-4496-3

PM 257: Lakshmibai, V. / Littelmann, P. / Seshadri, C.S. (Eds.)
Schubert Varieties (due 2007)
ISBN 978-0-8176-4153-5

PM 256: Borthwick, D.
Spectral Theory of Infinite-Volume Hyperbolic Surfaces (due 2007)
ISBN 978-0-8176-4524-3

PM 255: Kobayashi, T. / Schmid, W. / Yang, J.-H. (Eds.)
Representation Theory and Automorphic Forms (due 2007)
ISBN 978-0-8176-4505-2

PM 254: Ma, X. / Marinescu, G.
Holomorphic Morse Inequalities and Bergman Kernels (2007)
ISBN 978-3-7643-8096-0

PM 253: Ginzburg, V. (Ed.)
Algebraic Geometry and Number Theory. In Honor of Vladimir Drinfeld's 50th Birthday (2006)
ISBN 978-0-8176-4471-7

PM 252: Maeda, Y. / Michor, P. / Ochiai, T. / Yoshioka, A. (Eds.)
From Geometry to Quantum Mechanics. In Honor of Hideki Omori (2007)
ISBN 978-0-8176-4512-8

PM 251: Boutet de Monvel, A. / Buchholz, D. / Iagolnitzer, D. / Moschella, U. (Eds.)
Rigorous Quantum Field Theory. A Festschrift for Jacques Bros (2006)
ISBN 978-3-7643-7433-4

PM 250: Unterberger, A.
The Fourfold Way in Real Analysis. An Alternative to the Metaplectic Representation (2006)
ISBN 978-3-7643-7544-7

PM 249: Kock, J. / Vainsencher, I.
An Invitation to Quantum Cohomology. Kontsevich's Formula for Rational Plane Curves (2006).
ISBN 978-0-8176-4456-7

PM 248: Bartholdi, L. / Ceccherini-Silberstein, T. / Smirnova-Nagnibeda, T. / Zuk, A. (Eds.) Infinite Groups: Geometric, Combinatorial and Dynamical Aspects (2006)
ISBN 978-3-7643-7446-4

PM 247: Baues, H.-J.
The Algebra of Secondary Cohomology Operations (2006)
ISBN 978-3-7643-7448-8

PM 246: Dragomir, S. / Tomassini, G., Differential Geometry and Analysis on CR Manifolds (2006).
ISBN 978-0-8176-4388-1

PM 245: Fels, G. / Huckleberry, A.T. / J.A. Wolf
Cycle Spaces of Flag Domains. A Complex Geometric Viewpoint (2006). ISBN 978-0-8176-4391-1

PM 244: Etingof, P. / Retakh, V. / Singer, I.M. (Eds.)
The Unity of Mathematics. In Honor of the Ninetieth Birthday of I.M. Gelfand (2006)
ISBN 978-0-8176-4076-7

PM 243: Bernstein, J. / Hinich, V. / A. Melnikov (Eds.)
Studies in Lie Theory. A Joseph Festschrift (2006)
ISBN 0-8176-4342-3

PM 242: Dufour, J.-P. / Zung, N.T.
Poisson Structures and their Normal Forms (2005)
ISBN 978-3-7643-7334-4

PM 241: Seade, J.
On the Topology of Isolated Singularities in Analytic Spaces (2005). ISBN 978-3-7643-7322-1

PM 240: Ambrosetti, A. / Malchiodi, A.
Perturbation Methods and Semilinear Elliptic Problems in R^n (2005). ISBN 978-3-7643-7321-4

PM 239: van der Geer, G. / Moonen, B.J.J. / Schoof, R. (Eds.)
Number Fields and Function Fields – Two Parallel Worlds (2005)
ISBN 978-0-8176-4397-3

PM 238: Sabadini, I. / Struppa, D.C. / Walnut, D.F. (Eds.)
Harmonic Analysis, Signal Processing, and Complexity. Festschrift in Honor of the 60th Birthday of Carlos A. Berenstein (2005). ISBN 978-0-8176-4358-4

PM 237: Kulish, P.P. / Manojlovic, N. / Samtleben, H. (Eds.)
Infinite Dimensional Algebras and Quantum Integrable Systems (2005). ISBN 978-3-7643-7215-6

Advanced Courses in Mathematics CRM Barcelona

Edited by
Manuel Castellet

Since 1995 the Centre de Recerca Matemàtica (CRM) in Barcelona has conducted a number of annual Summer Schools at the post-doctoral or advanced graduate level. Sponsored mainly by the European Community, these Advanced Courses have usually been held at the CRM in Bellaterra.
The books in this series consist essentially of the expanded and embellished material presented by the authors in their lectures.